TODAY'S TECHNICIAN ™

Classroom Manual for

Automotive Brake Systems

Fourth Edition

TODAY'S TECHNICIAN ™

Classroom Manual for
Automotive Brake Systems

Fourth Edition

Clifton E. Owen
Griffin Technical College
Griffin, GA

DELMAR
CENGAGE Learning

Australia · Brazil · Japan · Korea · Mexico · Singapore · Spain · United Kingdom · United States

DELMAR
CENGAGE Learning™

**Today's Technician: Classroom Manual
for Automotive Brake Systems
Fourth Edition
Clifton E. Owen**

Vice President, Technology and
Trades ABU:
David Garza

Director of Learning Solutions:
Sandy Clark

Managing Editor: **Larry Main**

Senior Acquisitions Editor:
David Boelio

Product Manager: **Matthew Thouin**

Marketing Director:
Deborah S. Yarnell

Marketing Manager: **Erin Coffin**

Marketing Coordinator:
Patti Garrison

Director of Production:
Patty Stephan

Production Manager: **Stacy Masucci**

Content Project Manager:
Barbara L. Diaz

Content Project Manager:
Cheri Plasse

Technology Project Manager:
Kevin Smith

Editorial Assistant: **Lauren Stone**

For product information and technology assistance, contact us at
Professional & Career Group Customer Support, 1-800-648-7450

For permission to use material from this text or product, submit all requests online at **www.cengage.com/permissions**
Further permissions questions can be emailed to
permissionrequest@cengage.com

Library of Congress Control Number:

ISBN-13: 978-1-4180-6221-7

ISBN-10: 1-4180-6221-9

Delmar Cengage Learning
5 Maxwell Drive
Clifton Park, NY 12065-2919
USA

Cengage Learning products are represented in Canada by Nelson Education, Ltd.

For your lifelong learning solutions, visit **delmar.cengage.com**

Visit our corporate website at **www.cengage.com**

Notice to the Reader

Publisher does not warrant or guarantee any of the products described herein or perform any independent analysis in connection with any of the product information contained herein. Publisher does not assume, and expressly disclaims, any obligation to obtain and include information other than that provided to it by the manufacturer. The reader is expressly warned to consider and adopt all safety precautions that might be indicated by the activities described herein and to avoid all potential hazards. By following the instructions contained herein, the reader willingly assumes all risks in connection with such instructions. The publisher makes no representations or warranties of any kind, including but not limited to, the warranties of fitness for particular purpose or merchantability, nor are any such representations implied with respect to the material set forth herein, and the publisher takes no responsibility with respect to such material. The publisher shall not be liable for any special, consequential, or exemplary damages resulting, in whole or part, from the readers' use of, or reliance upon, this material.

Printed in the United States of America
3 4 5 XX 09

CONTENTS

Preface   ix

CHAPTER 1

Brake System Fundamentals   **1**

Introduction 1 ● Brake System Overview 1 ● Trailer
Brakes 11 ● Summary 15 ● Review Questions 15

CHAPTER 2

Principles and Theories of Operation   **17**

Introduction 17 ● Brake Operation 17 ● Brake System
Energy 18 ● Braking Dynamics 21 ● Friction Principles 22
 ● Energy and Works 28 ● Newton's Laws of Motion 29
 ● Hydraulic Principles 31 ● Vacuum and Air Pressure
Principles 36 ● Electrical Principles 36 ● Summary 38
 ● Review Questions 38

CHAPTER 3

**Related Systems: Tires, Wheels, Bearings, and
Suspensions**   **41**

Introduction 41 ● Tire Fundamentals 41 ● Run-Flat Tires 47
 ● Tire Pressure Monitoring System (TPMS) 47 ● Wheel
Fundamentals 48 ● Wheel Bearings 50 ● Wheel Alignment
Fundamentals 52 ● Effects on Braking Performance 56 ●
Performance Tires, Wheels, and Alignment 57 ● Summary 59
 ● Review Questions 60

CHAPTER 4

Master Cylinders and Brake Fluid   **63**

Introduction 63 ● Hydraulic Brake Fluid 63 ● Brake Pedal
and Pushrod 68 ● Dual-Piston Master Cylinder Construction
and Operation 72 ● Split Hydraulic Systems 84 ● Fast-Fill
and Quick Take-Up Master Cylinders 86 ● Central-Valve Master
Cylinders 89 ● Summary 91 ● Review Questions 91

CHAPTER 5

Hydraulic Lines, Valves, and Switches 95

Introduction 95 • Brake Lines and Hoses 95 • Brake Hydraulic Valves 106 • Brake Electrical Warning System 120 • Summary 128 • Review Questions 128

CHAPTER 6

Power Brake Systems 131

Introduction 131 • Increasing Brake Force Input 131 • Vacuum Principles 132 • Vacuum and Air Systems for Power Boosters 134 • Vacuum Power Boosters 137 • Tandem Boosters 145 • Hydraulically Assisted Power Brakes 149 • Electro-Hydraulic Brake (EHB) 157 • Summary 159 • Review Questions 160

CHAPTER 7

Disc Brakes 163

Introduction 163 • Disc Brake Advantages and Disadvantages 163 • Disc Brake Construction 169 • Caliper Construction and Operation 182 • Types of Disc Brakes 188 • Rear Wheel Disc Brakes 193 • Performance Disc Brakes 193 • Summary 195 • Review Questions 196

CHAPTER 8

Drum Brakes 199

Introduction 199 • Drum Brake Advantages and Disadvantages 199 • Drum Brake Construction and Operation 203 • Drum Brake Designs 219 • Summary 224 • Review Questions 225

CHAPTER 9

Parking Brakes 229

Introduction 229 • Parking Brake Operation 229 • Parking Brake Controls—Levers and Pedals 231 • Warning Lamps 234 • Parking Brake Linkage 234 • Rear Drum Parking Brakes 238 • Electrical Parking Brake Systems 239 • Rear Disc Parking Brakes 242 • Summary 244 • Review Questions 244

CHAPTER 10 | **Electrical Braking Systems** | **247**

Introduction 247 ● Common Components and Terms 247 ● Antilock Brake System (ABS) and Vehicle Control 252 ● ABS Types and General Operations 252 ● ABS Brands 254 ● ABS Components 254 ● Traction Control System (TCS) 260 ● System Integration 261 ● Continental Teves Mark 20 and Mark 20E 264 ● Delphi DBC-7 ABS 269 ● Active Brake Systems 272 ● Electronic Braking Systems and Future Brake Technology 273 ● Summary 274 ● Review Questions 276

Glossary   **279**

Index   **295**

PREFACE

Thanks to the support the Today's Technician™ series has received from those who teach automotive technology, Delmar Cengage Learning, the leader in automotive related textbooks, is able to live up to its promise to provide new editions of the series regularly. We have listened and responded to our critics and our fans and present this new updated and revised fourth edition. By revising our series regularly, we can and will respond to changes in the industry, changes in technology, changes in the certification process, and the ever-changing needs of those who teach automotive technology.

The Today's Technician™ series, by Delmar Cengage Learning, features textbooks that cover all mechanical and electrical systems of automobiles and light trucks (whereas the Heavy-duty Trucks portion of the series does the same for Heavy-duty vehicles). Principally, the individual titles correspond to the main areas of ASE (National Institute for Automotive Service Excellence) certification. Additional titles include remedial skills and theories common to all of the certification areas and advanced or specific subject areas that reflect the latest technological trends. Each text is divided into two volumes: a Classroom Manual and a Shop Manual.

Unlike yesterday's mechanic, the technician of today and for the future must know the underlying theory of all automotive systems and be able to service and maintain those systems. Dividing the material into two volumes provides the reader with the information needed to begin a successful career as an automotive technician without interrupting the learning process by mixing cognitive and performance learning objectives into one volume.

The design of Delmar Cengage Learning's Today's Technician™ series was based on features that are known to promote improved student learning. The design was further enhanced by a careful study of survey results, in which the respondents were asked to value particular features. Some of these features can be found in other textbooks, whereas others are unique to this series.

Each Classroom Manual contains the principles of operation for each system and subsystem. The Classroom Manual also contains discussions on design variations of key components used by the different vehicle manufacturers. This volume is organized to build on basic facts and theories. The primary objective of this volume is to allow the reader to gain an understanding of how each system and subsystem operates. This understanding is necessary to diagnose the complex automobiles of today and tomorrow. Although the basics contained in the Classroom Manual provide the knowledge needed for diagnostics, diagnostic procedures appear only in the Shop Manual. An understanding of the basics is also a requirement for competence in the skill areas covered in the Shop Manual.

A spiral-bound Shop Manual covers the "how-to's." This volume includes step-by-step instructions for diagnostic and repair procedures. Photo Sequences are used to illustrate some of the common service procedures. Other common procedures are listed and are accompanied by line drawings and photos that allow the reader to visualize and conceptualize the finest details of the procedure. This volume also contains the reasons for performing the procedures as well as when that particular service is appropriate.

The two volumes are designed to be used together and are arranged in corresponding chapters. Not only are the chapters in the volumes linked together, but also the contents of the chapters are linked. This linking of content is evidenced by marginal callouts that refer the reader to the chapter and page in which the same topic is addressed in the other volume. This feature is valuable to instructors. Without this feature, users of other two-volume textbooks must search the index or table of contents to locate supporting information in the other volume. This is not only cumbersome but also creates additional work for an instructor when planning the presentation of material and when making reading assignments. It is also valuable to the students; with page references, they also know exactly where to look for supportive information.

Both volumes contain clear and thoughtfully selected illustrations, many of which are original drawings or photos specially prepared for inclusion in this series. This means that the art is a vital part of each textbook and not merely inserted to increase the numbers of illustrations.

The page layout used in the series is designed to include information that would otherwise break up the flow of information presented to the reader. The main body of the text includes all of the "need-to-know" information and illustrations. In the wide side margins of each page are many of the special features of the series. Items that are truly "nice-to-know" information include simple examples of concepts just introduced in the text, explanations or definitions of terms that will not be defined in the Glossary, examples of common trade jargon used to describe a part or operation, and exceptions to the norm explained in the text. Many textbooks attempt to include this type of information and insert it in the main body of text; this tends to interrupt the thought process and cannot be pedagogically justified. By placing this information off to the side of the main text, the reader can select when to refer to it.

Highlights of This Edition—Classroom Manual

The text and figures of this edition are updated to show modern brake technology and applications. The Classroom Manual covers the complete mechanical-hydraulic automotive braking theories. It introduces the reader to the first generation of electric brake systems with a brief overview of brake-by-wire applications in the first chapter. The following chapters cover basic brake physics theories; discussion of newer components and materials, including a section on electric parking brakes; and any braking functions required for passenger cars and light trucks. The reader is introduced to fundamental information on trailer brakes and the DOT requirements for trailer brakes. In the trailer section electric, electrohydraulic, and surge-braking systems are discussed. A new Chapter 10 in this text entitled Electrical Braking Systems (EBS) simplifies the discussion on traditional antilock brake systems (ABS) while retaining the information for a complete understanding of ABS. That chapter guides the reader from ABS through traction control and the incorporation of those systems, to vehicle stability systems. Electrohydraulic brakes are discussed in detail along with straightforward information on vehicle stability fundamentals. The Classroom Manual guides the reader from traditional hydraulic brakes to the brake system of the future.

Highlights of This Edition—Shop Manual

Safety information has been moved from the Classroom Manual to the first chapter of the Shop Manual. This places this critical subject next to the work to be accomplished. Chapter 2, Brake Service Tools and Equipment, covers basic tools with more information on brake special tools and equipment. Some of the safety information that is pertinent to a particular piece of equipment has been moved to this chapter so safety issues are presented just before operation of the equipment. Another major improvement is that the related system information is presented in Chapter 3. This is more in accordance with standard brake operation and diagnosis procedures. Totally new information covers diagnosing electric parking brakes and electric braking systems. To clarify the diagnosis and repair procedures for electric braking two major ABS/TCS brands, Delphi DBC-7 and Teves Mark 20E, are used for discussion instead of the complete industry as in the last edition. This helps the reader better understand the technical diagnosing and repairing for all ABS/TCS. This edition of the Shop Manual will guide the student/technician through all the basic tasks in brake system repair.

Classroom Manual

Features of this manual include:

Cognitive Objectives

These objectives define the contents of the chapter and define what the student should have learned on completion of the chapter.

Each topic is divided into small units to promote easier understanding and learning.

Related Systems: Tires, Wheels, Bearings, and Suspensions

CHAPTER 3

Upon completion and review of this chapter, you should be able to:

❑ Describe the basic kinds of tire construction and identify the most common construction method for modern tires.

❑ Identify and explain the various letters and numbers used in tire size designations and other tire specifications.

❑ Explain the basic effects of tire tread design on vehicle handling and braking.

❑ Explain the most important effects of tire design and condition on brake performance.

❑ Explain how wheel and tire run out and wheel rim width and offset affect braking.

❑ Identify the common types of wheel and axle bearings used on cars and light trucks.

❑ Identify the basic wheel alignment and steering angles.

❑ Explain how wheel alignment and steering angles can affect braking.

❑ Explain how the condition of steering and suspension parts can affect braking.

Introduction

The brake shoes or pads apply friction to the wheels, but it is the friction between the tires and the road that actually stops the car. Tire design, condition, and inflation pressure can affect braking, and attention to these factors often can solve braking problems.

The tires are mounted on wheels, which ride on bearings on steering knuckle spindles and axles. Steering and axle components, in turn, are supported by suspension struts and springs. Any of these components can create braking problems if they are not in proper working order. This chapter outlines the key relationships between brake systems and the related systems of wheels, tires, wheel bearings, and suspensions.

The gross vehicle weight is the weight of the vehicle plus driver, passenger(s), full fuel tank, and the amount of other material loaded onto the vehicle.

Tire Fundamentals

Brake systems are engineered in relation to many vehicle factors of weight, size, and performance. Among these factors are the construction, size, and tread design of the tires and the amount of traction or friction expected to be available between the tires and the road. For the best and most reliable brake performance, tires at all four wheels should be identical in construction, size, and tread pattern.

Shop Manual pages 86–93

Carmakers' Recommendations

Most passenger cars and light trucks built since 1968 have a tire information placard on a door, a door pillar, or inside the glove compartment (Figure 3-1). The tire information placard lists the manufacturer's original equipment tire size and any recommended optional sizes. It also lists the recommended front and rear inflation pressures and maximum front and rear **gross vehicle weight rating (GVWR)**. Brake systems are engineered to work most efficiently with the tire sizes and pressures listed on the placard.

Gross vehicle weight rating (GVWR) is the total weight of a vehicle plus its maximum rated payload.

41

References to the Shop Manual

Reference to the appropriate page in the Shop Manual is given whenever necessary. Although the chapters of the two manuals are synchronized, material covered in other chapters of the Shop Manual may be fundamental to the topic discussed in the Classroom Manual.

Hydraulic Principles and Brake System Engineering

Engineers must consider these principles of force, pressure, and motion to design a brake system for any vehicle that will give maximum stopping efficiency but still be easy to control. If the engineer chooses a master cylinder with relatively small piston area, the brake system can develop very high hydraulic pressure, but the pedal travel will be extreme. Moreover, if the master cylinder piston travel is not long enough, this high-pressure system will not move enough fluid to apply the large-area caliper pistons regardless of pressure. If, however, the engineer selects a large-area master cylinder piston, it can move a large volume of fluid but may not develop enough pressure to exert adequate braking force at the wheels.

The overall size relationships of master cylinder pistons, caliper pistons, and wheel cylinder pistons are balanced to achieve maximum braking force without grabbing or fading. Most brake systems with front discs and rear drums have large-diameter master cylinders (large piston area) to move enough fluid and a power booster to increase the input force.

Vacuum and Air Pressure Principles

Vacuum is generally considered to be air pressure lower than atmospheric pressure (a true vacuum is a complete absence of air).

Vacuum is another force used in most brake systems. Most power brake systems use vacuum to provide a power assist for the driver. Because the most significant use of atmospheric pressure and vacuum in a brake system is in the operation of a power booster, these principles are covered in Chapter 6 of this manual on power brake systems.

Electrical Principles

Many of the brake system components you will work on are controlled or powered by electricity. Examples are brake system warning lamps, stoplamp switches, brake fluid level sensors, and ABS components. Therefore, a basic understanding of some of the electrical principles, including amperage, voltage, and resistance, is needed.

Amperage, Voltage, and Resistance

Ampere (A) is the unit for measuring electric current where 1 ampere equals a current flow of 6.28×10^{18} electrons per second.

Voltage is the electromotive force that causes current to flow; the potential force that exists between two points when one is positively charged and the other is negatively charged.

Think of electricity in terms of the same principles that work in hydraulic systems. The flow of electricity through a circuit is similar to the flow of fluid through a hydraulic line. Current is the movement of free electrons, under pressure, in a conductor. A flow of current through a conductor requires a source of free electrons to supply the demand, just as fluid in a tank or reservoir is a source of flow in a hydraulic system. The rate of fluid moving through a line often is measured in gallons per minute. The rate of current flowing in a conductor is measured in amperes (A). One **ampere (A)** equals 6.28×10^{18} electrons passing a given point in a circuit per second.

Just as pressure is necessary to move fluid through hydraulic lines, there must be pressure to move electrons through a conductor. The pressure pushing the electrons through an electrical circuit is called the **voltage**, measured in **volts (V)**.

Friction between the walls of a hydraulic line and the fluid will cause some resistance to the flow of fluid. Similarly, some resistance to electron flow through a circuit is offered by any material. Electrical resistance is measured in **ohms (Ω)**.

To summarize the comparison of electrical and hydraulic systems, we can say:

❑ Voltage is the pressure (or electrical force) that moves electrons (current or amperes) through a wire just as pressure moves fluid through a pipe.

❑ Amperage, or current, is similar to the fluid flowing in a line.

❑ Electrical resistance is a load on the moving current that must be present to do any useful work, just as a hydraulic system must have the load of an output piston or motor to do work.

36

Marginal Notes

New terms are pulled out and defined. Common trade jargon also appears in the margin, as well as some of the common terms used for components. These marginal notes allow the reader to speak and understand the language of the trade, especially when conversing with an experienced technician.

Figure 2-20 The output pistons' movement and their created force will be proportional to their size in relationship to the input piston size.

If the output piston is larger than the input piston, it exerts more force but travels a shorter distance. The opposite also is true. If the output piston is smaller than the input piston, it exerts less force but travels a longer distance. Apply the equation to the 5-square-inch output piston in Figure 2-20:

$$\frac{10 \text{ square inches (input piston)}}{5 \text{ square inches (output piston)}} = \frac{2}{1} \times 2 \text{ inches (input stroke} = 4.0 \text{ inches output motion)}$$

In this case, the smaller output piston applies only half the force of the input piston, but its stroke (motion) is twice as long.

This relationship of force, pressure, and motion in a brake system is shown when the force applied to the master cylinder pistons and the resulting brake force and piston movement at the wheels is considered. Wheel cylinder pistons move only a fraction of an inch to apply hundreds of pounds of force to the brake shoes, but the wheel cylinder piston travel is quite a bit less than the movement of the master cylinder piston. Disc brake caliper pistons move only a few thousandths of an inch but apply great force to the brake rotors.

A BIT OF HISTORY

All of the hydraulic principles that are applied in a brake system are based on the work of a seventeenth-century scientist named Blaise Pascal. Pascal's work is known as Pascal's law. Pascal's law says that pressure at any one point in a confined liquid is the same in every direction and applies equal force on equal areas.

One of the most important results of Pascal's work was the discovery that fluids can be used to increase force. Pascal was the first person to demonstrate the relationships of pressure, force, and motion and the inverse relationship of motion and force. On an automobile, Pascal's laws are applied not just to the brake system. These same hydraulic principles are at work in the hydraulic system of an automatic transmission and other systems. Pascal's laws are even at work in the movement of liquid fuel from a tank to the fuel-injection system on the engine.

A Bit of History

This feature gives the student a sense of the evolution of the automobile. This feature not only contains nice-to-know information, but also should spark some interest in the subject matter.

Summary

- ❑ Brake fluid specifications are defined by SAE Standard J1703 and FMVSS 116.
- ❑ Fluids are assigned DOT numbers: DOT 3, DOT 4, DOT 5, DOT 3/4, and DOT 5.1.
- ❑ Always use fluid with the DOT number recommended by any specific carmaker.
- ❑ Never use DOT 5 fluid in an ABS or mix with any other brake fluid.
- ❑ HSMO fluids are very rare and should never be used in brake systems designed for DOT fluids.
- ❑ The brake pedal assembly is a lever that increases pedal force to the master cylinder.
- ❑ The brake pedal lever is attached to a pushrod, which transmits force to the master cylinder pistons.
- ❑ The master cylinder has two main parts: a reservoir and a cylinder body.
- ❑ The reservoir can be a separate piece or cast as one piece with the cylinder.
- ❑ A dual-piston master cylinder has two separate pistons providing pressure for two independent hydraulic systems. Each of the two pistons in the master cylinder has a cup, a return spring, and a seal.
- ❑ During application, the piston and cup force fluid ahead of the piston to activate the brakes.
- ❑ During release, the return spring returns the piston.
- ❑ Fluid from the reservoir flows from the reservoir through the replenishing port around the piston cup.
- ❑ Excess fluid in front of the piston flows back into the reservoir through the vent ports.
- ❑ A front-to-rear split hydraulic system has two master cylinder circuits. One is connected to the front brakes and the other to the rear brakes.
- ❑ A diagonally split hydraulic system is one in which one master cylinder circuit is connected to the left front and right rear brakes and the other circuit is connected to the right front and left rear brakes.
- ❑ Quick take-up or fast-fill master cylinders have a step bore, which is a larger diameter bore for the rear section of the primary piston.
- ❑ Quick take-up master cylinders have a valve that provides rapid filling of the low-pressure spool area of the primary piston from the reservoir.
- ❑ Some ABS master cylinders have check valves in the heads of the pistons to reduce piston and pedal vibration and cup wear.

Terms to Know

Adjustable Pedal System (APS)
Cup seal
Diaphragm
Free play
Hydraulic system mineral oil (HSMO)
O-ring
Polyglycol
Quick take-up master cylinder
Quick take-up valve
Replenishing port
Reservoir
Residual pressure check valve
Vent port

Terms to Know

A list of new terms appears next to the Summary. Definitions for these terms can be found in the Glossary at the end of the manual.

Summaries

Each chapter concludes with summary statements that contain the important topics of the chapter. These are designed to help the reader review the contents.

Review Questions

Short-Answer Essays

1. Identify and explain the systems of tire size designations and other tire specifications.
2. Describe the basic kinds of tire construction and identify the most common construction method for modern tires.
3. Explain the use of unidirectional tread pattern and its disadvantage compared to standard production tire tread.
4. Discuss run-flat tires.
5. List the basic adjustable wheel alignment and steering angles.
6. List the nonadjustable alignment and steering angles.
7. Identify the common types of wheel bearings used on cars and light trucks.
8. Explain how the condition of related systems may affect braking.
9. Identify the principal dimensions used to specify wheel size.
10. Explain the operational differences between WSB and PSB systems.

Fill in the Blanks

1. The steel beads around the rim and layers of cords or plies that are bonded together to give a tire its shape and strength are called the _____.
2. The layer of tire rubber that contacts the road and that contains a distinctive pattern is called the _____.
3. A tire with the cords in the body plies of the carcass running at an angle of 90 degrees to the steel beads in the inner rim of the carcass is a _____ tire.
4. The _____ is the percentage of tire cross-sectional height to cross-sectional width.
5. The most common system used to identify tires for passenger cars and many light trucks today is the _____ system.
6. On wet pavement, a slick tire will _____ on a layer of water trapped between the tread and the pavement.
7. The distance between the centerline of a rim and the mounting plane of the wheel is _____.
8. The distance from the tire contact patch centerline to the point where the steering axis intersects the road is _____.
9. The inward or outward tilt of the wheel measured from top to bottom and viewed from the front of the car is _____.
10. The backward or forward angle of the steering axis viewed from the side of the car is _____.

60

Review Questions

Short-answer essay, fill in the blank, and multiple-choice questions follow each chapter. These questions are designed to accurately assess the student's competence in the stated objectives at the beginning of the chapter.

Shop Manual

To stress the importance of safe work habits, the Shop Manual dedicates one full chapter to safety. Other important features of this manual include:

Performance Objectives

These objectives define the contents of the chapter and define what the student should have learned on completion of the chapter.

Although this textbook is not designed simply to prepare someone for the certification exams, it is organized around the ASE task list. These tasks are defined generically when the procedure is commonly followed and specifically when the procedure is unique for specific vehicle models. Imported and domestic model automobiles and light trucks are included in the procedures.

Tools Lists

Each chapter begins with a list of the Basic Tools needed to perform the tasks included in the chapter. Whenever a Special Tool is required to complete a task, it is listed in the margin next to the procedure.

Marginal Notes

Page numbers for cross-referencing appear in the margin. Some of the common terms used for components, and other bits of information, also appear in the margin. These marginal notes provide an understanding of the language of the trade and help when conversing with an experienced technician.

Photo Sequences

Many procedures are illustrated in detailed Photo Sequences. These detailed photographs show the students what to expect when they perform particular procedures. They also can provide a student a familiarity with a system or type of equipment that the school may not have.

Service Tips

Whenever a special procedure is appropriate, it is described in the text. These tips are generally those things commonly done by experienced technicians.

Cautions and Warnings

Throughout the text, cautions are given to alert the reader to potentially hazardous materials or unsafe conditions. Warnings are also given to advise the student of things that can go wrong if instructions are not followed or if an unacceptable part or tool is used.

References to the Classroom Manual

Reference to the appropriate page in the Classroom Manual is given whenever necessary. Although the chapters of the two manuals are synchronized, material covered in other chapters of the Classroom Manual may be fundamental to the topic discussed in the Shop Manual.

Customer Care

This feature highlights those little things a technician can do or say to enhance customer relations.

Job Sheets

Located at the end of each chapter, the Job Sheets provide a format for students to perform procedures covered in the chapter. A reference to the ASE and NATEF Tasks addressed by the procedure is referenced on the Job Sheet.

Facsimile page 131:

Therefore, if a customer wants silicone fluid in his vehicle, all the polyglycol fluid must be completely flushed out. The best time to convert to silicone fluid is during a complete brake system overhaul.

Silicone fluid compresses slightly under pressure, which can cause a slightly spongy brake pedal feel. Silicone fluid also attracts and retains air more than polyglycol fluid does, which makes brake bleeding harder; it tends to outgas slightly just below its boiling point; and it tends to aerate from prolonged vibration. DOT 5 fluid has other problems with seal wear and water accumulation and separation in the system. All of these factors mean that DOT 5 silicone fluid should *never be used* in an ABS.

The best practice is to use a single, high-quality brand of brake fluid of the DOT type specified for a particular vehicle. Avoid mixing fluids whenever possible.

WARNING: DOT 5 should never be mixed with or used to replace DOT 3, DOT 4, or DOT 3/4, or DOT 5.1 fluids because of their chemical noncompatability. They will not mix and silicone can damage seals designed for polyglycol liquids.

CAUTION: DOT 5.1 and DOT 5.1 long-life brake fluids are not silicone-based fluids. Do not mix or replace DOT 5 fluid with DOT 5.1 or DOT 5.1 long-life fluids. Damage to the brake system and possible injury could occur due to damage to brake system components.

SERVICE TIP: Two good reasons for periodic flushing and refilling of the brake hydraulic system are: (1) Flushing the system and refilling with fresh fluid keeps sediment out of brake system; (2) Flushing and refilling the system also keeps that sediment out of the self-adjusting parking brake mechanisms on rear-wheel disc brakes.

Master Cylinder Fluid Service

Brake fluid service procedures are the most basic—but among the most important—brake system services. The following paragraphs provide instructions for checking the master cylinder fluid level and adding fluid to the system. Later sections of this chapter contain procedures for bench bleeding a master cylinder. Complete system bleeding instructions and fluid flushing and filling instructions are discussed later.

Checking Master Cylinder Fluid Level

Master cylinder fluid level and fluid condition should be inspected at least twice a year as part of a vehicle preventive maintenance schedule. If the car has a translucent fluid reservoir, general fluid level can be checked every time the motor oil is checked or changed.

Although normal brake lining wear causes a slight drop in fluid level, an abnormally low level in either chamber—especially an empty reservoir—usually means that there is a leak in the system. When you check the fluid in the master cylinder, you are checking two things. First, be sure the reservoir is filled to the correct level. A two-piece master cylinder with a plastic reservoir has graduated markings to indicate the correct fluid level (Figure 4-18). The markings may be on the inside of the reservoir if the reservoir is translucent, or they may be inside if the reservoir is opaque. Fill the reservoir to the FULL mark or equivalent.

If the master cylinder has a one-piece cast body with an integral reservoir, fluid level may be etched in the reservoir. In this case, fill the reservoir to ¼ inch (6 mm) from the top. If the reservoir is mounted at an angle on the master cylinder, measure fluid level at the point closest to the filler rim (Figure 4-19). Some composite master cylinders have opaque plastic reservoirs. Remove the reservoir caps or covers to check fluid level as you would for a one-piece reservoir.

Classroom Manual pages 63–68, 72–74

Special Tools
Brake fluid
Cloths

131

Facsimile page 169:

before disconnecting the injector. Repeat with each wheel brake until all show clear, air-free fluid from the cylinder/caliper. Fill the master cylinder to the proper level if needed. This setup would not work well in bench bleeding the master cylinder.

Brake Fluid Replacement: Flushing and Refilling the Hydraulic System

Manufacturers are about evenly divided on whether or not the brake system should be flushed periodically and refilled with fresh fluid. DOT 3 and DOT 4 fluids absorb moisture from the atmosphere. Most vehicles will contain about 2 percent water in their brake fluid after just one year of service. As little as 6 percent moisture in the brake fluid can cut the fluid boiling point in half. These facts may be very good arguments for periodically flushing and refilling the brake hydraulic system.

Currently, more than a dozen carmakers specify periodic brake fluid changes for some, or all, of their models built since 1985. Change intervals vary from as often as every 12 months or 15,000 miles to as infrequently as every 60,000 miles.

If this service is offered, however, there are a few general points to remember. All brake systems accumulate sludge over some period of time. Flushing the system can remove this sludge; but once it has been disturbed, make sure to get all of it out of the system. Stirring up sludge from the master cylinder reservoir may cause it to get into ABS valves and pumps if the sludge is not out of the system. The control valves for some rear-wheel ABS installations on some trucks may be particularly susceptible to sludge and dirt contamination.

Brake hoses for disc brakes usually enter the caliper near the top of the caliper body. The bleeder valve also is located at the top of the caliper bore. (It has to be because air rises.) If sludge accumulates in the caliper bore, it collects at the bottom. A quick, superficial bleeding of the caliper just passes a few ounces of fluid across the top of the bore. It does not flush out the sludge and all the old fluid. To flush a caliper thoroughly, pump several ounces of fluid through it. On some vehicles, it may be advisable to remove the caliper from its mounts and retract the piston to force out all the old fluid. Then reinstall it and thoroughly flush it with fresh fluid.

Flushing is done at each bleeder screw in the same manner as bleeding. Open the bleeder screw approximately one and a half turns and force fluid through the system until the fluid emerges clear and uncontaminated. Do this at each bleeder screw in the system. After all lines have been flushed, bleed the system using one of the bleeding procedures explained previously.

SERVICE TIP: Flush a brake system by draining the old fluid and adding denatured or isopropyl alcohol to the system. Continue to add alcohol until the system is clean. Flush out the alcohol with new brake fluid until all of the alcohol is removed.

CUSTOMER CARE: A little customer education can go a long way toward safer driving. Explain to your customers the importance of following their vehicle maintenance schedules for brake hydraulic system flushing and refilling. Refilling the system with fresh fluid is cheap insurance against hydraulic failure due to sludge, dirt, and moisture in the system.

CASE STUDY

A customer complained of excessive pedal travel on a 10-year-old domestic sedan. The master cylinder reservoir was full. A test drive confirmed the condition. The car was placed on the lift. Lines and hoses were in good condition and leak free at all connections. The wheels were pulled and the caliper and wheel cylinders were inspected for leakage. All checked out okay.

Classroom Manual pages 62–68, 84–86

Special Tools
Flushing requires the same special tools as bleeding.
The tools depend on the type of bleeding method used.

169

Facsimile Job Sheet 18 (page 235):

Job Sheet 18   18

Name: _____ Date: _____

Replace a Brake Hose

NATEF Correlation
This job sheet addresses the following NATEF tasks: Fabricate and/or install brake lines (double flare and ISO types); replace hoses, fittings, and supports as needed.

Objective
Upon completion of this job sheet, you should be able to replace a brake hose.

Tools and Materials
Basic hand tools

Procedure   Task Completed

1. Describe the vehicle being worked on:
   Year _____ Make _____ Model _____
   VIN _____ Engine type and size _____
   ABS _____ yes _____ No _____   ☐

2. Lift the vehicle, if necessary, and remove the left front tire and wheel assembly. Different wheels may be selected.   ☐

3. Explain why (or what circumstances would cause) this brake hose to be replaced.
   _____
                                                        ☐

   NOTE: This job sheet uses a front disc brake hose as an example. Other brake hoses are replaced in a similar manner.

4. Does this brake hose use banjo-type fittings at either end? If so, adjust your tool choice to perform the following steps.
   _____
                                                        ☐

5. Use flare-nut (line) wrenches to disconnect the hose from the steel line on or near the vehicle frame. Plug the steel line. What size wrench was used?
   _____
                                                        ☐

6. Use a small prybar or pliers to remove the hose retainer at the frame. What tool was used and how was the procedure performed?
   _____
                                                        ☐

7. Use a flare-nut (line) wrench to disconnect the hose from the caliper.   ☐

   _____

235

Case Studies

Case Studies concentrate on the ability to properly diagnose the systems. Chapters 3 through 10 end with a case study in which a vehicle has a problem, and the logic used by a technician to solve the problem is explained.

CASE STUDY

The problem with a 1995 imported SUV was that the brakes would barely work first thing in the morning. After the vehicle was warmed up for a few minutes and driven about a half mile, the brakes started working normally. The owner suspected a stuck master cylinder, an ABS problem, or some similar "high-tech" cause. The tech working on the car took a simpler approach.

After the vehicle sat outside overnight, the technician started the engine and removed the vacuum line from the power brake booster. Engine speed jumped up several hundred rpm. The tech then shook the hose and blew through it several times. The engine speed then dropped back to normal.

On many imported vehicles, the vacuum check valve for the power booster is in line in the hose, not in the fitting on the booster. On this Isuzu, the valve was sticking open when cold. A few minutes of driving and some engine heat would get it working again. Replacing the hose with a new OEM unit that contained a new check valve solved the problem.

Terms to Know

Pressure differential	Spool valve	Vacuum suspended

Terms to Know

Terms in this list can be found in the Glossary at the end of the manual.

ASE-Style Review Questions

1. *Technician A* says that if a vacuum booster is in good condition, starting the engine after the vacuum is exhausted should cause the brake pedal to drop slightly under foot pressure.
   Technician B says that the pedal should pulsate lightly after it drops.
   Who is correct?
   A. A only **C.** Both A and B
   B. B only **D.** Neither A nor B

2. The brake pedal of a high-mileage car is sluggish and moderately hard to apply.
   Technician A says that vacuum leaks or restrictions may exist between the engine and the booster.

3. *Technician B* says that a diesel-powered vehicle may have a hydraulic power brake booster.
   Who is correct?
   A. A only **C.** Both A and B
   B. B only **D.** Neither A nor B

4. *Technician A* says that a defective vacuum check valve in the power brake booster may cause a hard-pedal problem.
   Technician B says that a vacuum check valve may not be installed in a vacuum booster.
   Who is correct?
   A. A only **C.** Both A and B
   B. B only **D.** Neither A nor B

valve in the booster is pressure from leaking rakes are applied. valve is used to keep the booster when the

Both A and B
Neither A nor B

ASE-Style Review Questions

Each chapter contains ASE-style review questions that reflect the performance objectives listed at the beginning of the chapter. These questions can be used to review the chapter as well as to prepare for the ASE certification exam.

Terms to Know

Aqueous	Discard dimension	Parallelism
Bearing end play	Heat checking	Rotor lateral runout
Bedding in	Loaded caliper	Vernier caliper
Burnishing	Nondirectional finish	
Cross-feed		

ASE-Style Review Questions

1. While servicing the front disc brakes on a FWD vehicle:
   Technician A determines the right wheel pad is worn and replaces both the right and left wheel pads.
   Technician B determines the pads are not worn but rotates their position to ensure even pad wear.
   Who is correct?
   A. A only **C.** Both A and B
   B. B only **D.** Neither A nor B

2. After servicing the disc brakes on a vehicle:
   Technician A reinstalls the wheel nuts using an impact wrench to ensure a tight fit.
   Technician B refills the master cylinder reservoirs to the proper level.
   Who is correct?
   A. A only **C.** Both A and B
   B. B only **D.** Neither A nor B

3. *Technician A* says loaded calipers are replacement calipers that come with pads and hardware already installed.
   Technician B says loaded calipers should be installed in axle sets.
   Who is correct?
   A. A only **C.** Both A and B
   B. B only **D.** Neither A nor B

4. *Technician A* says it is very important that the rotor surface be made nondirectional during refinishing.
   Technician B says rotors should always be refinished as part of routine disc brake service.
   Who is correct?
   A. A only **C.** Both A and B
   B. B only **D.** Neither A nor B

5. When performing disc brake work:
   Technician A works on one wheel at a time to avoid popping pistons out of the other caliper and to allow the other caliper to be used as a guide.

Technician B hangs the calipers from the brake hoses as a convenience to speed up the job.
   Who is correct?
   A. A only **C.** Both A and B
   B. B only **D.** Neither A nor B

6. When measuring rotor runout:
   Technician A uses a micrometer.
   Technician B uses a dial indicator.
   Who is correct?
   A. A only **C.** Both A and B
   B. B only **D.** Neither A nor B

7. *Technician A* says that newly installed wheel bearings are self-adjusting and require no special adjustment.
   Technician B says that bearing assemblies are interchangeable with one another.
   Who is correct?
   A. A only **C.** Both A and B
   B. B only **D.** Neither A nor B

8. *Technician A* says that the minimum wear thickness of a rotor is the discard thickness of the rotor.
   Technician B says that the refinishing dimension is cast into a rotor.
   Who is correct?
   A. A only **C.** Both A and B
   B. B only **D.** Neither A nor B

9. When refinishing a rotor on a lathe, the rotor wobbles excessively.
   Technician A says that the lathe arbor may be bent.
   Technician B say that the mounting adapters of the lathe may be distorted.
   Who is correct?
   A. A only **C.** Both A and B
   B. B only **D.** Neither A nor B

ASE Practice Examination

A 50-question ASE practice exam, located in the Appendix, is included to test students on the content of the complete Shop Manual.

Appendix

ASE Practice Examination

1. A vehicle has a drift to the left as the brakes are applied:
   Technician A says that the left front disc brake is grabbing.
   Technician B says that the right rear drum brake is adjusted too loose.
   Who is correct?
   A. A only **C.** Both A and B
   B. B only **D.** Neither A nor B

2. The customer complains of rear wheel lockup during medium braking on a pickup truck with RWAL:
   Technician A says that the metering valve may be defective.
   Technician B says that a frame-mounted proportioning valve could be stuck.
   Who is correct?
   A. A only **C.** Both A and B
   B. B only **D.** Neither A nor B

3. Four-wheel, three-channel antilock brakes are being discussed:
   Technician A says that the system has four speed sensors.
   Technician B says that the two rear wheels are independently controlled.
   Who is correct?
   A. A only **C.** Both A and B
   B. B only **D.** Neither A nor B

4. Proportioning valves are being discussed:
   Technician A says that a bad valve may cause the rear brakes to engage too quickly during initial brake application.
   Technician B says that this valve prevents excessive pressure from reaching the front brakes during normal braking.
   Who is correct?
   A. A only **C.** Both A and B
   B. B only **D.** Neither A nor B

5. *Technician A* says that a jerk of the steering to the right only as the brakes are applied could indicate improper wheel alignment.
   Technician B says that brake fluid could cause the steering wheel to shake as the brakes are applied.
   Who is correct?
   A. A only **C.** Both A and B
   B. B only **D.** Neither A nor B

6. The brake pedal drops slowly to the floor after the vehicle has stopped and the brake is kept applied. The fluid level does not drop.
   Technician A says that the dump valve on the RABS could be leaking.
   Technician B says that the master cylinder could have an internal leak.
   Who is correct?
   A. A only **C.** Both A and B
   B. B only **D.** Neither A nor B

7. There is a heavy shuddering and vibration as the brakes are heavily applied on a dry road:
   Technician A says that the rotors are not parallel.
   Technician B says that the ABS is probably functioning.
   Who is correct?
   A. A only **C.** Both A and B
   B. B only **D.** Neither A nor B

8. ABSs are being discussed:
   Technician A says that a bent or damaged tone ring could cause the wheel to lock up.
   Technician B says that a damaged tone ring could cause the ABS to deactivate.
   Who is correct?
   A. A only **C.** Both A and B
   B. B only **D.** Neither A nor B

9. A vehicle with four-wheel antilock brakes has a problem with the right rear wheel locking:
   Technician A says that this could be caused by a bad speed sensor mounted at the wheel.
   Technician B says that the speed sensor mounted at the differential could cause this problem.
   Who is correct?
   A. A only **C.** Both A and B
   B. B only **D.** Neither A nor B

10. The brake pedal pulsates during braking:
    Technician A says that this could cause the ABS to deactivate.
    Technician B says that this could be caused by warped drums.
    Who is correct?
    A. A only **C.** Both A and B
    B. B only **D.** Neither A nor B

499

Reviewers

The author and publisher would like to extend special thanks to the following instructors for reviewing the draft manuscript:

Ron Chappell
Santa Fe Community College
Gainesville, FL

John Eichelberger
St. Philips College
San Antonio, TX

Raymond Karbowiak
University of Northwestern Ohio
Lima, OH

Jon D'Ambrosio
Mesa Community College
Mesa, AZ

Daniel Livingston
Jackson Community College
Jackson, MI

Darryl Malone
San Jacinto College
Pasadena, TX

Ruth Morrison
Southern Maine Community College
South Portland, ME

Les Peterson
Anoka Technical College
Anoka, MN

Anthony K. Rish
Gateway Community College
North Haven/New Haven, CT

Barry Stirn
University of Northwestern Ohio
Lima, OH

Christopher VanStavoren
Pennsylvania College of Technology
Williamsport, PA

Brake System Fundamentals

Upon completion and review of this chapter, you should be able to:

❏ List and describe the operation of the basic parts of a brake system.

❏ Describe the operation of the brake system during and after pedal application.

❏ Discuss the general operation of electronic braking systems.

❏ Explain how kinetic energy works in a brake system and the effects of mass and **weight.**

❏ Explain momentum and inertia as they apply to vehicle motion.

❏ Explain the effects of weight and mass as they relate to braking systems.

❏ Discuss the various dynamic factors involved in braking.

❏ Explain how **vacuum** and atmospheric pressure can be used to increase force.

Introduction

The brake system is one of the most important systems on a vehicle. It has four basic functions:

1. It must slow a moving vehicle.

2. It must bring a vehicle to a stop.

3. It must hold a vehicle stationary when stopped.

4. It allows directional control during maximum braking.

If the brake system does not operate properly, the driver and passengers could be injured or killed in an accident. Technicians who service the brake system must be highly skilled experts because the work they do can save lives. In this chapter, we start our study of the brake system by presenting the basic concepts and parts of all brake systems.

This chapter also highlights some of the dynamics associated with braking and controlling a vehicle. If all of the various dynamics are not considered during the design stage, most braking systems will under brake or over brake. When the brake system is not designed or operating correctly, it will be up to the driver to compensate, usually with poor results. In many cases, the human response is either too slow or too quick to react to a braking situation. In both cases, a loss of vehicle control is probably unavoidable.

Brake System Overview

The complete brake system consists of the major components shown in Figure 1-1. The complete brake system can be divided into the **service brakes,** which slow and stop the moving vehicle, and the **parking brakes,** which hold the vehicle stationary. On most late-model vehicles, the antilock brake system (ABS) is a third major subsystem; and many cars now also include traction control as part of the brake system functions.

Leverage and the Brake Pedal Design

 AUTHOR'S NOTE: A fulcrum is the point at which one lever pivots or sits to apply force to another lever or device. A seesaw pivots on a fulcrum.

Braking action on an automobile begins with the driver's foot on the brake pedal. The driver applies **force** to the pedal (which we learn more about later), and the pedal transfers that force to the master cylinder pistons. The brake pedal also multiplies the force of the driver's foot through leverage.

Figure 1-1 A typical automotive brake system comprises these major components and subsystems.

The brake pedal is mounted on a lever with a pivot near the top of the lever. The movement of the pedal causes a pushrod to move against a master cylinder. The master cylinder is mounted inside the engine compartment on the rear bulkhead. The master cylinder is a hydraulic pump that is operated by the driver through the brake pedal.

Most brake pedal installations are an example of what is called a second-class lever. In the science of physics, a second-class lever has a pivot point (or **fulcrum**) at one end and force applied to the other end. A second-class lever transfers the output force in the same direction as the input force, and multiplies the input force, depending on where the output load is placed. The brake pedal installation shown in Figure 1-2 has a 10-inch lever, and the load (the master cylinder pushrod) is 2 inches from the fulcrum (8 inches from the pedal). The pedal ratio, or the force multiplying factor, is the length of the lever divided by the distance of the load from the fulcrum. In this case, it is:

$$\frac{10}{2} = 5:1$$

If the driver applies 50 pounds of force to the pedal, the lever increases the force to 250 pounds at the master cylinder. When the driver applies 50 pounds of input force, the pedal may travel about 2.5 inches. When the lever applies 250 pounds of output force, the pushrod moves only 0.5 inch. Thus, as **leverage** in a second-class lever increases force, it reduces distance by the same factor:

$$\frac{2.5 \text{ inches}}{5} = 0.5 \text{ inch}$$

Leverage is the use of a lever and fulcrum to create a mechanical advantage, usually to increase force applied to an object.

Figure 1-2 The brake pedal assembly uses leverage to increase force applied to the master cylinder.

Service Brake History and Design

Modern automobile brakes evolved from the relatively crude brakes of horse-drawn vehicles. The earliest motor vehicle brakes were pads or blocks applied by levers and linkage to the outside of a solid tire on a wooden-spoked wheel. The same principles of leverage that work in modern brake pedal installations increased the force of the brake pad applied to the solid tire. These brakes worked well with speeds of 10 mph to 20 mph and little traffic. Higher performance (30 mph and beyond) and pneumatic tires meant that early wagon brakes were short-lived on automobiles.

By the end of the first decade of the twentieth century, automobiles were using either external-contracting band brakes or internal-expanding drum brakes. A few internal-expanding band brakes were tried on some early motor vehicles. External-contracting brakes have a band lined with **friction** material wrapped around a drum located on the driveline or on the wheels. The band is anchored at one end or at the center; levers and linkage tighten the band around the drum for braking force. The service brakes on Ford's famous Model T were a single contracting band applied to a drum inside the transmission.

Band brakes, either internal or external, lose their effectiveness when higher braking force is needed. When you study **drum brakes,** you will learn about the mechanical servo action of brake shoes. It is very difficult to develop servo action with an internal band brake, and higher brake force is thus needed. Servo action on an external band brake tends to make the brake grab at high brake forces and high drum speed. Other problems associated with band brakes included dirt and water damage and loss of friction with external bands and the tendency of these brakes to lock if the drum overheated and expanded too much. Internal band brakes also suffered from band and drum overheating and reduced braking force.

As drum brakes evolved, internal-expanding shoe-and-drum brakes became the standard. External-contracting band brakes were used as parking brakes until the late 1950s, but their days as service brakes were over by the late 1920s.

Drum Brakes. By the mid-1920s, drum brakes with internal-expanding shoes were the general rule. Early drum brakes were operated mechanically by levers and linkage (Figure 1-3). Expensive luxury cars such as the 1921 Duesenberg Model A were among the first to have hydraulic drum

Friction is the force that resists motion between the surfaces of two objects or forms of matter.

A **drum brake** is a brake in which friction is generated by brake shoes rubbing against the inside surface of a brake drum attached to the wheel.

Figure 1-3 Simple mechanical drum brake.

brakes. Hydraulic brakes started to appear on lower-priced cars in the mid-1920s with Chrysler's Light Six, which became the Plymouth. Ford Motor Company, however, used mechanical brakes through the 1938 model year.

The rigid brake shoes used with drum brakes could be made stronger than the flexible bands of earlier brake designs. This eliminated breakage problems that occurred with greater braking forces that were required as automobiles got more powerful and faster. With hydraulic actuation, four-wheel drum brakes remained the standard braking system for most cars into the middle and late 1960s. With the coming of Federal Motor Vehicle Safety Standards (FMVSS) 105 in 1967, brake systems had to pass specific performance tests that made front disc brakes the general rule in the 1970s. Even at the beginning of the twenty-first century, however, drum brakes are still used on the rear wheels of many cars and light trucks.

Disc Brakes. Modern automotive **disc brakes** were developed from aircraft brakes of World War II. Known originally as "spot" brakes, disc brakes work by applying pressure to two brake pads on opposite sides of a spinning rotor attached to the wheel hub (Figure 1-4). Disc brake pads are mounted in a **caliper** that sits above the spinning rotor. The caliper is either fixed or movable on its mounting. With a fixed caliper, hydraulic pressure is applied to pistons on both sides to force the pads against the rotor (Figure 1-5). With a movable caliper, pressure is applied to a piston on the inboard side only. This forces the inboard pad against the rotor, and the reaction force moves the outboard side of the caliper inward so that both pads grip the rotor (Figure 1-6).

All the friction components of a disc brake are exposed to the airstream, which helps to cool the brake parts and maintain braking effectiveness during repeated hard stops from high speeds. This, in turn, leads to longer pad life and faster recovery from brake fade. Disc brakes do not develop the mechanical servo action that you will learn about as you study drum brakes. Therefore, disc brakes require higher hydraulic pressure and greater force to achieve the same stopping power as a comparable drum brake. These pressure and force requirements for disc brakes are met easily, however, with large caliper pistons and power brake boosters. Because their advantages far outweigh any disadvantage, disc brakes have become the universal choice as the front brakes on all cars and light trucks built since the 1970s. Additionally, four-wheel disc brakes became standard equipment on high-performance automobiles, SUVs, and some light trucks.

A **caliper** is a major component of a disc brake system that houses the piston(s) and supports the brake pads.

Figure 1-4 Hydraulic pressure in the caliper forces the disc brake pads against the spinning motor.

Figure 1-5 Hydraulic pressure is applied equally to pistons on both sides of a fixed caliper.

Figure 1-6 Hydraulic pressure in a movable caliper forces the piston in one direction and the caliper body in the other. The resulting action and reaction force the pads against the rotor.

Hydraulic brakes were invented in 1918 in the California shop of Malcolm Loughead. He later changed the spelling of his name to Lockheed, and he and his brother founded the aircraft company of that name. The Lockheed hydraulic brake first appeared on the 1921 Duesenberg Model A.

Brake Hydraulic Systems

Hydraulic operation of service brakes has been the universal design for more than 60 years. The complete hydraulic system consists of the **master cylinder,** steel lines, rubber hoses, various pressure-control valves, and brake apply devices at each wheel.

The **master cylinder** is the liquid-filled cylinder in the hydraulic brake system or clutch in which hydraulic pressure is developed when the driver depresses a foot pedal.

Pressure is the force exerted on a given unit of surface area—force divided by area—measured in pounds per square inch (psi) or kilopascals (kPa).

Master Cylinder. The master cylinder is the start of the brake hydraulic system. It actually is a cylindrical pump. The cylinder is sealed at one end, and the movable pushrod extends from the other end (Figure 1-7). The pushrod moves a pair of in-line pistons that produce the pumping action. When the brake pedal lever moves the pushrod, it moves the pistons to draw fluid from a reservoir on top of the master cylinder. Piston action then forces the fluid under **pressure** through outlet ports to the brake lines.

All master cylinders for vehicles built since 1967 have two pistons and pumping chambers as shown in Figure 1-7. Motor vehicle safety standards require this dual-brake system to provide hydraulic system operation in case one hose, line, or wheel brake assembly loses fluid. Because the brake hydraulic system is sealed, all the lines and cylinders are full of fluid at all times. When the master cylinder develops system pressure, the amount of fluid moved is only a few ounces.

Split Systems. Modern-day vehicles have split brake systems. The pre-1970s vehicle had a single hydraulic system serving all four wheels. A leak anywhere in the system resulted in a complete braking failure. The split system was designed to prevent a total system failure. This required the use of a dual-piston master cylinder and the inclusion of various valves. A split system is fed by one piston

Figure 1-7 The master cylinder is a cylindrical pump with two pistons that develop pressure in the hydraulic lines to the front and rear brakes.

Figure 1-8 Each master cylinder piston feeds one system of a split hydraulic brake system. Shown is a diagonally split system.

in the master cylinder and feeds two wheel brakes of the vehicle. There are two types of split systems: diagonal and front/rear. The diagonal has one system feeding a front-wheel brake and the rear, opposing-side wheel brake, that is, left front and right rear (Figure 1-8). The second diagonal split is to the other wheel brakes. The front/rear split is exactly as it sounds. One side or split feeds the rear-wheel brakes and the other feeds the front wheels (Figure 1-9). Both types of split have advantages and disadvantages, but each prevents complete system failure from a single leak.

Figure 1-9 A front/rear split system.

Slot for mounting

Mounting clip

Caliper connection

Connection between steel and flexible hose

Flexible hose

Figure 1-10 A flexible hose provides a connection between the vehicle's rigid frame and the movement of the wheel and suspension assemblies.

Brake Lines and Hoses. The rigid lines or pipes of a brake hydraulic system are made of steel tubing for system safety. Flexible rubber hoses connect the wheel brakes to the rigid lines on the vehicle body or frame (Figure 1-10). The front brakes have a rubber hose at each wheel to allow for steering movement. Rear brakes may have separate hoses at each wheel or a single hose connected to a line on the body or frame if the vehicle has a rigid rear axle. Brake lines and hoses contain the high-pressure fluid, and the fluid acts as a solid rod to transmit force to the wheel cylinders and caliper pistons.

Pressure Control Valves. Almost all brake hydraulic systems built since 1967 have one or more valves to control system pressure. Metering and proportioning valves modulate hydraulic pressure to front disc or rear drum brakes to provide smooth brake application and reduce the tendency to lock the brakes. A pressure differential switch is a type of valve used in most systems to turn on the instrument panel warning lamp if half of the hydraulic system loses pressure.

The hydraulic system may have several individual valves or a single combination valve with multiple functions (Figure 1-11). You learn about pressure control valves in detail in subsequent chapters.

Although pressure control valves have been part of brake systems for more than 40 years, **antilock brake systems (ABSs)** may make some valves obsolete. An ABS electronic control module can modulate hydraulic pressure for normal braking better than metering and proportioning valves can. As ABS installations become more widespread, some older hydraulic functions may be given over to the electronic computer.

Wheel Cylinders and Caliper Pistons. Technically, the **wheel cylinders** of drum brakes and the caliper pistons of disc brakes are "slave" cylinders because they operate in response to the master cylinder. These hydraulic cylinders at the wheels change hydraulic pressure back into mechanical force to apply the brakes.

Most late-model systems with drum brakes have a single, two-piston cylinder at each wheel (Figure 1-12). Hydraulic pressure enters the cylinder between the two pistons and forces them outward to act on the brake shoes. As the shoes move outward, the lining contacts the drums to stop the car. Wheel cylinder construction and operation of drum brakes are covered in Chapter 8 of this manual.

The caliper pistons for disc brakes also act in response to hydraulic pressure that enters a fluid chamber in the caliper. Hydraulic pressure in a stationary caliper is applied to one or two pistons on each side of the caliper to force the pads against the rotor as shown in Figure 1-5.

An **antilock brake system (ABS)** is a service brake system that modulates hydraulic pressure to one or more wheels as needed to keep those wheels from locking during braking.

A **wheel cylinder** is the hydraulic slave cylinder mounted on the backing plate of a drum brake assembly.

Figure 1-11 This combination valve performs three hydraulic functions and switches on the brake warning light within a single housing.

Figure 1-12 Hydraulic pressure in the wheel cylinder moves the two pistons outward to force the shoes against the drum.

Pressure is applied to a single piston in a movable caliper on the inboard side to force the inboard pad against the rotor. As explained later in Chapter 2, hydraulic pressure is equal in all directions in a sealed chamber. This equal pressure creates a reaction force that moves the outboard side of the caliper inward so that both pads grip the rotor as shown in Figure 1-6. You learn more about caliper construction and operation of disc brakes in Chapter 7 of this manual.

Figure 1-13 The power brake booster increases the brake pedal force applied to the master cylinder.

Power Boosters

Almost all late-model brake systems have a power booster that increases the force of the driver's foot on the pedal (Figure 1-13). Most cars and light trucks use a vacuum booster that uses the combined effects of engine vacuum and atmospheric pressure to increase pedal force. Some vehicles have a hydraulic power booster that may be separate from the brake system and supplied with fluid by the power steering system, or may be a part of the brake system and driven by an electric motor. Chapter 6 in this *Classroom Manual* explains power boosters in detail.

Parking Brakes

After the service brakes stop the moving car, the parking brakes help to hold it stationary. Parking brakes are often mistakenly called "emergency" brakes, but their purpose is not to stop the vehicle in an emergency. The amount of potential stopping power available from parking brakes is much less than from the service brakes. Because the parking brakes work on only two wheels or on the driveline, much less friction surface is available for braking energy. In the rare case of total hydraulic failure, the parking brakes can be used to stop a moving vehicle, but their application requires careful attention and skill to keep the vehicle from skidding or spinning.

Electronic Braking Systems

ABS. Electronic braking covers all systems from antilock brake systems (ABSs) and Traction Control Systems (TCSs up to and including Vehicle Stability Systems [VSSs]). While there are different terms used by the various manufacturers for the same operation, all operate with electronic sensors and actuators with little or no input from the vehicle operator. Some systems like VSS are the first stages of development and application but all of the ones discussed in this text will soon be offered as standard equipment on most vehicles. They will only become more complex and more common over the next several years.

Whenever the brakes are applied with heavy pressure, the wheel may totally stop rotating. This condition is called wheel **lockup** or **negative wheel slip,** which does not help the car stop.

Lockup or **negative wheel slip** is a condition in which a wheel stops rotating and skids on the road surface.

Rather, the tire loses some frictional contact with the road and slides or skids. As the tires slide, the car is no longer stopping under control and the driver is in a dangerous situation. Experienced drivers try to prevent wheel lockup by pumping the brake pedal up and down rapidly. This stops and starts hydraulic pressure to the brakes and gives the driver control during hard braking.

Most late-model cars have an ABS. The ABS does the same thing as an experienced driver would, only it does it faster and more precisely. It senses when a wheel is about to lock up or skid. It then rapidly interrupts the braking pressure to the brakes at that wheel. Speed sensors at the wheels monitor the speed of the wheels and send this information to an on-board computer. The computer then directs the ABS unit to pulse the pressure going to the wheel that is starting to lock up.

Active Braking

Mass-produced **active braking** systems were first introduced by Ford in 1999. In the context here, "active" means the brake system will perform some functions without input from the operator. The ABS could be considered as the first active braking system, but it functioned only if the driver had applied the brakes and certain conditions, such as wheel skid, were present. The active braking systems of today take it one step further. The active braking system works with components from the ABS, **traction control system (TCS),** and **automatic ride control (ARC).** Most of those components are **wheel speed sensors, yaw** sensors, **actuators,** and the shared wiring. Each of the systems listed normally has its own controller or computer module. Before going further, it is best to understand what each of the noted systems does.

The TCS controls wheel spin or positive wheel spin during hard acceleration or slippery road conditions. This is accomplished by shifting the power from a "spinning" drive wheel to a wheel with more traction or by reducing engine power. Shared components among active braking, ABS, and TCS are the wheel speed sensors and hydraulic modulator. TCS is an outgrowth of the ABS.

ARC is an electronic suspension system designed to provide a more comfortable ride to the passengers and allow better control of the vehicle during cornering. To some extent it can command reduced engine power, thereby slowing the vehicle. The major component shared between ARC and active braking is the yaw sensor.

Active braking systems take the data from shared components, perform certain calculations, and apply the brakes to one or several wheels without any driver input at all. This, in effect, can help control the vehicle during some cornering and steering situations. It can be expected that, as electronic steering systems become more available, more cooperation between systems will increase and active braking will be more enhanced. There are more details on active braking in Chapter 10.

Wheel speed sensors are mounted at selected (or all) wheels to monitor wheel speed during vehicle operation.

Yaw is the deviation in the line of travel commonly referred to as roll or lean during cornering.

Trailer Brakes

Normally a textbook of this type does not deal with trailer brakes. However, with the advent of active braking systems and the increasing sales and usage of ¾- and 1-ton pickups, it is important that some information be provided. Consult local and state laws for specific requirements.

Older trucks used to tow trailers heavy enough to warrant a brake system always used an add-on system. This system was not efficient in some cases, and its efficiency in many cases was based on "you got what you paid for." If the trailer was lightly loaded or the road was slick, the trailer brakes could lock and the trailer would probably jack-knife. Heavy trucks over 1 ton did have a trailer brake system that was either "hard-wired" at the factory or at least had to meet stringent Department of Transportation (DOT) standards. The newest ¾- and 1-ton truck models offer a factory-installed trailer brake system. This system borrows from the ABS and active braking systems. Trailer brake systems are usually divided into two types: hydraulic surge and electric.

Hydraulic Surge

A hydraulic surge brake system is completely mounted on the trailer. It may be a disc or drum type and operates hydraulically and mechanically in a very similar manner to the drum brakes on the tow vehicle. There is a pressure differential valve mounted somewhere in the tow vehicle's rear brake system, usually at or near the vehicle's master cylinder. This valve operates so the pressure supplied to the trailer's master cylinder is in direct proportion to the pressure being directed to the tow vehicle's rear brake. The valve works very similarly to the regular proportioning valve found on almost every roadworthy vehicle. The pressure delivered to the trailer's master cylinder applies a mechanical pushrod, which, in turn, creates the correct pressure to apply the trailer brake. Trailer drum brakes may be uniservo with one wheel cylinder and one pushrod or duo servo with one wheel cylinder and two pushrods. The uniservo type is the most common because of cost. The disc-brake type operates almost exactly like the disc brakes on a standard vehicle.

Electric Brakes

Electric trailer brakes are most commonly used on utility and RV units, whereas boat trailers or others that are designed to be submerged underwater use the hydraulic surge. This is because water severely shortens the life expectation of electric brake components.

A trailer electric brake system requires a brake controller mounted in the tow vehicle within hand control of the vehicle driver. It is usually mounted on the dash near the steering column (Figure 1-14). The driver may reduce or increase the power of the trailer brakes or lock them down as a kind of parking brake. The controller is not readily switched between different tow vehicles, so each tow unit must be equipped with its own controller. There are three types of controllers, but each performs basically the same function. For detailed information on the controllers and troubleshooting trailer brakes, investigate Champion Trailers at its Web site at http://www.championtrailers.com/techsup.html. The regulated electric power from the controller is sent to magnets within the wheel drum. The energized magnets are pulled toward a specially machined flat surface of the drum. As the magnets move, they, in turn, move a lever(s) attached to the brake shoe. The brake shoe is applied against the rotating drum in direct proportion to the braking action of the rear brakes of the tow vehicle.

Trailer Breakaway Condition

The DOT requires that any trailer with its own braking system have a method to apply its brakes in case the trailer disconnects from its tow vehicle. On an electric system, an emergency power battery is mounted on the trailer. If a breakaway condition occurs, the pull pin at the trailer/vehicle hitch is pulled loose, triggering a switch that connects the full power of the emergency battery to the wheel magnets. This effectively locks the wheels. The hydraulic surge system uses a slightly different method to accomplish the same result. A chain or cable runs from the tow vehicle to a lever positioned to apply force to the piston in the trailer's master cylinder. When a disconnect situation occurs, the lever is moved by the chain/cable and applies the trailer brakes. In either case, both breakaway systems may be used as parking brakes when the trailer is stored or being loaded/unloaded.

Reverse Braking

Trailer brakes do not recognize reverse braking from forward braking. Usually it is best if the trailer does not have braking during reverse operation. To prevent this, the hydraulic surge brake

Figure 1-14 This is a typical add-on trailer brake controller.

Standard　　　　　　　　　**Free Backing**

Figure 1-15 The free backing brake on the right recognizes reverse movement and the brake cannot be aplied. (Courtesy of Champion Trailer Parts Supply)

may be standard or free backing (Figure 1-15). The standard type requires a reversing solenoid that is triggered when the tow vehicle backup lights are activated (Figure 1-16). The solenoid is plumbed into the system at the outlet of the trailer's master cylinder. With the solenoid energized, pressurized fluid from the master cylinder is blocked from the wheels and is directed back into the master cylinder's reservoir. Free backing operation recognizes that the wheels are rotating backward and deactivates the brakes. The brake system reactivates when the wheels rotate forward. This is the system of choice for most trailers because it can be manufactured into the system. It should be noted that disc brakes on trailers must have a free backing (reversing)

Figure 1-16 The reversing solenoid is needed with the standard brakes in Figure 1-15 to prevent braking in reverse. (Courtesy of Champion Trailer Parts Supply)

solenoid. Electric brake systems use only a reverse solenoid to prevent the trailer brakes from functioning in reverse. The solenoid is triggered by the backup lights circuit.

 AUTHOR'S NOTE: Technicians or do-it-yourselfers can get into difficulties if they do not understand how to connect an add-on trailer brake system. Also, if the add-on system is on the less expensive side, the rule "you get what you pay for" applies. There are some add-on units that meet DOT standards but have no leeway if the trailer is loaded or too heavy for the brake system.

AUTHOR'S NOTE: Go to http://www.ford.com for information on one of the first production factory-installed trailer brake systems.

Before installing an add-on system, read the specifications carefully and talk to the customer about the weight and number of axles on the trailer. On multiple-axle trailers, braking components may be installed at each wheel or only on one axle. The total weight of the loaded trailer, not just the payload of one axle, must be considered when selecting the capacity of the braking system.

AUTHOR'S NOTE: I bought a used 1999 Dodge Dakota in 2000 at very low cost. The tailgate and fender were crushed in on the left side. Because I am curious (and cheap), I convinced the salesperson to get me in contact with the previous owner. The owner had bought the Dakota because it pulled his 32-foot cabin cruiser easily on the road and was reasonable on fuel. However, one day he had to make an emergency downhill stop on the way to the boat ramp. Because the boat and trailer were too much for the little truck, the trailer lifted the back of the truck. The ABS on the truck was working, but there was no ABS on the trailer. In a classic example of lost friction, the boat and trailer tried to pass the truck. The boat made it to the truck's left rear fender before crushing a hole in its side and damaging the fender and tailgate. So the owner traded up to a Dodge B3500 with dual wheels and a Cummins diesel—one extreme to another. Moral of the story: If your customer is looking for a Dakota-size truck to pull the cabin cruiser, you would do well to brief him on the difference between towing power and stopping power.

Summary

- ❏ An automotive brake system consists of a master cylinder connected hydraulically through lines to disc and drum brake units that stop the wheels.

- ❏ A hydraulic or vacuum power assist is used on most cars to decrease the braking effort required from the driver.

- ❏ A mechanical brake, operated by levers and cables, is used for parking.

- ❏ Many cars have an ABS to improve brake operation during emergency stopping.

- ❏ Active braking systems may function with or without driver input.

- ❏ Active braking systems share components with the ABS, TCS, and ARC to provide better vehicle control and comfort.

- ❏ Trailer brakes may now be controlled or regulated by components similar to those used in active braking systems.

- ❏ Regardless of the type of braking system, mechanical/hydraulic or electronic, the final braking action occurs between the tire and the road.

Review Questions

Short-Answer Essays

1. List the main parts of a basic brake system and explain their purposes.

2. Discuss the general operation of drum brakes. Confine the answer to the components mounted at the wheel.

3. Discuss the operation of disc brakes. Confine the answer to the components mounted at the wheel.

4. Explain why a parking brake system is not viable for stopping a moving vehicle.

5. Explain the reasons for having a split brake system and the two types of split systems.

6. List some of the components shared among active braking systems, ABS, TCS, and ARC.

7. Explain the purpose of flexible hoses within the brake system.

8. Discuss the general interaction between a towing vehicle and its trailer brake.

9. Explain how the outboard disc brake pad is applied against the rotor with a typical movable caliper.

10. Explain why ABS is an important system in controlling a vehicle.

Fill in the Blanks

1. A brake system can have a _____ or _____ type of power assist.

2. An _____ prevents tires from skidding during braking.

3. The _____ _____ system is designed to hold the vehicle stationary.

Terms to Know

Active brakes
Actuators
Ampere (A)
Antilock brake system (ABS)
Automotive ride control (ARC)
Caliper
Disc brake
Drum brake
Force
Friction
Fulcrum
Leverage
Lockup
Master cylinder
Parking brakes
Pressure
Service brakes
Traction control system (TCS)
Vacuum
Weight
Wheel cylinder
Wheel speed sensors
Yaw

4. On a 2:1 ratio, the driver's input of 75 pounds of force is boosted to

   _____ pounds of force.

5. During braking, the driver's input mechanical force is converted into

   _____ pressure by the master cylinder.

6. The lines of a brake system are primarily designed for _____

   _____.

7. Brake valves are used to provide _____ _____

   _____ and reduce wheel lockup.

8. Active braking systems share components with ABS, ARC, and _____.

9. Trailer brakes do not recognize _____ from forward braking.

10. A breakaway condition exists when the _____ from the tow vehicle.

Multiple Choice

1. Parking brake systems may share components with which of the listed brake systems?
   A. driveline brake
   B. disc brakes
   C. drum brakes
   D. both B and C

2. The brake pedal mechanism is
   A. always mounted from the top.
   B. a fulcrum.
   C. a lever.
   D. both A and B.

3. On a pedal mechanism with a 4:1 ratio, applying 250 pounds of force and moving the brake pedal 2 inches results in a pushrod movement of
   A. 8 inches.
   B. 0.5 inch.
   C. 0.25 inch.
   D. 2.5 inches.

4. Disc brake calipers may (are)
   A. be fixed on their mounts.
   B. be moveable on their mounts.
   C. mechanically activated.
   D. both A and B.

5. The typical mechanical components of a hydraulic brake system include
   A. pedal, master cylinder, wheel cylinders.
   B. pedal, disc calipers, control valves.
   C. brake shoes, brake pads, master cylinder.
   D. pedal, brake pads, brake drums.

6. The brake caliper
   A. sits above/over the rotor.
   B. is fixed with a single pad.
   C. uses a moveable caliper support.
   D. all of the above.

7. Split brake systems require
   A. dual-piston master cylinders.
   B. separate lines from the master cylinder to each wheel.
   C. control valves.
   D. both A and C.

8. An Active Brake System normally works with components of the
   A. ABS.
   B. TCS.
   C. ARC.
   D. all of the above.

9. The two electronic systems that may restrict engine performance are
   A. ABS and TCS.
   B. ABS and ARC.
   C. TCS and ARC.
   D. none of the above.

10. The major hydraulic components of a typical light vehicle brake system include
    A. master cylinder.
    B. wheel cylinder.
    C. valves.
    D. all of the above.

Principles and Theories of Operation

Upon completion and review of this chapter, you should be able to:

❏ Discuss braking dynamics.

❏ Discuss the conversion of energy from one type to another.

❏ Explain how work is accomplished.

❏ Explain the importance of kinetic and static friction in a brake system.

❏ Describe the effects of pressure, surface area, and friction material on producing friction during braking.

❏ Discuss coefficient of friction.

❏ Explain why heat dissipation is important in a braking system.

❏ Explain how the laws of motion affect the design and operation of a vehicle.

❏ Discuss hydraulic principles and how they may be applied in a vehicle.

❏ Explain the method for calculating mechanical advantage.

❏ Discuss how hydraulics can be used to transmit force.

❏ Discuss how a hydraulic brake system can produce mechanical advantage.

❏ Define and explain the basic electrical terms: amperes, voltage, and resistance.

❏ Explain how to use Ohm's law to calculate the amount of resistance, voltage, and current in a circuit.

Introduction

No vehicle or any other device will function correctly if the laws of physics are not considered during design and manufacturing. On the other hand, there is no current technology to make *all* designs meet *all* factors of physics. The automotive industry has performed well in making compromises between different physics laws to design and manufacture vehicles that work and are affordable. In this chapter, various laws of physics are discussed so that the reader will have a basic understanding of the physical environment that the designers, the manufacturers, the driver, and even yourself must overcome to operate any type of vehicle.

Brake Operation

Before discussing the energy and dynamics associated with braking, a review of brake system operation is essential. This may help when trying to understand the next sections.

When the driver decides to slow or stop the vehicle, the first action is to remove force from the accelerator pedal. The slowing engine, in most cases, begins to slow the vehicle. The driver then applies force to the brake pedal.

The movement of the brake pedal activates the brake power booster, which, in turn, boosts and transmits the driver's input force to the master cylinder pistons via pushrods (Figure 2-1). The master cylinder pistons turn mechanical action into hydraulic action by compressing the brake fluid and forcing it out through the hoses and lines to the wheel braking components (see Chapter 5). Valves between the master cylinder and wheels control the pressure and volume of brake fluid reaching the wheels. It should be understood that these valves do not control actual braking force, but mainly divide the force between the front and rear brakes for a smooth stop.

Force converted to hydraulic pressure

Applied force

Hydraulic pressure converted to mechanical action (force) at each wheel

Figure 2-1 The energy exerted to brake a vehicle is converted from mechanical to hydraulic and back to mechanical.

At the wheels, hydraulic pressure is converted into mechanical action and applies the last segment of the braking mechanism. In drum brakes, this action is accomplished by the wheel cylinders, brakes shoes, and brake drums. On disc brakes, the same action is performed by the calipers, brake pads, and brake rotors. The antilock brake system (ABS) becomes involved only if it senses one (or more) wheels locking up and sliding. The ABS can then modulate hydraulic pressure to that wheel(s) to reduce braking effects. All told the braking system is a straightforward and simple concept and operation, but it is one of the more critical systems on a vehicle.

Brake System Energy

All brake systems work according to a few principles or "laws" of physics, and the concept of energy is a basic part of physical science. Energy is the ability to do work and comes in many familiar forms: chemical energy, mechanical energy, heat energy, and electrical energy are among the most obvious forms in all automotive systems.

A brake system converts one form of physical energy to another. To slow and stop a moving vehicle, the brakes change the **kinetic energy** of motion to heat energy through the application of friction. When the brakes change one form of energy to another, they are doing work. Work is the result of releasing or using energy.

Kinetic Energy, Mass, Weight, and Speed

Kinetic energy is the energy of mechanical work or motion. When an automobile starts, accelerates, decelerates, and stops, kinetic energy is at work. The amount of kinetic energy at work at any moment is determined by a vehicle's mass (weight) and speed and the rate at which speed is changing.

Kinetic energy is the energy of mechanical work or motion.

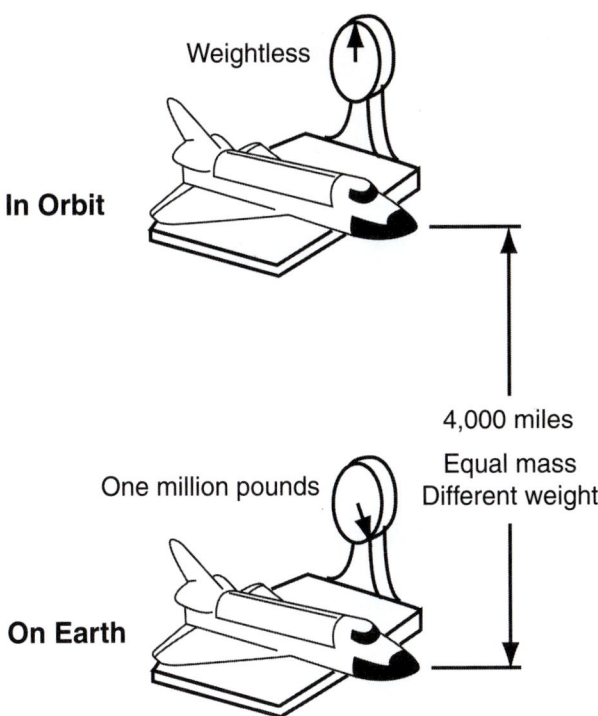

Figure 2-2 The space shuttle has equal mass but different weights on Earth and in orbit.

We can use the terms "mass" and "weight" interchangeably to describe objects on the surface of the Earth, but the two terms are not technically the same. **Mass** is a measurement of the number of molecules that make up an object. **Weight** is a measurement of the effect of gravity on that mass. All objects have mass, from a steel brake shoe to a quart of hydraulic fluid to the air in an air compressor. Without going too deeply into the science of physics, we can say that the greater the number of molecules in an object and the more complex the molecules are, the greater the mass of that object and the more dense it is. The effect of gravity on the mass of an object is that object's weight.

The basic difference between mass and weight can be understood by thinking of the space shuttle, which weighs about 1 million pounds on the launch pad, on Earth. When the shuttle is in orbit, outside Earth's gravity, it is weightless (Figure 2-2). Its mass stays the same, however.

The combined effects of weight and speed constitute kinetic energy, but speed has a much greater effect than weight. You can calculate the kinetic energy of any moving object with this formula, which is really quite simple:

$$\frac{mv^2}{29.9} = Ek$$

where m = mass (weight) in pounds
        v = velocity (speed) in miles per hour
       Ek = kinetic energy in foot-pounds

Consider two cars, both traveling at 30 miles per hour (mph). One weighs 2,000 pounds; the other weighs 4,000 pounds (Figure 2-3).

$$\frac{2,000 \times 30^2}{29.9} = 60,200 \text{ foot-pounds of kinetic energy for the lighter car}$$

$$\frac{4,000 \times 30^2}{29.9} = 120,400 \text{ foot-pounds of kinetic energy for the heavier car}$$

Mass is the measure of the inertia of an object or form of matter or its resistance to acceleration; it also is the molecular density of an object.

Weight is the measure of the Earth's gravitational force, or pull, on an object.

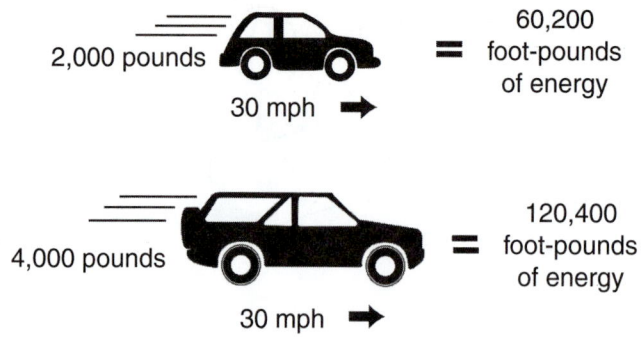

Figure 2-3 Kinetic energy increases proportionally with vehicle weight.

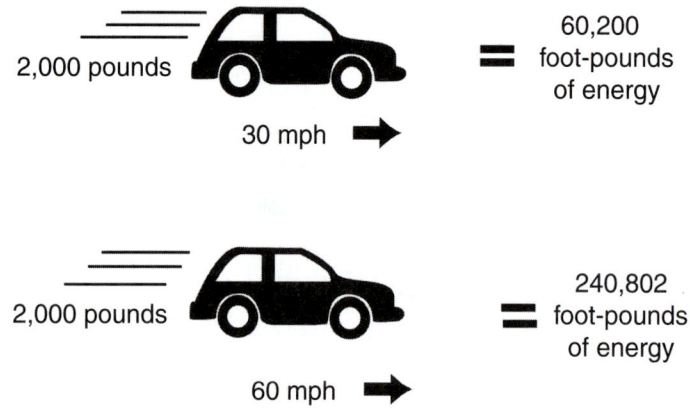

Figure 2-4 Kinetic energy increases exponentially with vehicle speed.

Doubling the car weight doubles the kinetic energy when the speeds are equal. Therefore, kinetic energy increases and decreases proportionally with weight. Now let us accelerate the lighter car to 60 mph (Figure 2-4).

$$\frac{2,000 \times 60^2}{29.9} = 240,802 \text{ foot-pounds of kinetic energy for the lighter car}$$

When we double the speed, kinetic energy increases not two times, but four times. Kinetic energy increases as the square of the speed. If we accelerate the same 2,000-pound car to 120 mph, its kinetic energy is 16 times greater than it was at 30 mph—almost 1,000,000 foot-pounds.

Remember the brake performance test described in Chapter 1 of the *Shop Manual,* in which the required stopping distance at 30 mph was 57 feet, but increased to 216 feet at 60 mph. That is a practical example of kinetic energy at work. The effects of kinetic energy are also why a high-performance car needs a brake system with much greater stopping power than an economy car with only modest performance.

Inertia and Momentum

Inertia is the tendency of an object in motion to keep moving and the tendency of an object at rest to remain at rest.

When we accelerate a car and then decelerate it and bring it to a stop, we are dealing with two forms of **inertia.** Inertia is simply the resistance to a change in motion. An object at rest tends to remain at rest; an object in motion tends to remain in motion. In both cases, the object resists any change in its motion.

Static inertia is the inertia of an object at rest; dynamic inertia is the inertia of an object in motion. When the brake system slows and stops a vehicle, it overcomes dynamic inertia and imposes static inertia. The brake system also must overcome the vehicle's momentum. **Momentum** is another way to view kinetic energy at work because it, too, is the mathematical product of an object's weight times its speed. Physical force starts an object in motion and gives it momentum. Another kind of force must overcome the momentum to bring the object to a stop. That force is friction.

Braking Dynamics

One important brake system dynamic is called weight transfer. You have probably had the experience of applying the brakes hard for an emergency stop. During such a stop, the front of the car lowered and you could feel yourself being thrown forward against your seat belt. This is caused by the vehicle weight being transferred from the rear to the front during braking, which means that more of the braking must be done by the front wheels and less by the rear wheels.

The weight of a car is not distributed evenly on all four wheels even when the vehicle is standing still. The position of the heavy engine and powertrain components determines weight distribution. During braking, the weight of these components transfers forward. On a rear-wheel-drive car, about 70 percent of the weight ends up on the front. On a front-wheel-drive car, as much as 90 percent of the car weight ends up on the front during braking. The vehicle's momentum and weight combine to cause the rear wheels to lift (have less down force applied) and the front wheels to be forced down (Figure 2-5). This is why front brake systems are larger than rear units.

Another dynamic to think about is braking power. As the previous section of this chapter about kinetic energy explained, the more the car weighs and the faster it is moving, the greater the braking power must be to stop the car. The brake system is actually designed with more power than the engine. Consider a typical car with a 200-horsepower engine. This car may be capable of accelerating from 0 to 60 mph in 10 seconds. The brake system must be able to decelerate the car from 60 to 0 mph in nearly one-fifth that time. This means that the comparable power to stop this car is about 1,000 horsepower.

This chapter focuses on the friction between the brake linings and the brake rotor or drum. We need to remember, however, that the vehicle is actually stopped by the friction between the tire and the road. Large, effective brake assemblies will stop the wheels from rotating, but the car will stop only if the tires maintain traction with the road.

When the brake system completely stops the tire so that it can no longer maintain traction with the road, the tire locks up or skids. A car with locked brakes takes much longer to stop because hot molten rubber can form between the tire and the road, causing a lubricating effect.

Figure 2-5 Weight transfer during braking can place as much as 90 percent of the vehicle weight on the front brakes.

Steering control can be lost, sending the car out of control. Even if the car stops before hitting something, lockup can damage the tires by causing flat spots on them.

Friction Principles

Service brakes use friction to stop the car. The parking brakes use friction to hold the car stationary. Friction is used in the wheel brake units to stop the wheels. The friction between the tires and the road stops the car. Because friction is so important to the brake system, it is necessary to understand some of its principles.

Kinetic and Static Friction

Two basic types of friction are at work in the brake system (Figure 2-6). The first is called kinetic, or moving, friction. The second is called static, or stationary, friction.

When the brakes are applied on a moving car, the frictional parts (brake shoes or pads) are forced against the rotating parts of the car (brake rotors or drums). The friction causes the rotating parts to slow down and stop. Just as the energy of the rotating parts is called kinetic energy, the friction used to stop them is called **kinetic friction.**

Kinetic friction changes kinetic energy into **thermal energy** (heat). For a simple example of kinetic friction changing kinetic energy to heat, rub the hands together. Rub them together fast enough and the palms will feel warm.

Brake parts and tires get very hot during braking because of kinetic friction. For example, the temperature of the brake friction material in a typical car going 60 mph (95 km/h) can be more than 450°F (230°C) during an emergency stop.

Static friction holds the car in place when it is stopped. The friction between the applied brake components and between the tire and the road resists movement. To move the car, the brake components must be released. Then, the power of the engine must be great enough to overcome the **rolling resistance** between the tires and the road. The car can then begin to move.

Friction and Pressure

One important factor in the amount of friction developed in a braking system is the amount of pressure used to force the friction material against the rotating brake part. The more pressure applied to two frictional surfaces, the harder they grip each other and the harder they resist any movement between them.

Figure 2-7 shows the basic frictional parts of a disc brake system. The rotor is the part connected to, and rotating with, the car wheel. The friction pads are forced against the frictional surfaces or sides of the rotor. The friction causes the rotor to slow and stop, which stops the wheel from turning.

Kinetic friction is the friction between two moving objects or between one moving object and a stationary surface. It is used to stop, control, or reduce kinetic energy.

Thermal energy is the energy of heat.

Static friction is friction between two stationary objects or surfaces.

Rolling resistance is a combination of inertia of the mass to moving, including the friction between all moving parts of the vehicle and the outside environment.

Figure 2-6 Both static and kinetic friction are at work during braking.

Figure 2-7 The greater the pressure against the friction pads, the greater the braking force.

Braking force in the brake system is achieved by using mechanical leverage, hydraulic pressure, and different kinds of power-assist systems. The more force used to push the friction pads against the rotor, the more friction developed. The more friction developed, the more braking action—and heat—achieved.

Friction and Surface Area

Another important factor affecting the amount of friction produced is the amount of frictional surface being contacted. Two hands will stop a revolving shaft faster than one hand. The larger of the two brake units shown in Figure 2-8 has more frictional contact area than the smaller unit. The larger unit will stop a car faster than the smaller unit, which is why large cars and trucks require larger brake components than do small cars.

Coefficient of Friction

It takes more force to move some materials over a surface than others, even though the applied pressure and the amount of surface in contact are the same. Different materials have different frictional characteristics or coefficients of friction. It is easier to slide an ice cube over a surface than it is a piece of sandpaper.

We can calculate the **coefficient of friction** by measuring the force required to slide an object over a surface and then dividing it by the weight of the object. The moving force that slides

Coefficient of friction is a numerical value that expresses the amount of friction between two objects, calculated by dividing tensile force (motion) by weight force; it can be either static or kinetic.

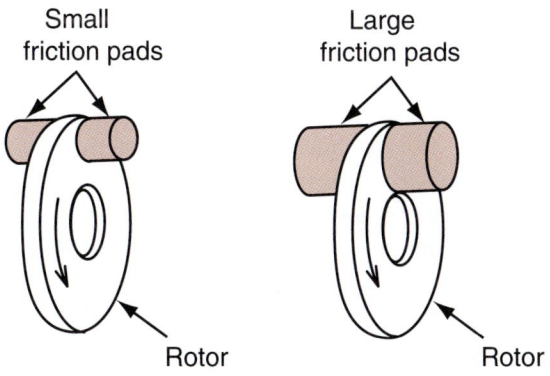

Figure 2-8 The larger the frictional contact area, the greater the braking force.

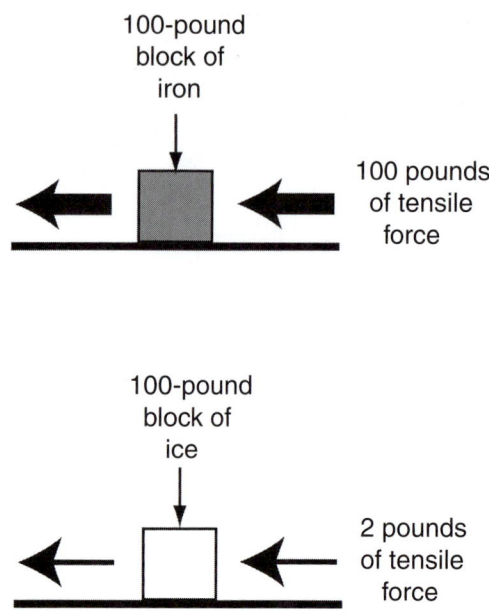

100-pound
block of
iron

100 pounds
of tensile
force

100-pound
block of
ice

2 pounds
of tensile
force

Figure 2-9 A high coefficient of friction requires greater force to move one surface against another.

an object over a surface is **tensile force.** The weight of the object is the force pushing the object against the surface.

As an example, assume it takes about 100 pounds of force to slide a 100-pound block of iron over a concrete floor, but only about 2 pounds of force to slide a 100-pound block of ice over the same surface (Figure 2-9). When the amount of force is divided by the weight of the object, we find that the coefficient of friction for the metal block is 1.0, whereas the coefficient for the ice block is 0.02.

$$\frac{\text{Tensile force (motion)}}{\text{Weight force of object}} = \text{coefficient of friction}$$

$$\frac{100 \text{ pounds tensile force}}{100\text{-pound iron block}} = \text{coefficient of friction: } 1.0$$

$$\frac{2 \text{ pounds tensile force}}{100\text{-pound ice block}} = \text{coefficient of friction: } 0.02$$

The results of these calculations provide the coefficient of friction between two materials. The coefficient of friction between brake friction components is always less than 1.0. A coefficient of friction greater than 1.0 means that material actually is transferred from one surface to another. Although brake friction surfaces wear, material is not transferred from pads to rotors or from shoe linings to drums. The coefficient of friction between a tire and the road can exceed 1.0, however. When material transfers from the tire to the road, we see a skid mark. That means that the coefficient of friction momentarily exceeded 1.0.

Designing or selecting brake materials in relation to the coefficient of friction is not a simple matter of picking the highest number available. If the coefficient of friction of a material is too high, the brakes may grab and cause the wheels to lock up and slide. If the coefficient of friction is too low, excessive pressure on the brake pedal would be required to stop the car.

Automotive engineers carefully determine the required coefficient of friction for the best braking performance on a particular vehicle. The friction materials selected for brake pads and rotors or brake shoes and drums are designed to give the best cold and hot temperature performance. The best materials have a coefficient that stays within narrow limits over a wide range of temperatures. If the selected material increases or decreases its coefficient of friction as temperature changes, then the brakes can fade or grab or work erratically. This illustrates an important reason to use approved replacement brake parts so that the brake performance will meet the system design requirements.

Three factors affect the coefficient of friction in a brake system:

1. The surface finish of both friction surfaces
2. The material: the metal of the rotors and drums and the friction material of the pads and linings
3. Temperature

Surface Finish. The 100-pound metal block previously used to calculate a basic coefficient of friction can be used to show how friction can be easily changed. By polishing the metal surface and waxing the concrete floor, the required tensile force to move the block may be changed from 100 pounds to 50 pounds. The metal block still weighs 100 pounds, but the coefficient of friction is now:

$$\frac{50 \text{ pounds tensile force}}{100\text{-pound metal block}} = 0.5 \text{ coefficient of friction}$$

In a brake system, the surface finishes of pads, rotors, linings, and drums must be a trade-off between smoothness for long life and roughness for good stopping power. Friction surfaces as rough as coarse sandpaper could provide excellent stopping action—for just a few brake applications.

Materials. The kinds of materials being rubbed together also have a big effect on the coefficient of friction, as the earlier comparison of a 100-pound block of metal with a 100-pound block of ice shows. Just as it is with selecting a surface finish, selecting the kinds of brake friction materials is a trade-off. Iron and steel work best as rotors and drums because of their combined characteristics of strength, their ability to hold a surface finish for a long time, and their ability to transfer heat without being harmed by high temperatures.

Friction materials for pads and linings can be allowed to wear but must have a reasonable lifespan. If pads and linings are soft and wear fast, the vehicle may have very good braking ability, but friction material life will be short. However, if friction material is hard and wears slowly for long life, it will create a low coefficient of friction and poor stopping ability.

Asbestos was used almost universally for decades as the basis of brake friction material. The health hazards of asbestos led to its removal from most pads and linings. Today, most friction materials are semimetallic or organic compounds. All have different coefficients of friction, and engineers select material that best matches the intended use of a vehicle and its braking requirements.

Temperature. Temperature also affects the coefficient of friction, but it affects different materials in different ways. A moderate amount of heat actually increases the coefficient of friction of most brakes. The semimetallic and carbon fiber materials of some racing brake linings must be heated quite a bit for maximum efficiency.

Too much heat, however, reduces the coefficient of friction (Figure 2-10), and as heat increases more, the coefficient of friction continues to drop. This leads to brake fade, discussed later in this chapter, and braking efficiency is reduced.

Figure 2-10 Initially, the coefficient of friction increases with heat, but very high temperatures cause it to drop off.

Heat Dissipation. The weight and speed of the car during braking determine how much frictional energy is required to stop the vehicle. As mentioned previously, heat is a natural result of friction. The more frictional energy needed to stop the car, the greater the amount of heat generated during braking.

The heat caused by friction must be removed or it will damage the brake system. Heat is removed from the friction surfaces as it passes through the friction material and metal of the brake components into the surrounding air. This heat removal is called heat dissipation.

Brake Fade

The buildup of heat from continuous braking can cause the friction material to become glazed or polished and the metal surface of the rotors or drums to become hardened. As a result, the driver will need to push harder on the brake pedal to stop the car.

Brake drums and rotors must absorb heat faster than they can dissipate it to the surrounding air. Drum or rotor temperature can rise 100°F (55°C) in just a couple of seconds during a hard stop. Dissipating that same heat to the air can take 30 seconds or more. Repeated hard stops in a short period of time can overheat brake components severely and reduce braking effectiveness. This is called **brake fade.**

A brake system is designed to provide the best possible heat dissipation. Parts are often vented to allow maximum airflow around the hot surfaces. The sizes of the friction surfaces are carefully designed with heat dissipation in mind. The larger the frictional contact area of the brake system, the better the heat dissipation and its resistance to brake fade. Severe heat causes three kinds of brake fade:

1. **Lining fade.** Lining fade occurs when the pad or lining material overheats to a point at which its coefficient of friction declines severely. Pressing harder on the brake pedal may further overheat the lining and increase—not decrease—the fade. Lining fade affects both disc and drum brakes but is usually less severe with discs. The rotor friction surfaces are exposed directly to cooling airflow, whereas drum surfaces are not. Ventilated rotors also aid the cooling process and counteract lining fade.

2. **Mechanical fade.** Mechanical fade occurs with drum brakes when the drums overheat and expand away from the shoes. More pedal travel is then needed to move the shoes farther and let the linings contact the drum. Extreme mechanical fade can cause the pedal to travel to the floor before braking action starts. Pumping the brakes rapidly sometimes can force more fluid into the lines and move the shoes farther.

Large and heavy drums and drums with cooling fins help to dissipate heat and reduce mechanical fade. Mechanical fade is not a factor in disc brake operation.

3. **Gas fade.** Gas fade is a rare condition that occurs under very hard or prolonged braking. A thin layer of hot gas develops between the pads or linings and the rotors or drums. The thin layer of gas—although only a few molecules thick—acts as a lubricant and an insulator and reduces friction. Gas fade is a greater problem with friction materials that have large surface areas that trap the gas. Slots cut in pads and linings provide airflow that aids cooling and a route for hot gases to escape. Gas fade is one of the reasons there are signs at the top of some hills requesting truck drivers to select a lower gear before heading down.

Gas fade results from **adsorption,** which is the condensation of a gas on the surface of a solid.

Brake Friction Materials

The friction material for a disc brake is bonded, riveted, or molded to a metal brake pad (Figure 2-11). The friction material for a drum brake is bonded or riveted to a metal brake shoe (Figure 2-12).

Several different friction materials are used for brake pads and shoes. Each of these has different characteristics. Five of the most important characteristics are:

1. The ability to resist fading when the brake system temperature increases
2. The ability to resist fading when the parts get wet
3. The ability to recover quickly from heat or water fading

Figure 2-11 The friction material of a disc brake pad is bonded or riveted to the metal pad.

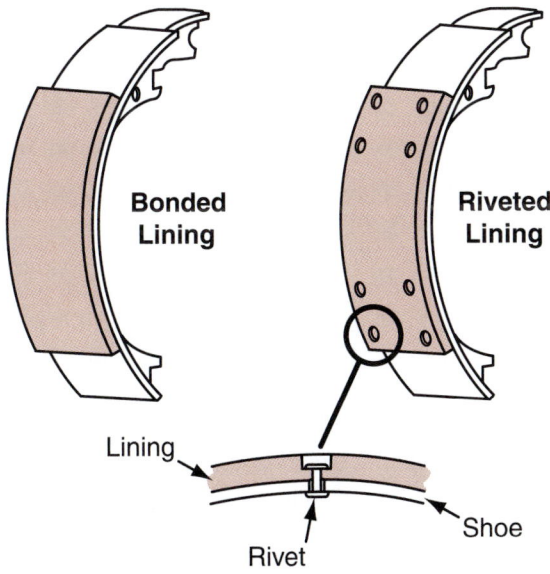

Figure 2-12 Drum brake lining also can be bonded or riveted to the metal shoe.

4. The ability to wear gradually without causing wear to brake rotors or drum surfaces

5. The ability to provide a quiet, smooth frictional contact with a rotor or drum

Asbestos was the most common brake lining friction material for many years. Asbestos has very good friction qualities and provides long wear but its health hazards caused its removal from most brake friction materials. The most common types of nonasbestos lining materials can be divided into four general categories:

1. *Nonmetallic materials* are made from synthetic fibers bonded together to form a composite lining.

2. *Semimetallic materials* are made from a mix of organic or synthetic fibers and metals molded together. Semimetallic linings are harder and more fade-resistant than nonmetallic materials but require higher brake pedal effort. They are often blamed for wear to the rotors and drums.

3. *Fully metallic materials* have been used for many years in racing. The lining is made from powdered metal that is formed by heat and pressure in a process called sintering. These materials provide the best resistance to brake fade but require high brake pedal pressure and create the most wear on rotors and drums.

4. **Ceramic** pads are based on clay or porcelain material blended with copper fibers. They provide suitable life, provide good stopping, and reduce chances of fading, but they retain heat. This heat can distort or damage other adjacent brake components.

Chapter 7 in this *Classroom Manual* provides more detailed information on friction materials used in disc and drum brake linings.

A BIT OF HISTORY

The first friction materials used in brake systems were wood, leather, and camel hair. These materials would burn or char easily, causing erratic brake action. In 1908, the Raymond Company of Bridgeport, Connecticut, developed an asbestos and copper wire mesh brake lining. This material worked much better than the other materials because it did not char and provided a smooth and grab-free pedal effort. The company later became known as the Raybestos Corporation.

Energy and Work

No matter how much a person may enjoy the task being performed, work is being done. Work is the expenditure of energy. Energy, in a loose definition, is the fuel for work. Energy cannot be created nor can it be destroyed. However, and this is the key, energy can be converted or changed from one form to another.

Many times we hear the phrase "that plant produces or makes electrical energy." Not really. The process is much simpler than making electrical energy, a task that no one can do at this time. The process is actually using the energy of falling water or heat to operate a mechanical device rotating within a magnetic field and converting that input energy into electrical energy. Although the mechanics and engineering required to perform that process are complex, the actual concept is very simple.

Let's move a little closer to home. The human body uses a lot of energy just to function without seemingly doing any work. The heart, lungs, and other organs work all the time. The body changes the chemical and heat energy of food into electrical energy for the brain and muscle control that, in turn, allows most organs to perform some type of chemical, electrical, or mechanical function. When the energy supply is low, the organs may become exhausted or unable to work. It is feeding time and the process continues. The same is true of vehicles.

The most common vehicle fuel of today is gasoline derived from crude oil. Crude oil is a by-product of all types of ancient animals that died, decayed, then heated under enormous pressure. This oil is recovered by the petroleum industry and eventually distilled into fuel for a vehicle. The original energy from the plants and animals has been converted into a chemical energy contained in the fuel. That chemical energy is compressed and burned in the engine and is converted into radiant heat and expanding gases. The gases push downward on a piston, and the last change in this process is the piston's mechanical action or energy. Run out of gas and everything stops.

There is one major drawback to all energy-conversion devices used today. It is a fact that a machine or device cannot produce more energy than it consumes, or, in other words, there is no **perpetual energy** device today. An example of a device that seems to contradict that statement is the alternating current generator or alternator on a vehicle. The magnetic fields in an alternator are both energized or charged by electricity from the vehicle battery or from the output current flow of the alternator. Typically a fully charged battery has about 13.2 volts of direct current, 13.2 volts to 13.8 volts with the engine operating. However, if the alternator's regulator fails and goes to "full field," it is possible for the alternator to produce up to 20 volts. This seems to prove that the alternator is a perpetual-energy device until the engine power needed to drive the alternator is considered. The end result is a large use of energy to produce a mere 20 volts of electrical current.

The preceding text is a simplistic overview of energy, but it shows that energy is neither destroyed nor created. It is just changed, and the changes can be used to do work. So whether you are engaging in a fun-filled activity or working the graveyard shift at the local factory, energy is being used and work is being performed. In the next sections, the discussion highlights how various physics or theories of operation must be considered and how the engineers may use those theories to accomplish work with the amount of energy available at any given time.

Perpetual energy is everlasting, continuous energy.

Newton's Laws of Motion

One of the biggest uses of energy in transportation is the energy to overcome the laws of motion. Every object on Earth has to deal with these basic theories of motion. Newton developed several basic laws while proving that the Earth revolved around the sun instead of the other way around. This section covers some of those that apply directly to the automobile. The first one deals with inertia.

This unchanging action applies to an object in motion or an object at rest. Without an outside force being applied to the object, its actions will not change. A stationary object will remain stationary, whereas a moving object will continue to move. And it will move in a straight line. Obviously this would not work well with a vehicle. Although natural forces like wind, gravity, and terrain will affect the vehicle, the driver may have to wait a while before the vehicle begins to move, and the driver still would not have control of it.

An engine and a driveline are used to power the vehicle, with steering and suspension providing some means of control. The brake system is used to overcome the inertia of motion and helps stop the vehicle. A typical vehicle uses large amounts of energy to move the vehicle from stationary and even more energy to stop it. The steering system also requires energy to move the vehicle out of a straight line motion so it can follow the road. But there are other factors involved.

For every action there is an opposite and equal reaction. This is generally the second law. The force applied to the drive wheels must create enough friction (see next section) against the road to produce that equal, opposite reaction (Figure 2-13). A spinning tire wastes that

Inertia is the tendency of an object to continue doing what it is already doing.

APPLIED ACTION

Horsepower (torque) + friction = driving power

OPPOSING ACTION

Inertia + mass + air resistance = resists driving force

Figure 2-13 The force needed to move and accelerate a vehicle must be greater than the opposing force. When the two forces are equal, the vehicle cannot achieve more speed or torque.

reaction and the vehicle does not move as it should. On the other end, the same wheels must grip the road so the applied braking force can provide the slowing action. The problem with stopping a vehicle is the speed at which it is traveling. Upon acceleration, the energy is used to move the mass of the vehicle. During braking the speeding mass must be overcome and still retain driver control.

Inherent in this law is a rule regarding acceleration of an object. The acceleration of a vehicle is directly proportional to the driving forces applied to it. This applied force must overcome the friction of the various moving components, the tire against the road, and the total mass of the vehicle and its load. Mass is not the weight of the object but the amount of matter in that object. A vehicle weighing 2 tons on Earth will weigh about 670 pounds on the moon, but its mass will be the same. When calculating the acceleration rate of a vehicle, other forces must be considered. As the drive wheels' force is applied against the road and begins to accelerate the vehicle, it is resisted by a force, equal to the vehicle's mass multiplied by its acceleration rate. Then aerodynamic forces begin to act simultaneously against the vehicle in the opposite direction of the driving force, reducing the acceleration rate. In general, this reduces the amount of force available to keep the acceleration rate high. At some point the driving forces equal the resisting forces, indicating that the vehicle has reached its maximum speed. Most light vehicles will accelerate quickly to a certain speed before the rate of acceleration drops off. This can be checked easily with a stopwatch and an open, clear stretch of road. Measure the time taken to accelerate from 0 to 25 mph. Record the time but continue running the clock until the speed reaches 50 mph. The acceleration time between 25 and 50 will be a little slower than the time between 0 and 25. The distance needed for that extra acceleration will also be a little greater. Even low-powered, four-cylinder vehicles can usually reach 25 mph quickly and in a short distance. The next 25 mph sometimes becomes a question mark.

AUTHOR'S NOTE: The so-called performance cars of today are usually truly considered performance only in a fairly narrow range of engine power. Either the vehicle can accelerate quickly to a certain speed with a definite top end, or it can accelerate reasonably well with a higher top speed or some combination of the two. Very few production vehicles can have a high acceleration rate *and* a very high top-end speed with good fuel mileage and comfort. Production vehicles require compromise among all possible designs to overcome the laws of motion. The example that proves this is the top fuel dragster that requires hundreds of thousand of dollars to operate for a quarter of a mile on 4 to 6 gallons of fuel for 3.5 seconds, then the engine is rebuilt for the next one. This would be hard to sell to the public.

Hydraulic Principles

Automotive brake systems use the force of hydraulic pressure to apply the brakes. Since automotive brakes use hydraulic pressure, we need to study some hydraulic principles used in brake systems. These include the facts that fluids cannot be compressed; fluids can be used to transmit movement and force; and fluids can be used to increase force.

Fluids Cannot Be Compressed

Hydraulic devices such as the brake system work because fluids, unlike gases, do not compress. If you fill a container with a top on it with a gas as shown in Figure 2-14 and put a weight on the top, the weight will push the top down. The top moves down because the gas can be compressed. If you fill the container with a fluid, however, the weight will not push the top down, because fluid in a sealed container cannot be compressed. Fluid resistance to compression allows it to transmit motion and pressure.

Fluids Can Transmit Movement

Fluids can be used to transmit movement. Two cylinders of the same diameter are filled with a fluid and connected by a pipe as shown in Figure 2-15. If you force piston A downward, fluid will

Fluid trapped in a sealed system such as the brake system will act like a steel rod when force is applied to it. That force will be transmitted to every interior surface of the system equally. It is measured in pounds per square inch or pressure.

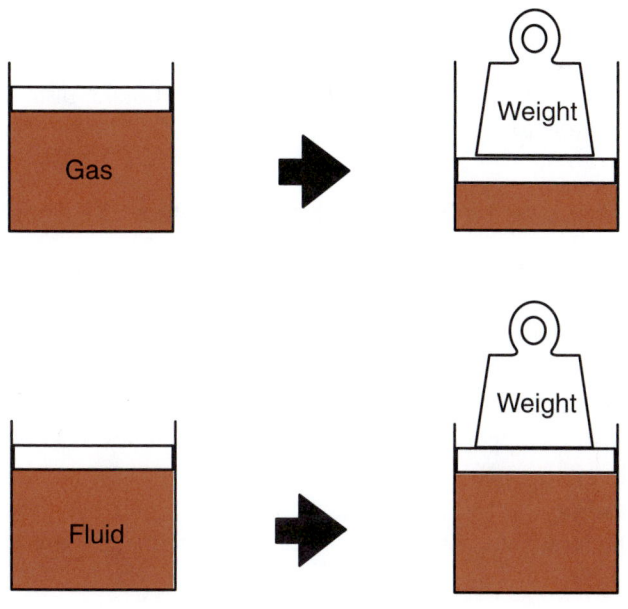

Figure 2-14 Gases are compressible, but fluids are not.

Figure 2-15 Fluid can transmit motion through a closed system.

Figure 2-16 The master cylinder is an apply piston, working as a pump, to provide hydraulic pressure to the output pistons at the wheel brakes.

push piston B upward. Because piston A starts the movement, it is called the apply piston. Piston B is called the output piston. If the apply piston moves 10 inches, the output piston also will move 10 inches (Figure 2-16).

The principle that motion can be transmitted by a liquid is used in hydraulic brake systems. A master cylinder piston is pushed when the driver applies the brakes. The master cylinder piston is the apply piston. The fluid in the master cylinder is connected by pipes to pistons in each of the car's front and rear wheel brake units. Each of the wheel brake pistons is an output piston. They move whenever the master cylinder input piston moves.

Fluids Apply Pressure to Transmit and Increase Force

Fluids are not just used to transmit movement; they are also used to transmit **force.** Force is power working against resistance; it is the amount of push or pull exerted on an object needed to cause motion. Force is usually measured in the same units that are used to measure weight: pounds or kilograms.

Piston A and piston B (Figure 2-17) are the same size. If a 100-pound force is applied to piston A, the same force will be applied to piston B. This demonstrates that force is transmitted from the apply piston to the output piston.

This principle can be used to increase the output force of a hydraulic system. To do this, change the sizes of the input and output pistons and calculate the amount of pressure developed in the system.

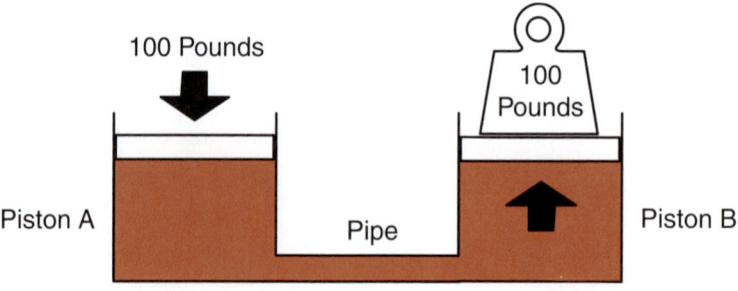

Figure 2-17 As hydraulic fluid transmits motion through a closed system, it also transmits force.

Output Force Creates System Pressure. Pressure is the force exerted on a given unit of surface area. Therefore, pressure equals force divided by surface area:

$$\frac{F}{A} = P$$

where
F = force in pounds (or kilograms)
A = area in square inches (or centimeters)
P = pressure

The pressure of a liquid in a closed system such as a brake hydraulic system is the force exerted against the inner surface of its container, which is the surface of all the lines, hoses, valves, and pistons in the system. In customary English units, pressure is measured in pounds per square inch (psi). In the metric system, pressure can be measured in kilograms per square centimeter, but the preferred metric pressure measurement unit is the pascal. Regardless of how it is measured, pressure applied to a liquid exerts force equally in all directions. When pressure is applied to a movable output piston, it creates output force.

In Figure 2-18, input piston A is smaller than output piston B. Piston A has an area of 20 square inches, and, in the example, 200 pounds of force are applied. Therefore,

$$\frac{200 \text{ pounds } (F)}{20 \text{ square inches } (A)} = 10 \text{ psi } (P)$$

If that same 200 pounds of force is applied to a piston of 10 square inches, system pressure is 20 psi, because

$$\frac{200 \text{ pounds } (F)}{10 \text{ square inches } (A)} = 20 \text{ psi } (P)$$

Therefore, pressure is inversely related to piston area. The smaller the piston, the greater the pressure that is developed with the same amount of input force.

Pressure and Output Piston Area Determine Force. Apply the 10 psi of pressure in the first example to an output piston with an area of 50 square inches. In this case, output force equals pressure times the surface area:

$$P \times A = F$$

where
F = force in pounds (or kilograms)
A = area in square inches (or centimeters)
P = pressure

Therefore, 10 psi of pressure on a 50-square-inch piston develops 500 pounds of output force: $10 \times 50 = 500$. Brake systems use hydraulics to increase force for brake application. This is called **mechanical advantage.**

Figure 2-18 A hydraulic system also can increase force.

Figure 2-19 Different-size output pistons produce different amounts of output force from the same hydraulic pressure.

Figure 2-19 shows a hydraulic system with an input piston of 10 square inches. A force of 500 pounds is pushing on the piston. The pressure throughout the system is 50 psi (force 500 ÷ area 10 = 50 psi). A pressure gauge in the system shows the 50-psi pressure.

There are two output pistons in the system. One has 100 square inches of area. The 50-psi pressure in the system is transmitted equally everywhere in the system. This means that the large output piston has 50 psi applied to 100 square inches to deliver an output force of 5,000 pounds (100 square inches × 50 psi = 5,000 pounds).

The other output piston in Figure 2-19 is smaller than the input piston with a 5-square-inch area. The 5-square-inch area of this piston has 50-psi pressure acting on it to develop an output force of 250 pounds (5 square inches × 50 psi = 250 pounds).

In a brake system, a small master cylinder piston is used to apply pressure to larger pistons at the wheel brake units to increase braking force. It is important that the pistons in the front brakes (now almost exclusively caliper pistons) have a larger surface area than the pistons in the rear brakes. This creates greater braking force at the front wheels to overcome the weight transfer that momentum creates during braking.

Hydraulic Pressure, Force, and Motion. Hydraulic pressure acting on pistons of different surface areas creates different output forces. As the output pistons move, another effect of pressure and force appears. When output force increases, output motion decreases. If the 10-square-inch input piston moves 2 inches as it applies 50 psi to the 100-square-inch output piston, that output piston will move only 0.2 inch as it applies 5,000 pounds of output force (Figure 2-20). The ratio of input motion to output motion is the ratio of the input piston area to the output piston area, and you can use this simple equation to calculate it:

$$\frac{A1}{A2} \times S = M$$

where $A1$ = input piston area
$A2$ = output piston area
S = input piston stroke (motion)
M = output piston stroke (motion)

$$\frac{10 \text{ square inches (input piston)}}{100 \text{ square inches (output piston)}} = \frac{1}{10} \times 2 \text{ inches (input stroke)} = 0.2 \text{ inch output motion}$$

Figure 2-20 The output pistons' movement and their created force will be proportional to their size in relationship to the input piston size.

If the output piston is larger than the input piston, it exerts more force but travels a shorter distance. The opposite also is true. If the output piston is smaller than the input piston, it exerts less force but travels a longer distance. Apply the equation to the 5-square-inch output piston in Figure 2-20:

$$\frac{10 \text{ square inches (input piston)}}{5 \text{ square inches (output piston)}} = \frac{2}{1} \times 2 \text{ inches (input stroke} = 4.0 \text{ inches output motion}$$

In this case, the smaller output piston applies only half the force of the input piston, but its stroke (motion) is twice as long.

This relationship of force, pressure, and motion in a brake system is shown when the force applied to the master cylinder pistons and the resulting brake force and piston movement at the wheels is considered. Wheel cylinder pistons move only a fraction of an inch to apply hundreds of pounds of force to the brake shoes, but the wheel cylinder piston travel is quite a bit less than the movement of the master cylinder piston. Disc brake caliper pistons move only a few thousandths of an inch but apply great force to the brake rotors.

A BIT OF HISTORY

All of the hydraulic principles that are applied in a brake system are based on the work of a seventeenth-century scientist named Blaise Pascal. Pascal's work is known as Pascal's law. Pascal's law says that pressure at any one point in a confined liquid is the same in every direction and applies equal force on equal areas.

One of the most important results of Pascal's work was the discovery that fluids can be used to increase force. Pascal was the first person to demonstrate the relationships of pressure, force, and motion and the inverse relationship of motion and force. On an automobile, Pascal's laws are applied not just to the brake system. These same hydraulic principles are at work in the hydraulic system of an automatic transmission and other systems. Pascal's laws are even at work in the movement of liquid fuel from a tank to the fuel-injection system on the engine.

Hydraulic Principles and Brake System Engineering

Engineers must consider these principles of force, pressure, and motion to design a brake system for any vehicle that will give maximum stopping efficiency but still be easy to control. If the engineer chooses a master cylinder with relatively small piston area, the brake system can develop very high hydraulic pressure, but the pedal travel will be extreme. Moreover, if the master cylinder piston travel is not long enough, this high-pressure system will not move enough fluid to apply the large-area caliper pistons regardless of pressure. If, however, the engineer selects a large-area master cylinder piston, it can move a large volume of fluid but may not develop enough pressure to exert adequate braking force at the wheels.

The overall size relationships of master cylinder pistons, caliper pistons, and wheel cylinder pistons are balanced to achieve maximum braking force without grabbing or fading. Most brake systems with front discs and rear drums have large-diameter master cylinders (large piston area) to move enough fluid and a power booster to increase the input force.

Vacuum and Air Pressure Principles

Vacuum is generally considered to be air pressure lower than atmospheric pressure (a true vacuum is a complete absence of air).

Vacuum is another force used in most brake systems. Most power brake systems use vacuum to provide a power assist for the driver. Because the most significant use of atmospheric pressure and vacuum in a brake system is in the operation of a power booster, these principles are covered in Chapter 6 of this manual on power brake systems.

Electrical Principles

Many of the brake system components you will work on are controlled or powered by electricity. Examples are brake system warning lamps, stoplamp switches, brake fluid level sensors, and ABS components. Therefore, a basic understanding of some of the electrical principles, including amperage, voltage, and resistance, is needed.

Amperage, Voltage, and Resistance

Ampere (A) is the unit for measuring electric current where 1 ampere equals a current flow of 6.28×10^{18} electrons per second.

Think of electricity in terms of the same principles that work in hydraulic systems. The flow of electricity through a circuit is similar to the flow of fluid through a hydraulic line. Current is the movement of free electrons, under pressure, in a conductor. A flow of current through a conductor requires a source of free electrons to supply the demand, just as fluid in a tank or reservoir is a source of flow in a hydraulic system. The rate of fluid moving through a line often is measured in gallons per minute. The rate of current flowing in a conductor is measured in amperes (A). One **ampere (A)** equals 6.28×10^{18} electrons passing a given point in a circuit per second.

Voltage is the electromotive force that causes current to flow; the potential force that exists between two points when one is positively charged and the other is negatively charged.

Just as pressure is necessary to move fluid through hydraulic lines, there must be pressure to move electrons through a conductor. The pressure pushing the electrons through an electrical circuit is called the **voltage,** measured in **volts (V).**

Friction between the walls of a hydraulic line and the fluid will cause some resistance to the flow of fluid. Similarly, some resistance to electron flow through a circuit is offered by any material. Electrical resistance is measured in **ohms (Ω).**

To summarize the comparison of electrical and hydraulic systems, we can say:

❑ Voltage is the pressure (or electrical force) that moves electrons (current or amperes) through a wire just as pressure moves fluid through a pipe.
❑ Amperage, or current, is similar to the fluid flowing in a line.
❑ Electrical resistance is a load on the moving current that must be present to do any useful work, just as a hydraulic system must have the load of an output piston or motor to do work.

Ohm's Law

The amount of fluid flowing through a pipe will increase if the fluid pressure is increased. Likewise, the amount of fluid flowing in the pipe will decrease if the resistance offered by the hydraulic system increases. In the same way, the amount of current flowing in an electrical circuit will increase if the voltage is increased and decrease if the resistance of the circuit increases.

The relationship of these electrical factors (voltage, current, and resistance) is summarized by a mathematical equation known as **Ohm's law** (Figure 2-21). Ohm's law states that it takes 1 volt of electrical pressure to push 1 ampere of current through a resistance of 1 ohm. Mathematically, Ohm's law is expressed by the following equations, using the symbols E for voltage, I for current, and R for resistance:

$E = I \times R$ to find voltage
$I = E \div R$ to find current
$R = E \div I$ to find resistance

For example, if a 12-volt circuit has 2 amperes of current flowing, you can use Ohm's law to determine the resistance as follows:

$R = E \div I$
$\quad = 12 \div 2$
$\quad = 6$ ohms

The resistance of the circuit is 6 ohms.

These equations can be used to calculate the voltage, current, or resistance of a circuit. In most cases, you can use electrical testing instruments to measure these values. There are times, however, when this is not possible. Then, if you know two of the values, you can find the third value by using the problem-solving wheel, which is based on Ohm's law.

Ohm is the unit used to measure the amount of electrical resistance in a circuit or an electrical device.

Georg Simon Ohm (1787–1854) wrote the law regarding electrical theory and measurement. It came to be known as Ohm's law.

A BIT OF HISTORY

Early experiments with electricity by Volta and Ampere led to names for the new phenomenon that remains today. The term *electromotive force* is now called volts. The term *intensity* is now called amps.

If E is unknown, cover it with a finger and the formula becomes:

$$I \times R$$

(I multiplied by R)
If I is covered with a finger, the formula reads:

$$\frac{E}{R}$$

(E divided by R)
If R is covered with a finger, the formula is:

$$\frac{E}{I}$$

(E divided by I)

Figure 2-21 If you know two of the three electrical factors of voltage, current, and resistance, you can calculate the third as shown here.

Summary

Terms to Know

Adsorption

Ampere (A)

Brake fade

Ceramic

Coefficient
of friction

Energy

Force

Gas fade

Inertia

Kinetic energy

Kinetic friction

Lining fade

Mass

Mechanical
advantage

Mechanical fade

Momentum

Ohms

Ohm's law

Perpetual energy

Rolling resistance

Static friction

Tensile force

Thermal energy

Vacuum

Voltage (volt)

Weight

❏ The energy used by the braking system is based on the kinetic energy, mass, weight, and speed of the vehicle and its braking system.

❏ Inertia is an object's resistance to change.

❏ Vehicle subsystems are used to overcome inertia.

❏ The acceleration rate of a vehicle is the force applied against the opposing forces.

❏ Kinetic and static friction are at work in the brake system.

❏ The amount of friction in a brake system depends on the pressure (force) exerted on the friction components and the surface area.

❏ Friction materials may be made from organic, metallic, or ceramic compounds.

❏ Energy provides the fuel for work.

❏ Work is accomplished any time energy is expended.

❏ The laws of motion affect the vehicle and all its subsystems.

❏ Hydraulic fluid can transmit motion and force because a fluid cannot be compressed.

❏ A small hydraulic input piston and a large output piston can increase force at the wheel braking components.

❏ Mechanical advantage in a hydraulic system is proportional to the size of the input piston versus the output piston.

❏ Electrical circuits have an electrical pressure (force) that is used to move electrons through a circuit.

❏ In Ohm's law, E = voltage, I = current (ampere), and R = resistance (ohm).

❏ In electrical terms, a change in $E, I,$ or R will cause a change in one or both of the other two.

❏ Resistance in a circuit is needed to change electrical energy into mechanical energy (action).

Review Questions

Short-Answer Essays

1. Define *kinetic friction* and explain how it is used in a brake system.

2. Define *static friction* and explain how it is used in a brake system.

3. Define *rolling resistance* and how it affects vehicle movement.

4. Why must heat from braking be dissipated?

5. Explain how inertia affects vehicle motion.

6. Discuss how hydraulics can be used to transmit force and motion.

7. Describe how the use of different-size pistons in a hydraulic system can be used to increase force.

8. Define *ampere, voltage,* and *resistance.*

9. Explain how the formula, Ohm's law, is used to calculate voltage when the current is 5 amperes and the resistance is 50 ohms.

10. Explain why a device cannot produce more energy than it uses.

Fill in the Blanks

1. The amount of friction developed in a brake system is the amount of _____ generated to _____ the friction materials together.

2. The friction used to stop the rotating parts of the brake system is called _____.

3. The major issue to moving a vehicle is its _____.

4. Friction causes _____, which must be removed or dissipated.

5. Because fluids cannot be _____, they can be used to transmit force.

6. The measurement _____ is used to measure the force against all internal components of a hydraulic system.

7. The flow of current in an electrical system is measured as _____.

8. The force or pressure needed to move electrons is called _____.

9. Without _____, electrical energy would not be able to perform work.

10. Ohm stated that it takes _____ volt to push _____ ampere through _____ of resistance.

Multiple Choice

1. Perpetual energy is
   A. everlasting.
   B. not possible using current knowledge.
   C. the ability to produce energy more than what is used.
   D. all of the above.

2. Inertia affects the
   A. rate of acceleration.
   B. mass of an object.
   C. movement of an object.
   D. slowing of an object.

3. The rate of acceleration of a vehicle is affected
   A. by the proportion of driving force applied.
   B. by the air resistance on the vehicle.
   C. by the number of people in the vehicle.
   D. all of the above.

4. A general definition of kinetic friction is
   A. stationary friction or inertia.
   B. moving friction or inertia.
   C. the amount of heat applied for stopping.
   D. the amount of force applied.

5. Rolling resistance can be caused by
   A. inflation of the tires.   C. static friction.
   B. kinetic friction.   D. kinetic energy.

6. Coefficient of friction is the
   A. tensile force to slide two objects over each other.
   B. tensile force multiplied by the mass of an object.
   C. tensile force divided by the weight of an object.
   D. sum of the tensile force and weight of an object.

7. Brake fade is most often caused by
   A. lack of heat dissipation.
   B. quick heating of the braking components.
   C. incorrect brake fluid.
   D. pumping the brakes.

8. Brake friction material must meet which of the following criteria?
   A. water repellency, heat absorption, and resistance to fading
   B. quick recovery from heat fading, resistance to water fading, and resistance to heat fading
   C. must be bonded, riveted, or stamped to the metal shoe or pad
   D. able to withstand a cold environment, resistance to cold fading, and resistance to gas fading

9. Hydraulic systems can be used to
   A. transfer force.
   B. increase force.
   C. convert force to pressure and pressure to force.
   D. all of the above.

10. According to hydraulic theory, a 0.5-inch (13 mm) input piston with a 2-inch (52 mm) output piston has a mechanical advantage of
   A. 2:1.   C. 4:1.
   B. 10:1.   D. 0.25:1.

Related Systems: Tires, Wheels, Bearings, and Suspensions

Upon completion and review of this chapter, you should be able to:

- ❏ Describe the basic kinds of tire construction and identify the most common construction method for modern tires.

- ❏ Identify and explain the various letters and numbers used in tire size designations and other tire specifications.

- ❏ Explain the basic effects of tire tread design on vehicle handling and braking.

- ❏ Explain the most important effects of tire design and condition on brake performance.

- ❏ Explain how wheel and tire run out and wheel rim width and offset affect braking.

- ❏ Identify the common types of wheel and axle bearings used on cars and light trucks.

- ❏ Identify the basic wheel alignment and steering angles.

- ❏ Explain how wheel alignment and steering angles can affect braking.

- ❏ Explain how the condition of steering and suspension parts can affect braking.

Introduction

The brake shoes or pads apply friction to the wheels, but it is the friction between the tires and the road that actually stops the car. Tire design, condition, and inflation pressure can affect braking, and attention to these factors often can solve braking problems.

The tires are mounted on wheels, which ride on bearings on steering knuckle spindles and axles. Steering and axle components, in turn, are supported by suspension struts and springs. Any of these components can create braking problems if they are not in proper working order. This chapter outlines the key relationships between brake systems and the related systems of wheels, tires, wheel bearings, and suspensions.

The gross vehicle weight is the weight of the vehicle plus driver, passenger(s), full fuel tank, and the amount of other material loaded onto the vehicle.

Tire Fundamentals

Brake systems are engineered in relation to many vehicle factors of weight, size, and performance. Among these factors are the construction, size, and tread design of the tires and the amount of traction or friction expected to be available between the tires and the road. For the best and most reliable brake performance, tires at all four wheels should be identical in construction, size, and tread pattern.

Shop Manual
pages 86–93

Carmakers' Recommendations

Most passenger cars and light trucks built since 1968 have a tire information placard on a door, a door pillar, or inside the glove compartment (Figure 3-1). The tire information placard lists the manufacturer's original equipment tire size and any recommended optional sizes. It also lists the recommended front and rear inflation pressures and maximum front and rear **gross vehicle weight rating (GVWR).** Brake systems are engineered to work most efficiently with the tire sizes and pressures listed on the placard.

Gross vehicle weight rating (GVWR) is the total weight of a vehicle plus its maximum rated payload.

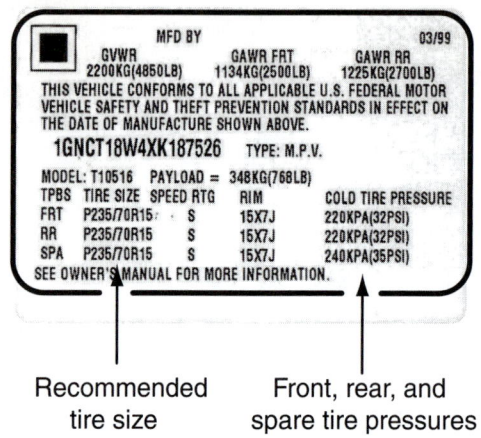

Recommended tire size

Front, rear, and spare tire pressures

Figure 3-1 This placard is located on the driver door and lists recommended tire size and cold inflation pressure.

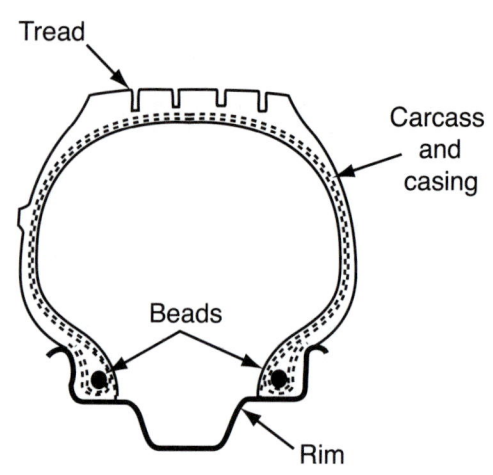

Figure 3-2 Major parts of a tire.

A few carmakers install different size wheels and tires at the front and rear of some vehicles, but this practice is reserved for a small percentage of high-performance sports cars like the Porsche 911. More than 99 percent of the vehicles on the road were originally fitted with wheels and tires of the same size at each corner. Although manufacturers may recommend one or two optional tire sizes at the rear that are larger than the front original equipment size, a large variation from the carmaker's recommendation can lead to braking problems, as well as problems with other vehicle systems.

For example, an extreme difference in tire diameters from front to rear may produce unequal speed signals from the wheel speed sensors of ABSs. Tires much larger than those recommended by the vehicle maker may produce inaccurate vehicle speed-sensor signals to the PCM or the ABS control module. This same problem exists if all four tires are larger or smaller than the manufacturer's recommendations.

Tire Construction

Every tire is constructed of these basic parts (Figure 3-2):

Carcass: Steel beads around the rim and layers of cords or plies that are bonded together to give the tire its shape and strength.

Casing: Additional layers of sidewall and undertread rubber added to the carcass.

Tread: The layer of rubber that contacts the road and that contains a distinctive pattern to provide the desired traction.

A BIT OF HISTORY

When radial tires were first introduced, there was a lot of resistance by drivers to using the new design. Complaints ranged from "feels funny when driving" to "they don't have enough air in them." Some drivers even went so far as to remove radial tires from a brand-new vehicle and to install bias tires. Two major characteristics of the radial tire overcame this die-hard resistance: a much smoother ride and increased fuel mileage.

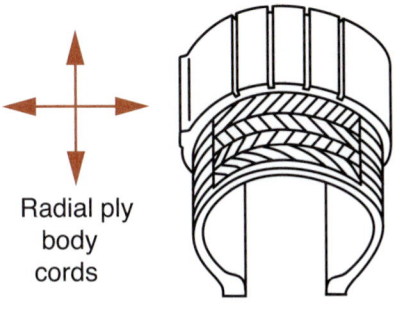
Radial ply
body
cords

Figure 3-3 Radial ply tire construction.

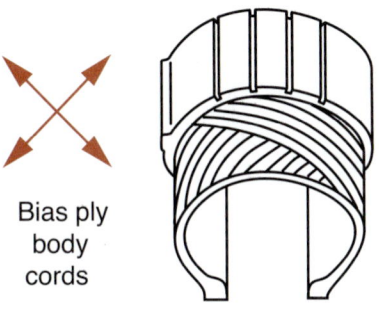
Bias ply
body
cords

Figure 3-4 Bias ply tire construction.

Bias ply
body
cords

Figure 3-5 Belted bias ply tire construction.

Since the mid-1970s, almost all original equipment and replacement tires for cars and light trucks have been the radial ply type. The cords in a **radial ply tire** run at 90 degrees to the tread, creating a more flexible sidewall and creating the impression that the tire is low on air pressure (Figure 3-3). Each cord is parallel to the radius of the tire circle, which gives the tire the name radial ply. Two or more additional belts of cords are added around the tire circumference between the radial plies and the tread for more strength and stability. These belts may be made of nylon, rayon, fiberglass, or steel. The radial plies provide maximum flexibility in the tire sidewall, and the belts ensure excellent tread stiffness for good road contact.

Before radial ply tires became the standard of the industry, tires were made with the plies laid at an angle or bias to the tire beads. This construction was called bias ply (Figure 3-4). The cords in a **bias ply tire** run at an angle closer to the tread, giving a much stiffer sidewall but creating a rougher ride than a radial tire. A second ply is added to the first so that its cords run in the opposite direction across the first ply. Additional plies can be added for more strength. Bias ply tires can carry more load and are still used on large trucks and most trailers. Most passenger car and light truck tires were made with two, four, six, or eight plies.

The cords of bias ply tires were usually made of nylon or polyester. This construction provided decent stiffness to prevent unnecessary twisting while still allowing enough flexibility to absorb some road shock and cushion the ride.

Belted bias ply tires incorporate the belts used in radial ply tires with the older bias ply construction. Two or more belts of nylon, fiberglass, or steel cords are wrapped around two or more bias plies (Figure 3-5). The belts keep the tire from deforming out of round at high speeds and thus improve tire stability. Belted bias ply tires were basically an interim design that combined some of the features of radial ply tires with the older bias ply construction. They are still used on some vehicles today.

Of the three types of tire construction, radial ply tires have the best braking performance. The flexible sidewalls of a radial ply tire provide a larger tread contact patch than a comparable size bias ply or belted bias ply tire. Radial ply construction also has greater flexibility on the wheel during cornering without decreasing tread contact with the road (Figure 3-6). The circumferential belts give a radial ply tire maximum stability and protect against deforming at high speed.

Tire Size

All tire sizes are based on three dimensions: the inside diameter, the section width, and the aspect ratio. The inside diameter is the same as the wheel size, or rim diameter, on which the tire fits: for example, 14 inches, 15 inches, or 16 inches. This diameter is measured across the tire beads or across the bead seat of the wheel rim. The **section width** is the width of the tire across the widest point of its cross section, usually measured in millimeters. The **aspect ratio** is the ratio of the cross-sectional height to the cross-sectional width expressed as a percentage. A high aspect ratio indicates a tall tire. Most modern tires have aspect ratios from 45 to 80.

RADIAL

BIAS

Vehicle in left turn (outward force)

Road surface

Inside edge of tread tends to remain flat

Inside edge of tread tends to lift

Figure 3-6 A radial tire tread tends to remain flat on the road, whereas the bias tread tends to lift the inner edge from contact with the road.

Most small cars use a tall tire, aspect ratio of 75 or 80, whereas larger cars use aspect ratios of 60 or 70. A fad that became popular in the late 1990s was to install tires with aspect ratios of 45, 50, and 60. This low profile lowered the vehicle drastically and gave a pleasing appearance until tire, wheel, or brake service was required. Extra labor costs may be charged on these "lowered" cars.

Over the history of the automobile, several different systems have been used to indicate the relative sizes of tires. When all tires were made with a simple bias ply construction and had a common aspect ratio of approximately 85, tire sizes were specified simply by the section width and wheel diameter. For example, an 8.20-15 tire had a section width of 8.2 inches and fit on a 15-inch wheel. Today, the most common system used to identify tires for passenger cars and many light trucks is the **P-metric system.**

In the P-metric system, the first character indicates the type of service for which the tire is designed (Figure 3-7). The most common designation is "P" for passenger car, which gives the size identification system its name. The numbers after the first letter indicate the tire section width in millimeters, and the numbers after the slash indicate the aspect ratio. An optional speed-rating letter may follow the aspect ratio, and the final letter indicates the construction type. The last numbers indicate the wheel diameter. The European metric variation of this size specification system omits the first letter, such as "P" or "C," but includes the speed-rating letter.

The alphanumeric tire size identification system was the most common system until the mid-1970s and is still found on a few tires today. In the alphanumeric system, the first letter identifies the approximate size and load capacity of the tire. Size letters range from "A" (the smallest) through "N" (the largest). Common passenger car sizes are "E," "F," "G," and "H." If the tire is a radial ply tire, the letter "R" follows the first letter. The letters are followed by a pair of numbers separated by a dash. The first number is the aspect ratio, such as 70, and the second number is the wheel diameter, such as 14.

Other Tire Specifications

Other tire specifications appear on tire information placards, in owner's manuals, and on the tire sidewall itself. The **tire load range** is indicated by the letters "A" through "L," which often appear

Figure 3-7 Typical tire markings.

after the size specification on the tire sidewall. The load range letters replace the older method of rating tire strength by the number of plies used in its construction. Load range "A" corresponds to a two-ply rating, and load range "L" corresponds to a twenty-ply rating. Most tires for cars and light trucks are load ranges "B," "C," and "D." As an alternative, tires may be marked with load index numbers that range from 65 to 104. These numbers indicate the load capacity of the tire from 639 pounds to 1,984 pounds. Most passenger car tires have a load index from 75 to 100. Many tires have a maximum inflation pressure molded into the sidewall. This is *not* the recommended operating pressure but the pressure that may cause the tire to burst if exceeded.

The maximum **cold inflation pressure** for any tire is molded into the tire sidewall (Figure 3-8). This is the inflation pressure after the tire has been standing for 3 hours or driven less than 1 mile after standing for 3 hours. The vehicle manufacturer's recommended inflation pressures are usually slightly less than the maximum cold inflation pressure.

Figure 3-8 The maximum cold inflation pressure specification is molded into the sidewall of every tire.

As tire aspect ratios or profiles have gotten lower, inflation pressures have gotten higher. Until the early 1980s, most passenger car tires were equivalent to the old load range "B" with a maximum pressure of 32 psi. Maximum pressures then moved upward to about 36 psi. Today, pressures in the low 40-psi range are common. Many light truck tires have their maximum load ratings at more than 50 psi. With the older maximum pressures of 32 psi to 36 psi, it was common to inflate tires to the maximum for best handling, tire life, and gas mileage. With today's higher pressures, the best practice is to follow the manufacturer's recommendations on a late-model vehicle's tire placard. If you go higher than the carmaker's specifications, hold the increase to no more than 4 psi and do not exceed the tire's maximum pressure rating.

AUTHOR'S NOTE: Starting in 1996, there were a lot of accidents involving a brand of SUVs equipped with a particular brand of tire. There were two opposing arguments concerning these accidents. First, the vehicle manufacturer claimed that the tires came apart during normal operation because of poor tire manufacturing. Second, the tire manufacturer claimed that the vehicle manufacturer's recommended tire pressure could critically damage the tire during operation. Both the SUVs and tires were recalled at a huge cost to both manufacturers and the public in general. Regardless of the root cause, it is obvious that tire pressure can affect overall operational durability, *particularly* if there is a problem with the tire's manufacturing process.

The Uniform Tire Quality Grading Standards were developed by an association of tire manufacturers, vehicle manufacturers, and governmental agencies in the early 1980s. The **uniform tire quality grading (UTQG) indicators** have been molded into the sidewall of every tire sold in the United States since 1985. The UTQG indicators are a number to indicate relative tread life, a letter to indicate wet weather traction, and a second letter to indicate heat resistance.

Tire Tread Design

The best possible traction and braking performance on dry pavement can be had with the treadless, slick tires used on race cars. Unfortunately, all of the great traction of a slick tire goes away on wet pavement. Therefore, all street tires have a tread pattern of ribs and grooves that lets water be displaced from under the tire while still maintaining good traction.

Tread designs vary widely, but most original equipment tires have some kind of mud-and-snow or all-weather tread pattern. In fact, the characters "M+S" or "M/S" often appear on a tire sidewall to indicate a mud-and-snow tread pattern. In addition to providing traction, the tread pattern acts as a radiator to expose more tire surface area to the air and improve tire cooling. Some high-performance tires have a **unidirectional tread pattern,** which means that each tire can be installed on only one side of the car.

Tread depth is measured from the top of the tread ribs to the bottom of the groove and usually is expressed in thirty-seconds of an inch. The tread on a new passenger car tire is usually in the range of $9/32$ inch to $15/32$ inch. The tread on truck tires and some mud-and-snow tires may be deeper.

All passenger car and light truck tires made since 1968 have **tread wear indicators** that appear as continuous bars across the tread when the tread wears down to the last $2/32$ ($1/16$) inch (Figure 3-9). Note that the tire should be replaced when the indicator is visible across two or more grooves. However, if the vehicle is extremely out of alignment, the tire may wear completely to the steel cords before wearing the indicator in the second groove. The two-groove indication assumes the vehicle is maintained and properly aligned.

Wear indicator
bars

Figure 3-9 When wear-indicator bars appear across any two tread grooves, the tire is ready for replacement.

Run-Flat Tires

Run-flat tires are not truly designed to be run flat. The actual design allows for one of two principal tire designs, both of which control the amount of air escaping through a puncture, thereby allowing the driver to locate and use facilities to change or repair the tire.

Some racing tires use race-proven technology by installing a tire-within-a-tire, but this is expensive and not suitable for mass-produced units. The most common design weaves the tire fabric in such a way that it will grip small penetrating objects in the tire. Another design weaves the fabric so it will attempt to close the hole if the penetrating object is withdrawn from the tire. In both cases, air inside the tire cannot escape quickly and, under the right conditions, will support the vehicle up to 50 miles at reasonable speeds and loads. Neither type will prevent the complete air loss due to a major incident such as a blowout. Leaking run-flat tires affect braking and steering of the vehicle the same as would any tire with insufficient air pressure. At least one federal government agency is looking at tire pressure.

Run-flat tires prevent quick air loss during punctures or minor damage to the tire.

Tire Pressure Monitoring System (TPMS)

Tires that are underinflated not only affect fuel mileage and tire life, they may also lead to problems in braking and steering. The TREAD Act of 2000 directed the **National Highway Transportation and Safety Agency (NHTSA)** to research and develop rules as a means of continuously monitoring tire pressure. After several years of testing, it was determined that two viable **tire pressure minitoring systems (TPMSs)** could perform the job, but each has it drawbacks. The **wheel-speed base (WSB)** system could use the wheel speed sensors that are used with ABS to detect the faster spinning or rotation caused by an underinflated tire. This would illuminate a light in the dash, warning the driver of the condition. A second option, **pressure-sensor base (PSB),** would have a pressure sensor located on each wheel and would read tire pressure directly. There were also questions posed about how much the tire pressure must drop (25 or 50 percent) before the driver alert was illuminated. A third issue with the PSB system was maintenance performed on the tires and wheels, road hazard, and reset procedures. This particular topic is discussed in Chapter 3 of the *Shop Manual.*

The NHTSA decided that vehicles manufactured after November 1, 2003, and through October 31, 2006, could meet either the 25 percent or the 50 percent limit with either the WSB or the

The **NHTSA** is a federal agency assigned to develop regulations for highway safety, including vehicle safety features.

Figure 3-10 A tire pressure sensor that is mounted on the valve stem in place of the stem cap.

PSB system. Any vehicle produced starting in October 2006 must have a TPMS (determined by the vehicle manufacturer) installed with a standard warning icon in the dash, to alert the driver if one or more tires have a pressure that is less than 25 percent of the cold inflation recommendation, and to be able to indicate which tire(s) is underinflated. During 2005 the NHTSA developed and issued the final ruling based on any new data collected during the 2003–2005 time period.

Based on current data, it appears that the PSB system is slightly more responsive, but advances in ABS technology may change that shortfall. The PSB may be factory installed on 2003–2005 models, or it could be an aftermarket kit. Both operate in a similar fashion. A wheel-mounted pressure sensor uses wireless communication to send the pressure signal to a PSB computer, which, in turn, illuminates the warning light or icon in the dash. Factory sensors are usually installed in place of the regular tire valve stem. Specific repair and reset procedures are mandated by the manufacturer. Aftermarket units usually have a pressure sensor installed on the valve stem in lieu of the valve cap (Figure 3-10). If a tire is replaced or repaired, or if the tires are rotated, both systems must be reset so the sensor can determine the rotational speed of the new/repaired/rotated tires and determine its own location on the vehicle. This reset and training procedure was one problem that the NHTSA found with the WSB and PSB systems. The final rule was issued in late 2006 and determines the standard warning icon, tire pressure cutoff, and standards for training and reset operations.

Wheel Fundamentals

Wheel offset and rim width are design factors of the wheel itself that affect brake performance. Their effects are most significant on the front wheels, which provide steering control as well as most of the braking force.

Rim Width

Rim width is measured between the inside surfaces of the rim flanges (Figure 3-11). Every tire has three or four rim widths on which it can be mounted. Similarly, every rim has three or four tire sizes that it can safely hold. Although there is not a rigid one-to-one correlation between tire sizes and rim sizes, there is a limited number of combinations for any given wheel or tire.

If a tire is mounted on a wheel rim that is too wide or too narrow, the beads will not seat properly and the contact patch area will be affected. The tire may shift excessively on the rim and cause uneven braking. In the worst cases, the tire may lose air and be pulled off the rim under hard cornering or braking.

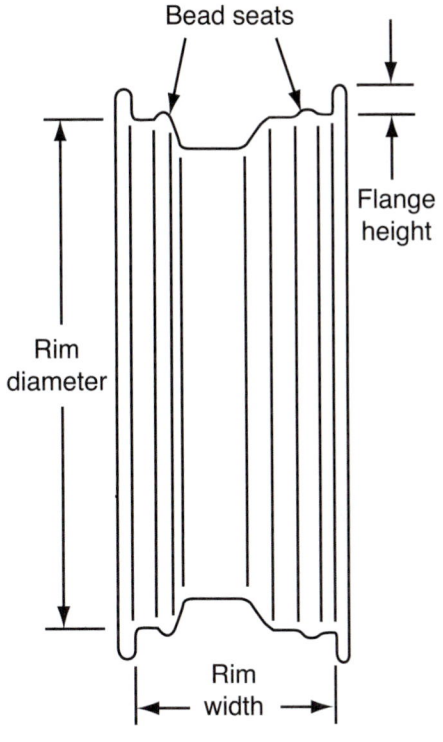

Figure 3-11 The principal wheel dimensions are diameter and width.

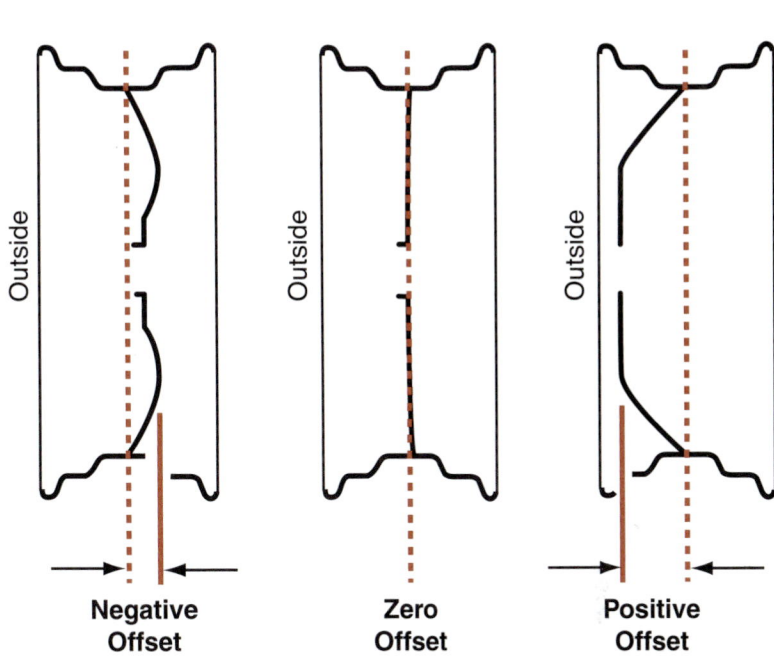

Figure 3-12 Wheel offset is the distance between the centerline of the rim and the mounting plane of the wheel.

Wheel Offset

Wheel offset is the distance between the centerline of the rim and the mounting plane of the wheel (Figure 3-12). If the mounting plane is outboard of the centerline, the wheel has positive offset. This is the more common design, particularly for front-wheel drive (FWD) cars. If the mounting plane is inboard of the centerline, the wheel has negative offset. Negative offset is common on many aftermarket wheels, particularly for trucks and SUVs. Along with low-profile tires, reverse center rims were also used to change the appearance of the vehicle. This type of rim pushed the tire away from the centerline of the body. Some tires were almost completely out of the wheel well. Reverse center rims and low-profile tires completely change the steering and suspension geometry and affect the braking characteristics of the vehicle. This involves extra cost over the price of the tires and rims.

The amount and direction of wheel offset are important design factors because they affect the load-carrying capacity of axles, spindles, hubs, and bearings. Wheel offset also affects the steering scrub radius and overstresses the wheel bearings.

The **scrub radius** is the distance from the tire contact patch centerline to the point where the steering axis intersects the road. If the tire contact patch centerline is outboard of the steering axis at the road, the car has positive scrub radius (Figure 3-13). If the tire contact patch centerline is inboard of the steering axis at the road, the car has negative scrub radius (Figure 3-14).

Scrub radius is a design factor that is engineered for each vehicle. Negative scrub radius is the more common design because it tends to push the tires inward as they roll. In case of a blowout or a failure of half of a diagonally split brake system, negative scrub radius helps to maintain vehicle stability. If scrub radius is changed from negative to positive by the installation of extremely offset wheels, the vehicle become unstable in case of tire or brake failure.

Scrub radius is the distance from the tire contact patch centerline to the point where the steering axis intersects the road.

Figure 3-13 When the pivot point is inside the tire contact point, the scrub radius is positive.

Figure 3-14 When the pivot point is outside the tire contact point, the scrub radius is negative.

 AUTHOR'S NOTE: Scrub is not the amount of rubbing an oversize tire does on a vehicle component. Scrub is an engineering term involving the movement of the tire on the road and must be considered during steering, suspension, and other system designing.

Wheel Bearings

Shop Manual
pages 93–106

Three basic kinds of wheel bearings are used on late-model cars and light trucks:

1. Tapered roller bearings
2. Straight roller bearings
3. Ball bearings

Tapered roller bearings are commonly used on the front or rear nondriving wheels of a vehicle. A complete tapered roller bearing assembly consists of an outer race (bearing cup), an inner race (cone), tapered steel rollers, and a cage that holds the rollers in place (Figure 3-15). The inner race, the rollers, and the cage are a one-piece assembly and cannot be separated from one another. The outer race is normally pressed into the wheel hub. Tapered roller bearings are installed in pairs with one large bearing set on the inboard side of the hub and a smaller bearing set on the outboard side. Tapered roller bearings must be cleaned and repacked with grease

Figure 3-15 Typical tapered roller bearing installation.

periodically and then have the end play adjusted. These bearing services are usually part of a complete brake job.

Straight roller bearings are used on the drive axles of most rear-wheel drive (RWD) vehicles (Figure 3-16). These bearing assemblies consist of an outer race pressed into the axle housing, straight steel rollers, and a cage to hold the rollers in place. The axle shaft acts as the inner bearing race. Because straight roller bearings are installed in the drive axle housing, they are lubricated by the final-drive lubricant. Periodic adjustment and repacking are unnecessary.

Ball bearings may be used on rear drive axles, and double-row ball bearings are used on the front hubs of many FWD cars. A typical ball bearing assembly consists of inner and outer races, steel balls, and a cage to hold the balls in place (Figure 3-17). Although ball bearings were used on the nondriven front wheels of many cars before 1960 and required periodic service, most modern ball bearing assemblies are sealed and permanently lubricated.

Some late-model cars have sealed double-row ball bearings or tapered roller bearings in an assembly that includes the wheel hub, the bearing races, and a mounting flange (Figure 3-18). These are nonserviceable, nonadjustable assemblies that must be replaced if damaged or defective.

Figure 3-16 Straight roller bearings are used on most RWD axles.

Figure 3-17 Typical ball bearing construction.

Figure 3-18 A typical sealed bearing may be pressed into the hub or may be an integrated part of the hub.

Wheel Alignment Fundamentals

Shop Manual
pages 106–109

Wheel alignment service is based on maintaining the geometric relationships designed into the front and rear suspension and steering systems of a vehicle. These relationships are a series of angles in the suspension and steering. Some of these angles affect braking more than others. Traditionally, the five basic steering and suspension angles are:

1. Camber
2. Caster
3. Toe
4. Steering axis inclination (SAI)
5. Toe-out on turns, or turning radius

Two other steering and suspension angles indicate the relationship among all four wheels. They are:

1. Thrust angle
2. Setback

Improper camber will wear the tire on one edge and cause steering problems based on the camber error.

Camber

Camber is the inward or outward tilt of the wheel measured from top to bottom and viewed from the front of the car. If the wheel tilts outward at the top, it has positive camber. If it tilts inward, camber is negative (Figure 3-19). Each front and rear wheel has its individual camber angle. Camber is measured in degrees.

Improper caster usually will not cause abnormal tire wear but may cause steering problems.

Caster

Caster is the backward or forward angle of the steering axis viewed from the side of the car. If the steering axis tilts backward at the top, the wheel has positive caster. If the steering axis tilts forward at the top, caster is negative (Figure 3-20). Each front wheel has its individual caster angle. Because the rear wheels are not steering wheels, rear caster is always zero. Caster is measured in degrees.

Figure 3-19 Camber is the inward or outward tilt of the wheel measured from top to bottom and viewed from the front.

Figure 3-20 Caster is the backward or forward angle of the steering axis viewed from the side.

Figure 3-21 The toe angle is the difference in the distance between the centerlines of the tires measured at the front and rear of the tires.

Toe

Toe angle is the difference in the distance between the centerlines of the tires on either axle (front or rear) measured at the front and rear of the tires and at spindle height. If the centerlines of the tires are closer together at the front of the tires than at the rear, the wheels are toed in (Figure 3-21). If the centerlines of the tires are farther apart at the front of the tires than at the rear, the wheels are toed out. If the centerline distances are equal from front to rear on the tires, the toe angle is zero.

Improper toe will feather wear the tread and may cause steering problems.

Steering Axis Inclination

Steering axis inclination (SAI) is the angle formed by the steering axis of a front wheel and a vertical line through the wheel when viewed from the front with the wheels straight ahead (Figure 3-22). SAI is measured in degrees independently for each wheel. SAI works with camber to maintain the

Included angle

True vertical

Steering pivot axis

Camber and centerline

SAI angle

Positive (+)
scrub radius

Figure 3-22 Steering axis inclination is the angle formed by the steering axis of a front wheel and a vertical line through the wheel.

vehicle load inboard on the larger front-wheel bearing. SAI is not adjustable, but it may be measured to identify suspension component damage. SAI, toe-out on turns, thrust angle, and setback are sometimes measured by alignment shops to determine why the wheel alignment cannot be brought to specifications. The technician uses measurements to determine if the vehicle has suffered accident damage or perhaps the steering wheels were bounced off a road curb.

Toe-Out on Turns

Toe-out on turns (turning radius) is the difference between the angles of the front wheels in a turn. Because of suspension and steering design, the inside front wheel turns at a greater angle during a turn (Figure 3-23). Thus, it toes out.

Toe-out on turns is measured in degrees by turning one wheel a specified amount and then measuring the angle of the other wheel. The difference between the two is the turning angle or the toe-out on turns. Toe-out on turns prevents tire scuffing and tire squeal during cornering. Toe-out on turns is not adjustable, but it may be measured to identify suspension or steering component damage.

Thrust Angle

To understand the **thrust angle,** you must understand the geometric centerline of the vehicle and the thrust line of the vehicle. The thrust angle is the angle between the geometric centerline and the thrust line (Figure 3-24).

The **geometric centerline** is a static dimension represented by a line through the center of the vehicle from front to rear. It can be used as a reference point for individual front- or rear-wheel toe, but it does not represent the direction in which the rear wheels are steering the vehicle. The **thrust line** is the bisector of total toe on the rear wheels. Put more simply, the thrust line is the direction in which the rear wheels are pointing.

Figure 3-23 Toe-out on turns is the difference between the angles of the front wheels in a turn.

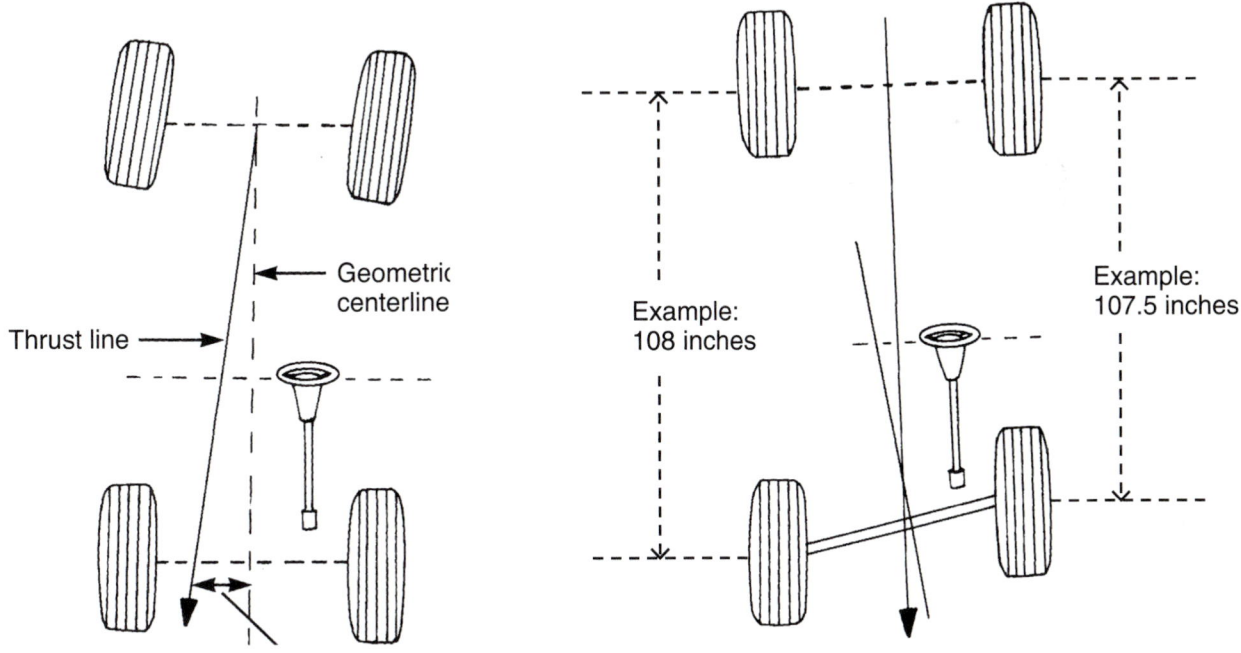

Figure 3-24 The thrust angle is the angle between the geometric centerline and the thrust line.

Figure 3-25 Setback is the difference in wheelbase from one side of the car to the other (exaggerated examples).

Setback

Setback refers to a difference in wheelbase from one side of the car to the other (Figure 3-25). Some vehicles are built with setback as a design feature. Ford trucks with twin I-beam front suspension are some of the best-known examples. Setback also can be present in a vehicle where it is not supposed to be. This is usually the result of a collision in which the frame is bent or suspension components are severely knocked out of position. Accidental setback also can change the vehicle thrust angle and cause handling problems.

Effects on Braking Performance

Shop Manual
pages 85–86

The bottom line of the braking system performance is how the tire contacts and holds the pavement, or the **tread contact patch** (Figure 3-26). The previous sections outline the overall requirements for tires, suspension, and related components. Major deviations from these requirements not only can result in poor ride and control through the steering, but may and will in some cases affect the braking ability of the vehicle.

The tread contact patch's friction ability is determined not only by the tire size, but also by how the tires contact the road. One primary failure is over- or underinflated tires. This causes a larger or smaller contact patch and a corresponding reduction or increase of braking friction. The angle at which the tire contacts the road may also affect braking.

Camber, caster, or toe angles that are excessively negative or positive will affect tire wear and the angle, and, to some extent, the size of the contact patch. The "perfect" contact patch has all the treads on the bottom of the tire in contact with the road. Excessive alignment angles may cause the tire to push toward one side during driving. This push or pull can be greatly increased as the tire tries to push but the vehicle is trying to go straight ahead. Even if the alignment is correct, worn or broken steering and suspension components may cause the same problem.

Wheels and tires that do not meet the vehicle manufacturer's specifications can, and do, cause the steering and suspension angles to change. Lifting or lowering the vehicle will also change those angles. Worn or damaged wheel or axle bearings can cause the wheel and tire assembly to wobble during operation or cause a drag that may initially be interpreted as brake drag.

Technicians must pay attention to what their customers are saying when they complain of a brake problem. Chapter 4 of the *Shop Manual* discusses ways and means for the technicians to isolate apparent braking problems from reduced performance of related systems and components. Remember that the engineers designed each system to work together. If one system fails to

Figure 3-26 Tire diameter and contact patch area are important factors in troubleshooting brake performance.

perform properly, then the other systems will not perform properly. This is especially true of brake, suspension, and steering systems because of their shared components and overlapping operation.

Performance Tires, Wheels, and Alignment

 AUTHOR'S NOTE: This section should give the reader an idea of differences between production vehicles and race vehicles. There are two classifications of "performance" vehicles. One classification consists of production lines like the Viper, Corvette, and Mustang, whereas the other consists of actual racing vehicles. The theory of operation for all performance vehicles is the same as that of a straight production product. However, race vehicles are usually "hand-engineered" and "hand-built." In other words, they work well on a controlled track, but not very well in downtown Atlanta traffic. A short section on the major differences between production and race products is included at the end of appropriate chapters.

Tires

Racing tires are engineered and manufactured for one reason: to provide traction for a particular track with a specific surface for speed and control. Most oval track (NASCAR) tires have no tread, but they provide much better traction than a typical treaded tire. However, there is one very distinct drawback: they do not work well on a damp track. Have you ever seen a NASCAR race conducted in the rain? Road course racers (Formula 1, Gran Prix) do use treaded tires because they run the race regardless of rain, but many times the rubber and chemical compound that these tires are made of is very similar to that of oval track tires.

AUTHOR'S NOTE: NASCAR tested a rain tire in Japan a few years back with some interesting results (full information can be found at http://www.nascar.com).

Race tires cost about $800 to $1,200 each. A set of racing tires could cost more than some cars on the road. Goodyear is the primary tire supplier for NASCAR and develops a tire specifically for a certain track. Other tire manufacturers like Bridgestone and Michelin supply tires for road course competitors like Formula 1 and Gran Prix. The tire may be of a softer compound with a short life and better grip or a hard compound with long life but not quite as much traction. The tire used on the short track at Martinsville, Virginia, with top speeds of about 130 mph is very different from the tire for Atlanta with top speeds in the 200-mph range. Race vehicles are usually equipped with an "inner" tire to help prevent blowouts. This is basically a little smaller tire mounted inside the visible tire.

Racing tires provide an excellent proving ground for tire engineers to test rubber compounds for wear and traction. Most of today's production tire designs and chemical compounds originated from research data collected during races. In fact, some of the ideas behind run-flat tires came from the oval track.

Wheels

Like the tires, the basic design of a racing wheel is very similar to that of a production wheel. Depending on the track and expected speeds, it may have more positive or more negative offset. Also, the bead-holding area of the wheel is usually a little deeper for a better grip on the tire bead. They are almost never made of anything other than high-quality steel. The side forces occurring during any race will destroy a production wheel in a fairly short period of driving. Typically racing wheels are a little wider than production wheels with the same size diameter mounted. Other than the extremely high cost to purchase wheels of this type, they could be used on production vehicles. Generally speaking, racing wheels are bland in appearance so few people want to pay an extra $5,000 to get wheels that would never wear out but do not look any better or racier.

Wheel Alignment

This area contains probably the biggest and most noticeable difference between production vehicles and racers. The theory behind wheel alignment is the same, but the application of that basic theory to racers is extreme. Road course racers use an alignment setup close to that of production, but oval track vehicles definitely appear to be completely abnormal to a nonexperienced passerby.

If a vehicle is set up for Atlanta, Daytona, or Talladega, it would appear that all four wheels are laid in (negative camber) (Figure 3-27). The camber is set very negative compared to production vehicles. This odd setup also has a very positive effect on vehicle handling. Due to the downforce on the vehicle racing on high-speed tracks, the wheels are forced out at the top, creating zero or positive camber. This is true of all vehicles when they are loaded or in some other condition in which the body is forced downward. However, routine operation on the public roads does not create the speed and resulting downforce required to change the camber setting. Also most oval track race vehicles are set up with a more positive camber on the left wheels than on the right. During that

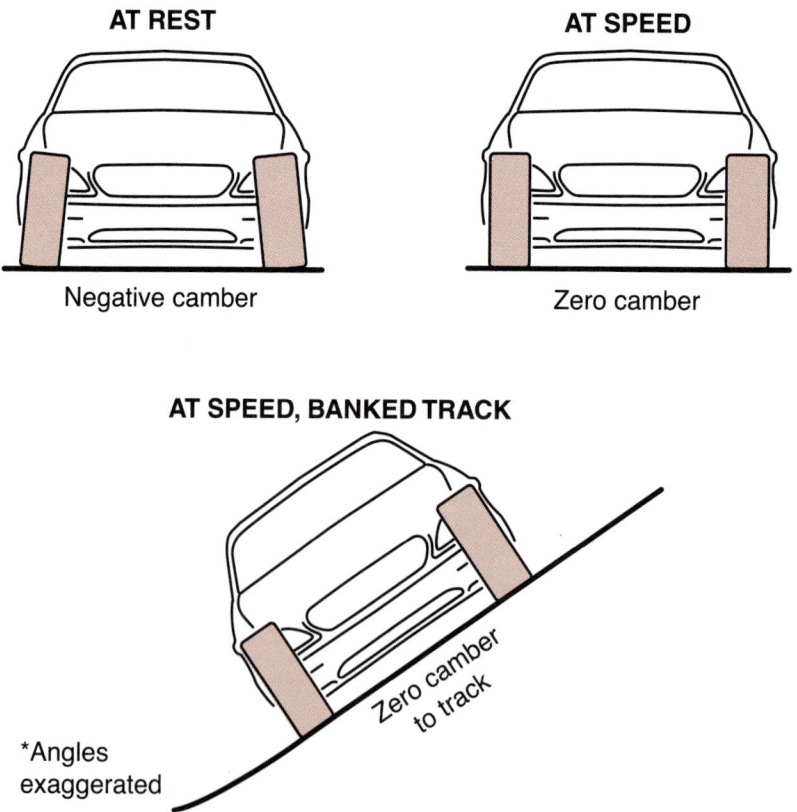

Figure 3-27 A simplistic race setup has the wheels set at negative camber (top left), and the downward force of the airflow and vehicle mass cause the wheels to move to zero or positive camber (bottom).

200-mph left turn, the vehicle body is forced downward and leans or tilts hard right. This changes the camber of the right-side tires to more positive than the left. If done correctly, all four tires try to stand straight up and the tread becomes flat on the pavement during those high-speed turns; hence, more and better traction is present when it is needed the most. Short-track ovals require a different camber setup because the speeds are much slower and the track surface is flatter.

Toe on a racer is not quite as important as camber, but it is set up with some differences from production vehicles. Because of body lean, excessive camber, and the construction of the suspension system, toe measurements are set up not only to control the vehicle but also to help keep the tread straight on the track surface. Of more importance to a race vehicle is toe-out on turn. Like production vehicles, this angle is determined by the construction and initial service on the suspension and steering systems and is not adjustable.

Another factor influencing racing setup is the banking angle of the track surface when compared to horizon. The high-speed tracks have up to 36 degrees of banking in the turns. This banking is one of the main factors for selecting the camber and toe settings. A high-bank track does not require as much offset camber as a short track (1 mile or less in length) with little banking. This is because the centrifugal forces at a low-bank track affect the sideways movement of the vehicle much more. The vehicle tends to slide sideways because the centrifugal force is about 90 degrees to the surface of the track or horizon. The camber is set to account for this force and body lean. A vehicle on a high-bank track tends to be pushed into the pavement because the pavement is at an angle to the horizon. Although the centrifugal force may be much greater, the track surface is flatter with regard to the vehicle chassis, forcing the vehicle into the track instead of across it. All of this means nothing, however, unless the suspension and steering systems are also matched to the track and vehicle.

It would not be a good idea to set up a production vehicle like a racer with regard to tires, wheels, suspension, and steering. Setting the high costs aside, the tires would wear out very quickly and the vehicle would ride like a horse-drawn wagon. That is assuming the driver has the strength to fight for control during straight-ahead and right-turn driving because most racing vehicles do not steer easily until operating at high speeds. It is suspected that the driver would wear out before the tires.

Summary

❑ If a vehicle has identical tires at all four wheels of the size and type recommended by the car manufacturer, brake performance should be at its best.

❑ Radial ply tires are used almost universally on late-model cars and light trucks.

❑ Tire size identification, maximum inflation pressure, and other specifications are molded into tire sidewalls.

❑ Tire and wheel runout, as well as wheel rim width and offset, can contribute to brake problems if they deviate from the car manufacturer's specifications.

❑ Late-model cars and light trucks use tapered roller bearings, straight roller bearings, and ball bearings at their wheels and axles.

❑ If wheel bearings have too much end play (looseness), wheel runout may be excessive and cause uneven braking and a pull to one side. Excessive runout also can contribute to pedal pulsation.

❑ The traditional wheel alignment angles of caster, camber, and toe can contribute to brake pull problems if they are out of specification or vary greatly from side to side.

❑ Loose or damaged steering and suspension parts also can cause brake pull problems.

Terms to Know
Aspect ratio
Belted bias ply tire
Bias ply tire
Camber
Carcass
Casing
Caster
Cold inflation
  pressure
Geometric
  centerline
Gross vehicle
  weight rating
  (GVWR)

Review Questions

Terms to Know (continued)

National Highway Transportation and Safety Agency (NHTSA)

P-metric system

Pressure-sensor base (PSB)

Radial ply tire

Run flat

Scrub radius

Section width

Setback

Steering axis inclination (SAI)

Thrust angle

Thrust line

Tire load range

Tire pressure monitoring system (TPMS)

Toe angle

Toe-out on turns (turning radius)

Tread

Tread contact patch

Tread wear indicator

Unidirectional tread pattern

Uniform tire quality grading (UTQG) indicators

Wheel offset

Wheel-speed base (WSB)

Short-Answer Essays

1. Identify and explain the systems of tire size designations and other tire specifications.

2. Describe the basic kinds of tire construction and identify the most common construction method for modern tires.

3. Explain the use of unidirectional tread pattern and its disadvantage compared to standard production tire tread.

4. Discuss run-flat tires.

5. List the basic adjustable wheel alignment and steering angles.

6. List the nonadjustable alignment and stearing angles.

7. Identify the common types of wheel bearings used on cars and light trucks.

8. Explain how the condition of related systems may affect braking.

9. Identify the principal dimensions used to specify wheel size.

10. Explain the operational differences between WSB and PSB systems.

Fill in the Blanks

1. The steel beads around the rim and layers of cords or plies that are bonded together to give a tire its shape and strength are called the _____.

2. The layer of tire rubber that contacts the road and that contains a distinctive pattern is called the _____.

3. A tire with the cords in the body plies of the carcass running at an angle of 90 degrees to the steel beads in the inner rim of the carcass is a _____ _____ tire.

4. The _____ _____ is the percentage of tire cross-sectional height to cross-sectional width.

5. The most common system used to identify tires for passenger cars and many light trucks today is the _____ system.

6. On wet pavement, a slick tire will _____ on a layer of water trapped between the tread and the pavement.

7. The distance between the centerline of a rim and the mounting plane of the wheel is _____ _____.

8. The distance from the tire contact patch centerline to the point where the steering axis intersects the road is _____ _____.

9. The inward or outward tilt of the wheel measured from top to bottom and viewed from the front of the car is _____.

10. The backward or forward angle of the steering axis viewed from the side of the car is _____.

Multiple Choice

1. Which of the following are listed on the tire placard?
   - **A.** cold tire pressure
   - **B.** load range
   - **C.** ply rating
   - **D.** all of the above

2. The casing of a tire is the portion
   - **A.** with the steel beads molded into it.
   - **B.** that provides the traction with the road.
   - **C.** with the additional layers of sidewall and undertread rubber.
   - **D.** none of the above.

3. The most common tread wear problem is
   - **A.** overinflation.
   - **B.** underinflation.
   - **C.** poor alignment.
   - **D.** incorrect tire size for the vehicle.

4. TPMS may use any of the following EXCEPT
   - **A.** ABS.
   - **B.** PSB.
   - **C.** WSB.
   - **D.** NHTSA.

5. Radial runout is the
   - **A.** side-to-side movement of a tire or wheel.
   - **B.** up-and-down movement of a tire or wheel.
   - **C.** backward and forward motion of the suspension.
   - **D.** up-and-down motion of the suspension.

6. When adjusting tire pressure, it should be adjusted to
   - **A.** the cold inflation pressure shown on the placard mounted on the underside of the hood.
   - **B.** the cold inflation pressure shown on the placard located at the driver's door.
   - **C.** the maximum cold inflation pressure shown on the sidewall.
   - **D.** either A or B.

7. Wheel offset will affect
   - **A.** wheel bearing wear.
   - **B.** the load-carrying capacity of the axles and spindles.
   - **C.** steering scrub radius.
   - **D.** all of the above.

8. FWD vehicles usually have which of the following type of bearing on the drive wheels?
   - **A.** sealed ball bearing
   - **B.** roller bearing
   - **C.** tapered roller bearing
   - **D.** ball bearing

9. Which of the following adjustable steering angles will NOT normally cause abnormal tread wear?
   - **A.** camber
   - **B.** caster
   - **C.** toe
   - **D.** toe-out on turns

10. Improper alignment angles can cause
    - **A.** pull to one side during braking.
    - **B.** pull toward the most negative caster wheel.
    - **C.** pull toward the most positive camber wheel.
    - **D.** all of the above.

Master Cylinders and Brake Fluid

Upon completion and review of this chapter, you should be able to:

❏ Explain the differences between different DOT brake fluid specifications.

❏ Explain proper brake fluid handling procedures.

❏ Identify the parts and explain the operation of a brake pedal and pushrod.

❏ Describe the mechanical advantage provided by the brake pedal and linkage.

❏ Describe the purpose of the master cylinder.

❏ Identify the main parts of a master cylinder.

❏ Explain the operation of a basic master cylinder.

❏ Describe the parts and operation of a dual-piston master cylinder.

❏ Explain the purpose and operation of the front-to-rear and diagonally split hydraulic systems.

❏ Describe the parts and explain the operation of a quick take-up master cylinder with a fast-fill valve.

❏ Describe the parts and explain the operation of a quick take-up master cylinder.

Introduction

This chapter covers hydraulic fluid, which makes the brake system work, as well as the master cylinder construction and operation. The master cylinder converts the force on the brake pedal to each of the four wheel brakes to stop the car. The master cylinder changes the driver's mechanical force on the pedal to hydraulic pressure, which is changed back to mechanical force at the wheel brakes. The master cylinder uses the fact that fluids are not compressible to transmit the pedal movement to the wheel brake units. The master cylinder also uses the principles of hydraulics to increase the pedal force applied by the driver. The brake pedal's construction and operation are also discussed in this chapter as part of the master cylinder operation.

Hydraulic Brake Fluid

The specifications for all automotive brake fluids are defined by Society of Automotive Engineers (SAE) Standard J1703 and Federal Motor Vehicle Safety Standard (FMVSS) 116. Fluids classified according to FMVSS 116 are assigned DOT numbers: DOT 3, DOT 4, DOT 3/4, DOT 5, and DOT 5.1. Basically, the higher the DOT number, the more rigorous the specifications for the fluid.

These specifications list the qualities that brake fluid must have, such as:

❏ Free flowing at low and high temperatures
❏ A high boiling point (over 400°F or 204°C)
❏ A low freezing point
❏ Noncorrosive to metal or rubber brake parts
❏ Ability to lubricate metal and rubber parts
❏ Ability to absorb moisture that enters the hydraulic system

Choosing the right fluid for a specific vehicle is not based on the simple idea that if DOT 3 is good, DOT 4 must be better, and DOT 5 better still. Almost all carmakers specify DOT 3 fluid for their vehicles; but Ford calls for a heavy-duty variation, which meets the basic specifications for

Shop Manual
pages 129–139

DOT 3 but has the higher boiling point of DOT 4. Import manufacturers are about equally divided between DOT 3 and DOT 4.

Polyglycol stands for polyalkylene-glycol-ether brake fluids that meet specifications for DOT 3 and DOT 4 brake fluids.

DOT 3 and DOT 4 fluids are polyalkylene-glycol-ether mixtures, called **polyglycol** for short. The color of both DOT 3 and DOT 4 fluid ranges from clear to light amber. DOT 5 fluids are all silicone based because only silicone fluid—so far—can meet the DOT 5 specifications. No vehicle manufacturer, however, recommends DOT 5 fluid for use in its latest brake systems. Although all three fluid grades are compatible in certain aspects, they do not combine if mixed together in a system. DOT 5, a silicone-based fluid, in particular should never be mixed with or used to replace other types of brake fluids. Therefore, the best general rule is to use the fluid type recommended by the carmaker and to not mix fluid types in a system.

Brake Fluid Boiling Point

The most apparent differences among the five fluid grades are the minimum boiling points as listed in Table 4-1.

The boiling point is important because heat generated by braking can enter the hydraulic system. If the temperature rises too high, the fluid can boil and form vapor in the brake lines. The stopping power of the system then will be reduced. The pedal will have to be pumped repeatedly to compress the vaporized fluid and build up pressure at the brakes to stop the vehicle.

The dry boiling point is the minimum boiling point of new, uncontaminated fluid. After brake fluid has been in service for some time, however, the boiling point drops because of water contamination. Polyglycol fluids are hygroscopic, which means that they readily absorb water vapor from the air. Brake systems are not completely sealed, and some exposure of the fluid to air is inevitable. SAE field evaluation program R11 showed that the average 1-year-old car has about 2 percent water in its brake fluid. Even this small amount is enough to lower the boiling point of DOT 3 fluid from 401°F to less than 320°F. DOT 4 fluid absorbs less moisture and maintains a higher boiling point than DOT 3 fluid.

Because both DOT 3 and DOT 4 fluids absorb moisture from the air, always keep containers tightly capped. Reseal brake fluid containers immediately after use. Although brake fluid is available in containers as small as 12 ounces to as large as 5 gallons (Figure 4-1), It is often wise to buy small containers and keep them sealed until needed. Taking these steps to minimize water in the brake fluid helps reduce corrosion of metal parts and deterioration of rubber components. It also keeps the brake fluid boiling point high to minimize the chance of vapor formation.

AUTHOR'S NOTE: Although brake fluid may be purchased in quantities larger than 1 quart, it will be a special-order item. Very few public parts vendors or repair shops keep anything larger than a 1-quart container of brake fluid. Even a "brake-only" repair shop would seldom need brake fluid in quantities that would justify storing 5-gallon containers of it. Plus when you consider that brake fluid must be placed in a smaller container for ease of pouring and labeled accordingly, it is not worth the trouble. Brake fluid should always be purchased in 1-quart or smaller containers.

Table 4-1 MINIMUM BOILING POINTS FOR THE SIX BRAKE FLUID GRADES

Boiling Point	DOT 3	DOT 4	DOT 5	DOT 3/4	DOT 5.1
Dry	401°F (205°C)	446°F (230°C)	500°F (260°C)	500°F (260°C)	585°F (307°C)
Wet*	284°F (140°C)	311°F (155°C)	356°F (180°C)	347°F (175°C)	365°F (185°C)

DOT 5.1 long-life has the same dry and wet boiling point of 424°F (218°C)

*WET means the fluid is saturated with water.

Figure 4-1 Brake fluid shoud be purchased only in the quantities needed for the job.

The DOT rating is found on the container of brake fluid (Figure 4-2). The vehicle service manual and owner's manual specify what rating is correct for the car. Do not use a brake fluid with a lower DOT rating than specified by the vehicle manufacturer.

DOT 3 fluid will last longer than DOT 4 under ideal conditions; that is, the system is clean, sealed, and kept sealed or "perfect." DOT 3 is considered to be a lifetime fluid under those conditions, whereas DOT 4 still must be flushed out and replaced every 1–3 years under the same conditions. It is best to completely flush light vehicle brake systems about every 2–3 years because no typical brake system is "perfect."

The synthetic fluids have higher wet and dry boiling points. This makes them, at least in theory, better for braking and durability while in the braking system. The service life of DOT 3/4 is about 24 months (2 years), whereas DOT 5.1 is rated for about 60 months (5 years). Long-life DOT 5.1 is rated at a service life of up to 10 years.

Figure 4-2 The DOT number is always on the label of a brake fluid container.

Other Brake Fluid Requirements

A high-temperature boiling point is not the only requirement that brake fluid must meet. Brake fluid must remain stable throughout a broad range of temperatures, and it must retain a high boiling point after repeated exposures to high temperatures. Brake fluid also must resist freezing and evaporation and must pass specific viscosity tests at low temperatures. If the fluid thickens and flows poorly when cold, brake operation would suffer at low temperatures.

In addition to temperature requirements, brake fluid must pass corrosion tests. It also must not contribute to deterioration of rubber parts and must pass oxidation-resistance tests. Finally, brake fluid must lubricate cylinder pistons and bores and other moving parts of the hydraulic system.

DOT 5 Silicone Fluid

DOT 5 Silicone fluid does not absorb water. This purple fluid has a very high boiling point, is non-corrosive to hydraulic system components, and does not damage paint as does ordinary fluid. However, DOT 5 fluid has other characteristics that are not so beneficial.

Silicone fluid compresses slightly under pressure, which can cause a slightly spongy brake pedal feel. Silicone fluid also attracts and retains air more than polyglycol fluid does, which makes brake bleeding harder; it tends to out gas slightly just below its boiling point; and it tends to aerate from prolonged vibration. DOT 5 fluid has other problems with seal wear and water accumulation and separation in the system. All of these factors mean that DOT 5 silicone fluid should *never be used* in an ABS.

Synthetic Fluids

As mentioned earlier, synthetic brake fluids are now available on the market. DOT 3/4 and DOT 5.1 are compatible with DOT 3 and DOT 4 fluids and can be mixed with them. DOT 5.1 is based on polyglycol chemistry. Both can be used to flush and replace DOT 3 or DOT 4 fluid in a brake system and can be used in an ABS. They are distinguished from each other by color: DOT 3 and DOT 4 are clear and amber, DOT 3/4 is clear and pale blue, DOT 5.1 is crystal clear, and the long-life version of DOT 5.1 is clear and yellow in color. DOT 5.1 has extra solvency to put gum and sludge deposits into suspension with additional anticorrosion additives in the long-life version.

DOT 5.1 and DOT 5.1 long-life brake fluids are *not* silicone-based fluids. Do not *mix* or replace DOT 5 fluid with DOT 5.1 or DOT 5.1 long-life fluids. Damage to the brake system and possible injury could occur due to damage to brake system components.

Hydraulic System Mineral Oil Fluids

Hydraulic system mineral oil (HSMO) is the rarest kind of brake fluid, used by only three car-makers: Citroen, Rolls Royce, and Audi, which uses it in some Audi models in the brake booster system. HSMO is not a polyglycol or silicone fluid; rather it is made from a mineral oil base. It has a very high boiling point, it is not hygroscopic, it is a very good lubricant, and it actively prevents rust and corrosion. HSMO fluid can be identified by its green color.

Because HSMO is petroleum based, systems designed for its use also require seals made of special rubber. If polyglycol or silicone fluid is used in a system designed for HSMO, these fluids will destroy the HSMO system seals. Similarly, if HSMO is used in a system designed for polyglycol or silicone fluid, it will destroy the seals of those systems. HSMO is not covered by the DOT classifications of FMVSS 116 and is not compatible with DOT fluids.

Fluid Compatibility

Although the performance requirements of DOT 3, DOT 4, DOT 5, and synthetic fluids are different, FMVSS 116 requires that the four grades of fluid must be compatible with each other in a system. Mixing different types of fluid in a system is not recommended, but it can be done without damaging the system or creating a damaging reaction between two types of fluid.

It is important to remember that if DOT 3 and DOT 4 fluids are mixed in a system, for example, the boiling point of the DOT 4 fluid will be reduced by the same percentage as the percentage of DOT 3 fluid in the mixture. Thus, overall system performance may be compromised by mixing fluids.

Although FMVSS 116 requires DOT fluids to be compatible, it does not require them to be homogeneous. They are not required to blend into a single solution unless the fluids are of a single type, such as two DOT 3 fluids or two DOT 4 fluids.

Silicone DOT 5 fluid has a lower specific gravity than polyglycol fluid. If the two types are mixed, they do not blend; the silicone fluid separates and floats on top of the polyglycol fluid. Therefore, if a customer wants silicone fluid in his or her vehicle, all the polyglycol fluid must be completely flushed out. The best time to convert to silicone fluid is during a complete brake system overhaul.

Although the synthetic fluids may be compatible with the DOT 3 and 4 fluids, they also should not be mixed except in an emergency situation. They will blend, but the performance level of the synthetics will decrease in proportion to the amount of polyglycol-based fluid remaining in the system. If a synthetic fluid is requested by the customer, then a complete flushing and removal of all polyglycol fluids must be accomplished. This would also be a good time to check for any damage to the brake seals. If the fluid has been in the system for several years, it would benefit the customer to replace or rebuild those brake components housing flexible or fixed seals.

The best practice is to use a single, high-quality brand of brake fluid of the DOT type specified for a particular vehicle. Avoid mixing fluids whenever possible.

Brake Fluid Storage and Handling

Polyglycol fluids have a very short shelf life. As soon as a container of DOT 3 or DOT 4 polyglycol fluid is opened, it should be used completely because it immediately starts to absorb moisture from the air. DOT 5 silicone fluids and HSMO fluids are not hygroscopic and can be stored for long periods of time.

All brake fluids must be stored in clean containers in clean, dry locations. Preferably, brake fluid should be stored in its original container with all labeling intact. Containers must be kept tightly closed when not in use.

When working with brake fluid, do not contaminate it with petroleum-based fluids, water, or any other liquid. Also keep dirt, dust, or any other solid contaminant away from the fluid. When filling a system with polyglycol fluid, keep it off painted surfaces and off your skin.

If possible, buy brake fluid in metal containers if it is to be stored. Plastic containers could allow air and/or moisture to seep through the container material, contaminating the fluid even when the container has not been opened. This is particularly true when the container is sitting in a shop or home garage where it is exposed to the environment to some extent. Contamination is not as great a concern at the vendor where the stock is stored and sold in a short period of time.

Other Brake Fluid Precautions

Brake fluid is considered a toxic and hazardous material. Used brake fluid must be disposed of in accordance with local regulations and EPA guidelines. Do not pour used brake fluid down a wastewater drain or mix it with other chemicals awaiting disposal. Observe these general precautions when working with brake fluid:

- ❑ Never mix polyglycol and silicone fluids because the mixture could cause a loss of brake efficiency and possible injury.
- ❑ Brake fluid can cause permanent eye damage. Always wear eye protection when handling brake fluid. If you get fluid in your eyes, rinse them thoroughly with water, then see a doctor immediately.
- ❑ Brake fluid may also irritate your skin. Wear chemical-resistant gloves. If fluid gets on your skin, wash the area thoroughly with soap and water.

❏ Always store brake fluid in clean, dry containers. Protect brake fluid from contamination by oil, grease, or other petroleum products, as well as contamination by water. Never reuse brake fluid.

❏ Do not spill polyglycol brake fluid on painted surfaces. Polyglycol fluid will damage a painted surface. Always flush any spilled fluid immediately with cold water.

Contaminated Fluid Problems

Always use the type of brake fluid recommended in the vehicle service manual or owner's manual. Using the wrong kind of fluid or using fluid that is contaminated with any other liquid can cause the brake fluid to boil or rubber parts in the hydraulic system to deteriorate.

Swollen master cylinder piston seals are the best indicator of contamination in the brake fluid. Deterioration also may be indicated by swollen wheel cylinder boots, caliper boots, or a damaged master cylinder cover diaphragm.

If you find water or other contaminants in the brake system and the master cylinder piston seals have been damaged, replace all sealing parts in the system, including the brake hoses and any valves or switches with rubber seals. Also check for brake fluid on the brake linings. If you find any, replace the linings.

 AUTHOR'S NOTE: There is a quick, easy test to check the brake fluid for contamination. It is covered in the *Shop Manual*.

If water or contaminants are in the system, but the master cylinder seals appear undamaged, check for leakage throughout the system or signs of heat damage to hoses or components. Replace all damaged components found. After repairs are made, or if no leaks or heat damage are found, drain the brake fluid from the system, flush the system with new brake fluid, refill, and bleed the system.

Brake Pedal and Pushrod

Shop Manual
pages 120–129

Braking action begins when the driver pushes on the brake pedal. The brake pedal (Figure 4-3) is a lever that is pivoted at one end, with the master cylinder pushrod attached to the pedal lever near the pivot point. This section explains how this lever arrangement multiplies the force applied at the brake pedal several times as it is applied to the master cylinder pushrod.

One end of the pushrod engages the master cylinder piston, and the other end is connected to the pedal linkage. The pushrod often is adjustable, with many made in two parts so they can be lengthened or shortened. Adjusting pushrod length is explained in the *Shop Manual*.

 AUTHOR'S NOTE: Vehicles with power brakes will have two pushrods, one between the pedal and power booster and another between the power booster and master cylinder. Both may be adjustable.

The brake pedal and master cylinder must be mounted close to each other so the pedal pushrod can operate the master cylinder pistons. There are two basic types of brake pedal mountings.

Older vehicles may have the pedal mounted to a pivot in the floorboard with the master cylinder under the floorboard. This exposed the working mechanism and the master cylinder to road debris and the usual culmination of trash and dirt on the floorboard.

Most late-model cars use a suspended pedal assembly (Figure 4-4). The pedal assembly is mounted to a support bracket that is attached to the inside of the engine compartment cowl or bulkhead. The pushrod that connects the pedal linkage to the master cylinder goes through a hole in the bulkhead.

The master cylinder is mounted on the opposite side of the bulkhead in the engine compartment. If the car has manual brakes, the cylinder is mounted directly to the bulkhead. If a vacuum power brake booster is used, the booster is mounted to the bulkhead and the cylinder is

Figure 4-3 The brake pedal is suspended under the dash on modern vehicles. Its lever action will increase the driver input braking force depending on the length and curve of the pedal arm.

Figure 4-4 Almost all brake pedal and master cylinder installations since the 1950s have been a suspended pedal with the master cylinder mounted on the engine compartment bulkhead.

mounted to the booster (Figure 4-5). With this arrangement, the master cylinder is easy to get to for checking or service.

Brake Linkage Free Play

All brake pedal linkage must provide some amount of **free play** between the master cylinder pistons and the pushrod. This free play is necessary to let the master cylinder pistons retract completely in their bores. Free play at the master cylinder is usually very slight: about $\frac{1}{16}$ inch (1.5 mm to 2.0 mm). At the

Figure 4-5 A suspended pedal installation with a vacuum power booster.

brake pedal, the free play is multiplied by the pedal ratio. Thus, if free play at the master cylinder is $\frac{1}{16}$ inch (1.5 mm) and the pedal ratio is 4:1, free play at the pedal will be $\frac{1}{4}$ inch (6 mm) (Figure 4-6).

Free play as measured at the pedal is usually specified by the vehicle manufacturer, at least as an inspection point. Free play is adjustable on some installations and not adjustable on others.

Free play at
pushrod
$\frac{1}{2}$ inch
(1.5 mm)

Pedal free play
$\frac{1}{8}$ to $\frac{1}{4}$ inch
(3 to 6 mm)

Figure 4-6 Free play at the master cylinder pushrod is multiplied by the pedal ratio, which increases free play when measured at the pedal.

Most adjustments are made by loosening a locknut at the pushrod clevis and turning the pushrod to lengthen or shorten it.

Vacuum power brake boosters have a second pushrod that transmits motion from the booster to the master cylinder. The booster pushrod may require adjustment separately from the brake pedal pushrod. The *Shop Manual* explains all necessary free play and pushrod adjustments.

Adjustable Pedal Systems (APS)

Beginning in 1999, Ford offered an **adjustable pedal system (APS)** in two of its models. Since then, other manufacturers have installed or will install adjustable pedal systems in their vehicles. Usually this applies only to the brake and accelerator pedals. The DaimlerChrysler Viper is the first to offer an adjustable pedal system for use on a manual transmission vehicle.

The APS is used to move the accelerator and brake pedals forward or back (up and down) for a more comfortable position for those drivers who are shorter or taller than the "average" driver. The APS has a second advantage that is a safety feature for many drivers. For some short drivers the chest area is too close to the driver's air bag, which could result in serious injury caused by air bag deployment. The same applies to taller drivers who may have to move the seat for comfort and get too far from the air bag to gain the protection offered.

There is only one motor. The motor drives the brake pedal adjustment directly and uses a flexible shaft to turn the adjusting screw on the accelerator pedal (Figure 4-7). The switch is dash-mounted and may have a memory saver in which one or more drivers can program their optimum pedal position and then recall it when needed. The pedal assembly is replaced as a unit. The pedals and the motor are mounted on a pedal sled.

Figure 4-7 The components of an adjustable pedal system.

Dual-Piston Master Cylinder Construction and Operation

Shop Manual
pages 140–151

Figure 4-8 is a simplified illustration of how the master cylinder and the hydraulic system work. The master cylinder pushrod is connected to a piston inside the cylinder, and hydraulic fluid is in front of the piston. When the pedal is pressed, the master cylinder piston is pushed forward. The fluid in the master cylinder and the entire system is noncompressible and immediately transmits the force of the master cylinder piston to all the inner surfaces of the system. Only the pistons in the drum brake wheel cylinders or disc brake calipers can move, and they move outward to force the brake shoes or pads against the brake drums or rotors.

Master Cylinder Reservoir

The master cylinder has two main parts: a **reservoir** and a body (Figure 4-9). The reservoir supplies brake fluid for cylinder operation. All master cylinders and reservoirs built since 1967 are split or dual-chamber designs, meaning they have two separate storage areas for two separate piston assemblies. The split design separates the hydraulic system into two independent sections providing reserve braking operation in case one section fails.

The reservoir may be cast as one piece with the cylinder body or it may be a separate molded nylon or plastic container (Figure 4-10). All reservoirs have a removable cover so that brake fluid can be added to the system. The one-piece reservoirs typically have a single cover that is held on the reservoir with a retainer bail. Nylon or plastic reservoirs may have a single cover or two screw caps on top of the reservoir. Separate reservoirs may be clamped or bolted to the cylinder body, or they may be pressed into holes in the top of the body and sealed with grommets or **O-rings.**

O-rings are circular rubber seals shaped like the letter "O."

All master cylinder caps or covers are vented to prevent a vacuum lock as the fluid level drops in the reservoir. A flexible rubber **diaphragm** at the top of the master cylinder reservoir is incorporated in the screw caps or the cover. The reservoir diaphragm separates the brake fluid from the air above it while remaining free to move up and down with changes in fluid level. The diaphragm keeps the moisture in the air from entering the brake fluid in the reservoir. Moisture in the brake fluid will lower the fluid boiling point.

If a vehicle with front disc and rear drum brakes has the brake system split so that the front brakes are on one circuit and the rear brakes on the other circuit, the reservoir chamber for the disc brakes is larger than the chamber for the drum brakes. As disc pads wear, the caliper pistons move

Figure 4-8 Simplified diagram of master cylinder and hydraulic system operation.

Figure 4-9 This master cylinder is viewed from the passenger side of the engine compartment. Note the power booster between the master cylinder and bulkhead. Also note that this master cylinder has a detachable reservoir.

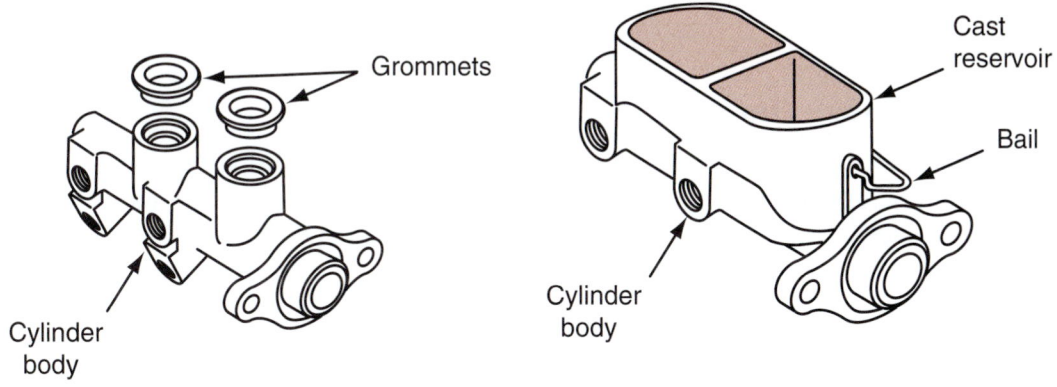

Figure 4-10 The master cylinder assembly to the left is the most common one in use today. However, there are still many vehicles on the road that are equipped with the one-piece unit shown to the right.

out farther in their bores. More fluid is then required to keep the system full from the master cylinder to the calipers. Drum brake wheel cylinder pistons always retract fully into the cylinders regardless of brake lining wear so the volume of fluid does not increase with lining wear. Vehicles with four-wheel disc brakes or diagonally split hydraulic systems usually have master cylinders with equal-size reservoirs because each circuit of the hydraulic system requires the same volume of fluid.

Plastic reservoirs often are translucent so that fluid level can be seen without removing the cover. Although this feature allows a quick check of fluid level without opening the system to the air, you should not rely on it for thorough brake fluid inspection. Stains inside the reservoir

The piston bore in a disc caliper is usually much larger than the one in a wheel cylinder, which is an additional reason for the disc side to have a larger reservoir chamber.

can give a false indication of fluid level, and contamination cannot be seen without removing the reservoir caps or cover.

American Motors Corporation introduced the dual hydraulic system on its 1960 cars. In 1963, several manufacturers added dual tandem in-line master cylinders to their cars. By 1967, DOT established FMVSS 105 that requires that all cars sold in the United States have split brake hydraulic systems. In 1978, Chrysler Corporation, later DaimlerChrysler, introduced the first mass-produced cars with a diagonally split hydraulic system.

Master Cylinder Ports

Many different names have been used for the ports in the master cylinder, and often the same name has been applied to each of the ports. For the sake of uniformity, this text refers to the forward port as the "vent" port and the rearward port as the "replenishing" port, which are the names established by Society of Automatic Engineers (SAE) Standard J1153.

The **vent port** has been called the compensating port and the replenishing port by some manufacturers. To further confuse the nomenclature, the **replenishing port** has been called the compensating port by some manufacturers, as well as the vent port, the bypass port or hole, the filler port, or the intake port by many manufacturers.

However, it is not so important what you call these ports (or holes) as it is that you understand their purposes and operations. Figure 4-11 shows the different names for these master cylinder ports. Although the vehicle and parts manufacturers have not always observed the nomenclature of SAE Standard J1153, its use in this manual maintains consistency and understanding.

Master Cylinder Construction

Figure 4-12 shows the parts of the dual-piston master cylinder in cross section. Figure 4-13 is a disassembled view. A single-piston bore contains two piston assemblies. The piston assembly at the rear of the cylinder is the primary piston, and the one at the front of the cylinder is the secondary

> The **vent port** is the forward port in the master cylinder bore.
>
> The **replenishing port** is the rearward port in the master cylinder bore.
>
> The vent port is also called replenishing port, compensating port, bypass hole.
>
> The replenishing port is also called compensating port, vent port, bypass port (hole), filler port, intake port.

Figure 4-11 The vent port and the replenishing port have been called many different names, but the terms used here are established by the SAE standard for master cylinder terminology.

Vent ports

Replenishing ports

Pushrod

Secondary piston

Primary piston

Figure 4-12 Cross section of a dual master cylinder.

Reservoir cover

Diaphragm

Separate reservoir

Grommets

Cylinder body

Spring

Spring retainer

Seal

Secondary piston

Seal

Primary piston

Lockring

Figure 4-13 Note how the two pistons and related components slide into the master cylinder bore. The primary piston will be the one closest to the driver and will serve the front wheels in most cases.

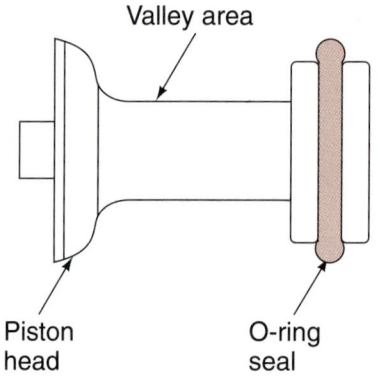

Figure 4-14 Simplified spool-shaped piston for master cylinder.

The **cup seal** is the circular rubber seal with a depressed center section that is surrounded by a raised sealing lip to form a cup; cup seals are often used on the front ends of hydraulic cylinder pistons because they seal high pressure in the forward direction of travel but not in the reverse.

piston. Each piston has a return spring in front of it. The primary piston has a long extension screw in front of it that is used if there is a hydraulic failure. There is a **cup seal** in front of each piston and a cup or seal at the rear of each piston. The seals retain fluid in the cylinder chambers and prevent seepage between the cylinders.

Two ports are at the bottom of each reservoir section. One set of ports is called the vent ports; the others are replenishing ports. The vent ports and replenishing ports let fluid pass between each pressure chamber and its fluid reservoir during operation.

Two spool-shaped pistons are inside the cylinder. Figure 4-14 is a simplified illustration of a spool-shaped piston. The piston has a head on one end and a groove for an O-ring seal on the other end. The seal seats against the cylinder wall and keeps fluid from leaking past the piston. The smaller diameter center of the piston is the valley area, which lets fluid get behind the head of the piston. This prevents low pressures developing here as the piston moves forward.

Each master cylinder piston works with a rubber cup seal that fits in front of the piston head (Figure 4-15). The cup has flexible lips that fit against the cylinder walls to seal fluid pressure ahead of the piston head. The cup lip also can bend to let fluid get around the cup from behind. When the brakes are applied, pressure in front of the cup forces the lip tightly against the cylinder wall and lets the seal hold very high hydraulic pressure. The lip of a cup seal is always installed toward the pressure to be contained or away from the body of the piston. The cup seals only in

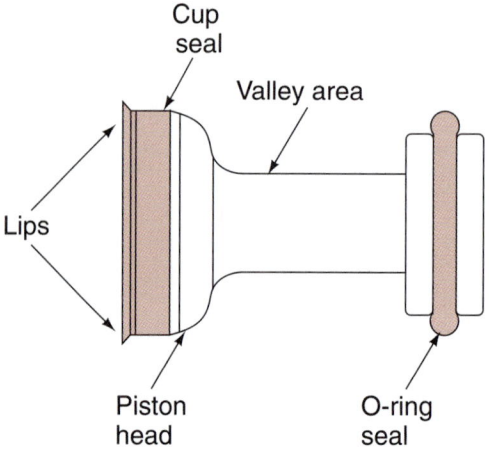

Figure 4-15 A cup seal and O-ring seal the piston in the cylinder bore.

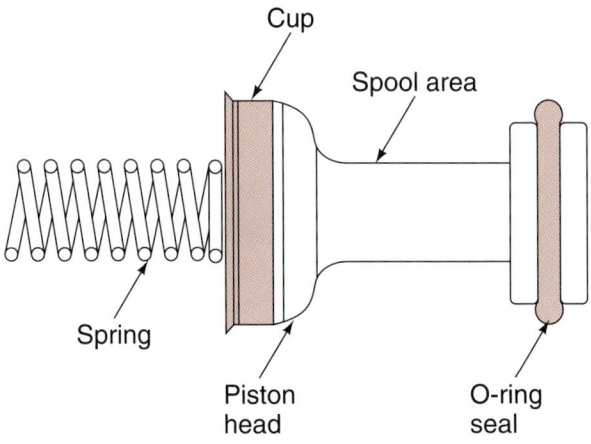

Figure 4-16 Each piston has a spring to return it to the released position.

one direction. If pressure behind the lip exceeds pressure in front of it, the higher pressure will force the lip away from the cylinder wall and let fluid bypass the cup.

Pistons have small coil springs (Figure 4-16) that return the pistons to the proper position when the brake pedal is released. Sometimes the springs are attached to the pistons; sometimes they are separate parts.

A dual master cylinder has two piston assemblies. The piston nearest the pushrod is the primary piston; the other is the secondary piston. Each piston provides a separate hydraulic system for the front and rear brakes or, on a diagonally split system, between one set of front brakes and one set of rear brakes. Two outlet holes provide the connection for the hydraulic lines. A snapring holds the components inside the cylinder, and a flexible boot fits around the rear of the cylinder and pushrod to keep dirt from entering the cylinder.

Figure 4-17 shows the major parts of the two-piece master cylinder. This master cylinder has an aluminum body to lower the weight of the assembly. Aluminum offers the advantage of being much lighter than cast iron, but has large pores in the metal that can damage the lip seals of the bore pistons. The interior walls of the bore are rolled, or bearingized, to close these pores. The aluminum master cylinder bore should not be honed during service because the pores will be reopened. Replacement is required if the interior bore is worn or damaged. The removable nylon or plastic reservoir is also much lighter than a cast-iron unit. Because the master cylinder is made from two materials, it is often called a composite master flexible. The pistons, cups, and springs used in a composite master cylinder are essentially the same and work the same way as those in a one-piece master cylinder.

Master Cylinder Operation

The primary and secondary pistons of the master cylinder operate in the same way during normal braking. Figures 4-18 through 4-23 show the operation of the primary piston. The secondary piston is moved forward by the pressure created ahead of the primary piston. Each of the two pistons has a stop at its front end. This provides a means to allow one piston to build pressure if there is an external leak in the brake system. As the noted figures are examined, understand that the secondary piston is performing the same action as the primary.

Figure 4-18 shows a piston assembly and reservoir with the piston in the released position. The reservoir is full of brake fluid. The vent port in the bottom of the reservoir is located just ahead of the piston cup. Fluid flows from the reservoir through the vent port into the pressure chamber in front of the piston cup. The replenishing port is located above the valley area of the piston, behind the piston head. The O-ring seal on the piston keeps the fluid from leaking out the rear of the cylinder. The return spring in front of the piston and cup returns the piston when the brakes are released.

The low-pressure area is created by the fast rearward movement of the piston and the slow-moving brake fluid. This sudden increase of empty volume causes the pressure to drop.

Figure 4-17 A two-piece composite master cylinder has a plastic reservoir mounted on an aluminum body.

Figure 4-18 Basic operating parts of a single piston within a dual-piston master cylinder.

Figure 4-19 As the piston moves forward, the cup seal closes the vent port and lets the pressure develop ahead of the piston.

When the driver presses on the brake pedal, the pushrod pushes the master cylinder piston forward (Figure 4-19). As the piston moves forward, the piston pushes the cup past the vent port. As soon as the vent port is covered, fluid is trapped ahead of the cup. The fluid, under pressure, goes through the outlet lines to the wheel brake units to apply the brakes.

When the driver releases the brake pedal, the return spring forces the piston back to its released position. As the piston moves back, it pulls away from the fluid faster than the fluid can flow back from the brake lines to the pressure chamber. When this happens, low pressure is created ahead of the piston.

The piston must move back to the released position rapidly so it can be ready for another forward stroke, if necessary. The low-pressure area must be filled with fluid as the piston moves back. A path for fluid flow is provided from the valley area (Figure 4-20), past the primary cup protector washer and through several small holes in the head of the piston, or by having enough clearance between the piston head and the cylinder bore. Fluid flows through the piston or

Figure 4-20 As the piston moves backward on the return stroke, fluid flows around the piston head and cup from the low-pressure area in the piston valley.

Figure 4-21 The low-pressure valley area is refilled through the replenishing port.

around the lip of the cup and into the chamber ahead of the piston. This flow quickly relieves the low-pressure condition.

The fluid that flows from the valley area to the pressure chamber must be replaced. Figure 4-21 shows how the replenishing port lets fluid from the reservoir flow into this area.

When the piston is fully returned to its released position, the space in front of it is full of fluid. The piston cup again seals off the head of the piston. In the meantime, the fluid from the rest of the system has begun to flow back to the high-pressure chamber. If we could not release this fluid pressure, the brakes would not release. Figure 4-22 shows how the returning fluid flows back to the reservoir through the vent port. The vent port is covered by the piston cup at all times, except when the piston is released.

The replenishing port also has another important job. There are times when the amount of fluid in the wheel brake units and lines must be increased. When disc brake pads wear, there is more space in the brake calipers for hydraulic fluid. When drum brake lining wears, before the automatic adjusters work, more fluid is needed in the wheel cylinders. When the brake system is

Figure 4-22 On the return stroke, fluid flows from the high-pressure chamber back to the reservoir through the vent port.

Figure 4-23 Fluid can flow around the cup seal when more fluid is required.

serviced and air is in the system taking up space, more fluid is needed in the system to force the air out.

This is accomplished in the same way low pressure ahead of the piston is relieved. On the return stroke, fluid flows through the head of the piston and around the lip of the cup (Figure 4-23). When the piston returns, the vent port is open. There is not as much fluid returning from the wheel brake units and lines, so less fluid flows through the vent port and back into the reservoir. If the drum brakes are adjusted or new pads are installed on disc brakes, the system will automatically compensate for the amount of fluid needed on the next piston cycle.

Residual Pressure Check Valve

The pressure chamber for a master cylinder for some drum brake systems may have an additional part called a **residual pressure check valve.** The check valve (Figure 4-24) can be installed in the pressure chamber or the outlet line of the master cylinder. A residual pressure check valve is the oldest kind of pressure control valve used in a brake system. It can be found in older four-wheel drum brake systems, as well as in late-model systems.

Figure 4-24 A residual pressure check valve maintains slight fluid pressure in a drum brake master cylinder when the brakes are released.

This valve usually is installed in the master cylinder outlet port to drum brakes. It maintains a residual pressure of 6 psi to 25 psi in the brake lines when the pedal is released. This residual line pressure maintains slight pressure on the wheel cylinder pistons to keep the sealing lips of the piston cups forced outward against the cylinder walls. When the pedal is released during braking operation, the retracting master cylinder piston creates a pressure drop in the lines. If pressure were to drop low enough, the piston cups could be pulled away from the wheel cylinder walls and draw air into the system. The slight residual pressure prevents this from happening but is not high enough to overcome brake shoe spring tension. Maintaining pressure on the piston cups also maintains their fluid-sealing integrity and helps to prevent fluid leakage. This residual pressure also reduces the application time needed when the brakes are applied. Disc brake systems do not use this valve because residual pressure would cause the pads to drag on the rotor when the brakes are released.

The spring-loaded residual pressure check valve may be installed under the tubing seat in the master cylinder outlet port or inside the cylinder bore. When the brakes are applied, master cylinder pressure opens the valve and allows fluid flow to the wheel cylinders. When the brakes are released, pressure in the lines unseats the valve in the reverse direction to allow fluid return flow to the master cylinder. When line pressure drops below the pressure of the check valve spring, the valve closes to hold residual pressure in the lines. With the redesign of wheel cylinders, many late-model master cylinders do not have a residual pressure check valve in their drum brake lines.

Residual pressure check valves once were used in almost all brake systems, but their use has decreased since the late 1980s. Piston cup expanders were developed for wheel cylinders to hold the cups against the cylinder walls and to keep air from being drawn into the wheel cylinders (Figure 4-25). Cup expanders are simpler, cheaper, and more reliable than check valves.

Diagonally split brake systems are another reason for the elimination of residual pressure check valves. A diagonally split system usually pairs one disc brake with one drum brake for half of the hydraulic system. Disc brakes cannot operate properly with residual pressures. Any residual pressure at all would cause brake drag.

Figure 4-25 The cup expander, centered and attached to the spring, has eliminated the need for a residual check valve in the master cylinder.

Although residual pressure check valves have been eliminated from many systems, they are still used in others. Many master cylinders are assembled from a standard casting, and the finished cylinder may or may not have a check valve depending on the intended vehicle installation. When a master cylinder is replaced, it is very important to verify whether or not the vehicle requires a residual pressure check valve. Installing the wrong cylinder will cause improper brake operation and possible system failure.

A BIT OF HISTORY

Single-piston master cylinders used before 1967 had one piston and one hydraulic circuit for all four wheel brake units. The problem with this system was that a hydraulic failure in the master cylinder could cause a complete loss of brakes to all four wheels. Most times they were used on four-wheel drum brakes without power boosters.

The accompanying illustration shows the parts of a single-piston master cylinder. These units were typically made from cast iron. The reservoir and cylinder body were one piece. The filler cap was usually a metal screw cap.

One piston assembly had a primary cup in front of the piston head. The rear of the piston was often sealed with a secondary cup. A coil return spring and residual pressure check valve were located in front of the piston. A snap ring at the rear of the cylinder held the components in the cylinder. An adjustable pushrod and boot were located at the rear of the cylinder. A single outlet at the front of the cylinder provided pressure to all the wheel brake units. A replenishing vent port was located in the bottom of the reservoir. A single-piston master cylinder operates exactly like one piston in a dual master cylinder.

A few General Motors cars of the mid-1980s had a single-piston master cylinder when the vehicle was equipped with an ABS. These systems rely on a proportioning valve that has a safety bypass feature. If there is a pressure failure in the rear brake system, the proportioning valve closes, stopping pressure from being sent to the rear brakes and allowing safe stopping with the front brakes.

Split Hydraulic Systems

The dual master cylinder has two independent hydraulic systems. These two hydraulic systems can be connected to the wheel brake units in either a front-to-rear or diagonal arrangement.

The front-to-rear hydraulic split system is the oldest split system. This system has two master cylinder outlets (Figure 4-26). One is connected to a line going to the two rear brakes and the other to a line going to the two front brakes.

Although this system provides two independent hydraulic systems, it has some disadvantages. The front brake system does 60 percent to 70 percent of the braking. If the front hydraulic system were to fail, all of the braking would have to be done by the rear wheels. This means that the car would have to be stopped with only 30 percent to 40 percent of its brake power.

Most late-model cars have a diagonally split hydraulic system. The diagonally split system (Figure 4-27) has one of the master cylinder circuits connected to the left front and right rear brakes. The other circuit is connected to the right front and left rear brakes. The connection can be made directly to the master cylinder outlets or externally with a valve. The advantage of this system is that if one system fails, the vehicle will have less than 50 percent of the braking action from one front and one rear brake. The driver may experience a left or right brake pull.

Split Hydraulic System Operation

In the released position, the cups of the primary and secondary pistons uncover the vent ports as was shown in Figure 4-18. Under normal conditions, when the brakes are applied, the primary piston moves forward, and a combination of hydraulic pressure and the force of the primary piston spring moves the secondary piston forward. When the pistons have moved forward so that their

Figure 4-26 Typical front-to-rear split hydraulic system.

Figure 4-27 Typical diagonally split hydraulic system.

② Secondary piston bottoms

① Primary piston applies front brakes and secondary piston

No pressure →

← High pressure

③ Pressure difference turns on warning light

Rear Brakes

Front Brakes

Figure 4-28 Dual master cylinder operation with a primary (front) circuit failure.

cups cover the vent ports, hydraulic pressure is built up and transmitted to the front and rear wheels. The replenishing ports work as described previously.

If there is a hydraulic failure in the brake lines served by the secondary piston (Figure 4-28), both pistons will move forward when the brakes are applied, as under normal conditions, but there is nothing to resist piston travel except the secondary piston spring. This lets the primary piston build up only a small amount of pressure until the secondary piston bottoms in the cylinder

Figure 4-29 Dual master cylinder operation with a secondary (rear) circuit failure.

bore. Then, the primary piston will build enough hydraulic pressure to operate the brakes served by this half of the system.

In case of a hydraulic failure in the brake lines served by the primary piston (Figure 4-29), the primary piston will move forward when the brakes are applied but will not build up hydraulic pressure. Very little force is transferred to the secondary piston through the primary piston spring until the piston stem comes in contact with the secondary piston. Then, pushrod force is transmitted directly to the secondary piston and enough pressure is built up to operate its brakes.

Fast-Fill and Quick Take-Up Master Cylinders

Several carmakers use fast-fill or quick take-up master cylinders. These cylinders fill the hydraulic system quickly to take up the slack in the caliper pistons of low-drag disc brakes. Low-drag calipers retract the pistons and pads farther from the rotor than do traditional calipers. This reduces friction and brake drag and improves fuel mileage.

If a conventional single-bore dual master cylinder were used with low-drag calipers, excessive pedal travel would be needed on the first stroke to fill the lines and calipers with fluid and take up the slack in the pads. To overcome this problem, fast-fill and **quick take-up master cylinders** provide a large volume of fluid on the first stroke of the brake pedal.

A fast-fill or quick take-up master cylinder is identified by the dual bore design that creates a bulge or stepped outside diameter of the casting (Figure 4-30). The cylinder has a larger diameter bore for the rear of the primary piston than for the front of the primary piston. Inside the cylinder, a fast-fill or quick take-up valve replaces the conventional vent and replenishing ports for the primary

Figure 4-30 A typical quick take-up master cylinder.

Figure 4-31 The quick take-up valve serves the primary piston for low-drag calipers.

piston (Figure 4-31). Some master cylinders for four-wheel disc brakes also have a quick take-up valve for the secondary piston.

The **quick take-up valve** contains a spring-loaded check ball that has a small bypass groove cut in the edge of its seat. The outer circumference of the quick take-up valve is sealed to the cylinder body with a lip seal. Several holes around the edge of the hole let fluid bypass the lip seal under certain conditions. Some valves (those more often called "fast-fill" valves) are pressed into the cylinder body and sealed tightly by an O-ring. A rubber flapper-type check valve under the fast-fill valve performs the same bypass functions as a lip seal of a quick take-up valve. The following sections explain the operation of a quick take-up or fast-fill valve.

Brakes Not Applied. When the brakes are off, both master cylinder pistons are retracted, and all vent and replenishing ports are open. Fluid to both ports of the primary piston must flow through the groove in the check ball seat, however.

Brakes Applied. As the brakes are applied, the primary piston moves forward in its bore. Remember that the diameter of the primary valley area is larger than the diameter of the rest of the cylinder. As the primary piston moves forward into the smaller diameter, the volume of the valley area is reduced. This causes hydraulic pressure to rise instantly in the low-pressure chamber. The higher pressure forces the large volume of fluid in the valley area past the cup seal of the primary piston (Figure 4-32), providing the extra volume of fluid to take up the slack in the caliper pistons.

The lip seal of the quick take-up valve keeps fluid from flowing from the valley area back to the reservoir. Initially a small amount of fluid bypasses the check ball through the bypass groove, but this small amount is not enough to affect quick take-up operation.

As brake application continues, pressure in the valley area rises to about 70 psi to 100 psi. The check ball in the quick take-up valve then opens to let excess fluid return to the reservoir (Figure 4-33). Pressures in front and behind the primary piston equalize, and the piston moves forward to actuate the secondary piston. These actions all occur in a fraction of a second.

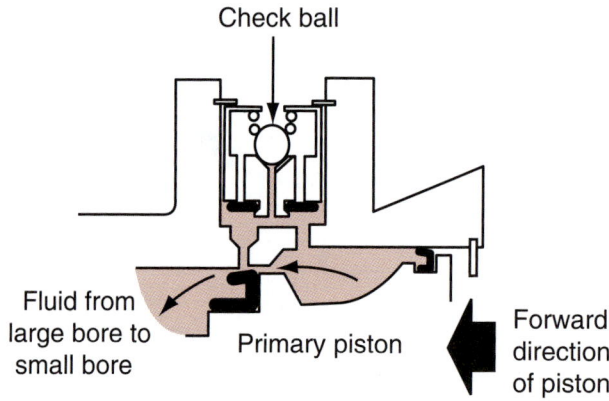

Figure 4-32 As the brakes are applied, fluid flows from the large bore to the small bore for the primary piston.

Figure 4-33 When pressures equalize in both chambers, fluid returns to the reservoir.

Seal opens

Primary piston

Reverse direction of piston

Figure 4-34 When the piston returns, low pressure draws fluid into the valley area.

All of the actions described above apply to the primary piston if it is serving front disc brakes and the secondary piston is serving drum brakes. If the hydraulic system is split diagonally, or if the car has four-wheel, low-drag discs, the quick take-up fluid volume must be available to both pistons. Some master cylinders have a second quick take-up valve for the secondary piston. Others provide the needed fluid volume through the design of the cylinder itself. As long as the primary quick take-up valve stays closed, the fluid bypassing the primary piston cup causes the secondary piston to move farther. This provides equal fluid displacement from both pistons and maintains equal pressure in the system. When the quick take-up valve opens, both pistons move together just as in any other master cylinder.

Brakes Released. When the driver releases the brake pedal, the return springs force the primary and secondary pistons to move back. Pressure drops in the pressure chambers, and fluid bypasses the piston cup seals from the valley area. Low pressure is created in the valley area, which lets atmospheric pressure in the reservoir force fluid past the seal of the quick take-up valve (Figure 4-34). Fluid from the reservoir then flows through both the vent and replenishing ports to equalize pressure in the pressure chambers and valley areas.

On the return stroke, fluid flow to the secondary piston is through a normal replenishing port unless the secondary piston also has a quick take-up valve. If a secondary quick take-up valve is installed, it works as described for a primary quick take-up valve.

Central-Valve Master Cylinders

Some ABSs use master cylinders that have central check valves in the heads of the pistons. If the master cylinder provides pressure during antilock operation (a so-called open system) and the system also has a motor-driven pump, the master cylinder pistons may shift back and forth rapidly during antilock operation. This could cause excessive pedal vibration and, more important, wear on the piston cups where they pass over the vent ports.

To prevent seal damage and pedal vibration, spring-loaded check valves are installed in the piston heads (Figure 4-35). When the brakes are released, fluid flows from the replenishing ports to the valley areas, through the open central check valves, and into the pressure chambers. As the brakes are applied, the central valves close to hold fluid in the pressure chambers. When the brakes are released again, the central check valves open to let fluid flow back through the pistons to the valley areas and the reservoir.

The central check valves in this type of master cylinder provide supplementary fluid passages to let fluid move rapidly back and forth between the pressure chambers and the valley areas during antilock operations. This is not much different in principle from non-ABS fluid flow, but

Secondary chamber

Central valves

Primary chamber

Figure 4-35 This ABS master cylinder has central valves in the piston heads in place of vent ports.

the extra passages reduce piston and pedal vibration and cup seal wear. Chapter 10 in this *Classroom Manual* provides more information on ABS master cylinders.

Performance Fluids and Master Cylinders

The pedal assembly, master cylinder, and brake fluid for racing vehicles are almost the same as those on a production vehicle. There are some differences in weight and increased performance, however.

Most race vehicles are equipped with low-drag disc brakes at all four wheels. To meet the initial braking requirement, a quick take-up master cylinder is needed. Racing master cylinders can be purchased with different bore sizes, resulting in more fluid pressure with reduced force by the driver.

In most instances, only one dual-piston cylinder is used with some type of split system. However, some race crews opt for two identical single-piston master cylinders. The two master cylinders act like a split hydraulic system in that one master cylinder serves the front wheels, wheras the other serves the rear wheels. The master cylinders are applied by one brake pedal acting through a balance bar between the pedal lever and the two pushrods. Some race units are equipped with a brake power booster, and others are not. In this case, it is more an issue of weight than of driver endurance.

Of primary importance to race vehicle braking is the type of brake fluid used. On short tracks with a lot of braking, the boiling point of the fluid can be reached quickly and may be sustained for long periods. Brake fluids developed for racing purposes have the same chemical properties as conventional fluids, generally, but they have much higher boiling points. Castrol offers a blend of polyglycol ester of dimethyl silane, ethylene polyglycols, and oxidation inhibitors. This blend has a dry boiling point of 450°F (232°C) and helps prevent fluid contamination during operation. Another brand, GS610, offers a fluid with a dry boiling point of 610°F (321°C). Racing brake fluids sell for about $45 per quart. One class of brake fluid would be needed for short-track competition and another for long-track. There are several manufacturers and suppliers of racing brake components. Brembo is one of the larger manufacturers of racing components, and some of its products are now being installed on some production performance vehicles. Visit http://www.brembo.com for additional information.

Summary

- Brake fluid specifications are defined by SAE Standard J1703 and FMVSS 116.
- Fluids are assigned DOT numbers: DOT 3, DOT 4, DOT 5, DOT 3/4, and DOT 5.1.
- Always use fluid with the DOT number recommended by the specific carmaker.
- Never use DOT 5 fluid in an ABS or mix with any other brake fluid.
- HSMO fluids are very rare and should never be used in brake systems designed for DOT fluids.
- The brake pedal assembly is a lever that increases pedal force to the master cylinder.
- The brake pedal lever is attached to a pushrod, which transmits force to the master cylinder pistons.
- The master cylinder has two main parts: a reservoir and a cylinder body.
- The reservoir can be a separate piece or cast as one piece with the cylinder.
- A dual-piston master cylinder has two separate pistons providing pressure for two independent hydraulic systems. Each of the two pistons in the master cylinder has a cup, a return spring, and a seal.
- During application, the piston and cup force fluid ahead of the piston to activate the brakes.
- During release, the return spring returns the piston.
- Fluid from the reservoir flows from the reservoir through the replenishing port around the piston cup.
- Excess fluid in front of the piston flows back into the reservoir through the vent ports.
- A front-to-rear split hydraulic system has two master cylinder circuits. One is connected to the front brakes and the other to the rear brakes.
- A diagonally split hydraulic system is one in which one master cylinder circuit is connected to the left front and right rear brakes and the other circuit is connected to the right front and left rear brakes.
- Quick take-up or fast-fill master cylinders have a step bore, which is a larger diameter bore for the rear section of the primary piston.
- Quick take-up master cylinders have a valve that provides rapid filling of the low-pressure spool area of the primary piston from the reservoir.
- Some ABS master cylinders have check valves in the heads of the pistons to reduce piston and pedal vibration and cup wear.

Terms to Know

Adjustable Pedal System (APS)

Cup seal

Diaphragm

Free play

Hydraulic system mineral oil (HSMO)

O-ring

Polyglycol

Quick take-up master cylinder

Quick take-up valve

Replenishing port

Reservoir

Residual pressure check valve

Vent port

Review Questions

Short-Answer Essays

1. List the main parts of a master cylinder piston assembly.
2. What is the purpose of the master cylinder piston primary cup?
3. What is the purpose of the master cylinder vent port?
4. What is the purpose of the master cylinder replenishing port?
5. Explain the path of fluid flow in the master cylinder when the piston is returning (brake released).

6. Explain the difference between single- and dual-piston master cylinders.

7. Explain the difference between front-to-rear and diagonally split hydraulic systems.

8. Explain the purpose and operation of a quick take-up master cylinder.

9. Explain why DOT 5 brake fluid should not be used in an ABS.

10. Describe how the master cylinder primary piston builds pressure when there is an external load in the secondary piston side.

Fill in the Blanks

1. Brake fluid specifications are developed by the U.S. _____ and the _____.

2. The two main parts of a master cylinder are the _____ and _____.

3. When the master cylinder piston is on a return stroke, fluid flow around the piston valley area comes through the _____ port.

4. After the master cylinder piston returns, excess fluid in front of the piston returns to the reservoir through the _____ port.

5. The pedal pushrod on a dual-piston master cylinder pushes on the _____ piston.

6. A dual-piston master cylinder can be hydraulically split either _____ or _____.

7. The master cylinder piston is returned on brake release by a _____.

8. The _____ in front of the piston traps fluid to build pressure in the pressure chamber.

9. Older drum brake systems keep pressure in the hydraulic lines with a master cylinder _____ check valve.

10. The use of low-drag calipers led to the development of _____ master cylinders.

Multiple Choice

1. *Technician A* says that the master cylinder primary cup seals pressure on brake application.
Technician B says that the master cylinder primary cup allows fluid flow during brake release.
Who is correct?
   A. A only
   B. B only
   C. Both A and B
   D. Neither A nor B

2. The operation of a master cylinder on the return stroke is being discussed:
Technician A says that the replenishing port lets fluid flow from the reservoir into the low-pressure valley area.
Technician B says that the replenishing port allows fluid to return to the master cylinder reservoir.
Who is correct?
   A. A only
   B. B only
   C. Both A and B
   D. Neither A nor B

3. The brake pads on a disc brake system are worn, and more fluid is required in front of the master cylinder piston:
Technician A says that this is the purpose of the vent port.
Technician B says that this is the purpose of the replenishing port.
Who is correct?

A. A only **C.** Both A and B
B. B only **D.** Neither A nor B

4. *Technician A* says that the primary piston is moved by the pedal pushrod in a dual-piston master cylinder. *Technician B* says that during normal braking, the secondary piston is moved by hydraulic pressure from the primary piston.
Who is correct?
A. A only **C.** Both A and B
B. B only **D.** Neither A nor B

5. The operation of a dual-piston master cylinder with a primary system leak is being discussed:
Technician A says that the secondary piston is moved mechanically by the primary piston.
Technician B says that the secondary piston is moved by hydraulic pressure from the primary piston.
Who is correct?
A. A only **C.** Both A and B
B. B only **D.** Neither A nor B

6. The operation of a dual-piston master cylinder with a secondary system leak is being discussed:
Technician A says that the secondary piston is moved mechanically by the primary piston.
Technician B says that the secondary piston is moved by hydraulic pressure from the primary piston.
Who is correct?
A. A only **C.** Both A and B
B. B only **D.** Neither A nor B

7. The construction of a quick take-up master cylinder is being discussed:
Technician A says that the primary piston fits in two bore sizes.

Technician B says that both ends of the secondary piston are the same size.
Who is correct?
A. A only **C.** Both A and B
B. B only **D.** Neither A nor B

8. *Technician A* says that the quick take-up valve in a quick take-up master cylinder allows flow from the reservoir to the primary piston spool area.
Technician B says that the quick take-up valve allows flow from the primary piston spool area to the reservoir.
Who is correct?
A. A only **C.** Both A and B
B. B only **D.** Neither A nor B

9. *Technician A* says that DOT 3 brake fluid has a higher boiling point than DOT 5.
Technician B says that DOT 4 brake fluid has a lower boiling point than DOT 5.1.
Who is correct?
A. A only **C.** Both A and B
B. B only **D.** Neither A nor B

10. *Technician A* says that silicone-based brake fluids are best suited for ABS use.
Technician B says that silicone-based brake fluids should never be mixed with other fluid in a brake system.
Who is correct?
A. A only **C.** Both A and B
B. B only **D.** Neither A nor B

Hydraulic Lines, Valves, and Switches

Upon completion and review of this chapter, you should be able to:

- ❏ Describe the purpose and types of hydraulic brake lines.

- ❏ Identify the two types of flares used on brake line tubing.

- ❏ Explain the purpose and identify the types of brake line fittings.

- ❏ Describe the purpose and explain the mounting of flexible brake line hoses.

- ❏ Explain the reasons for strength requirements for tubing and hoses and identify the construction features that fulfill these requirements.

- ❏ List the general precautions for working with brake tubing and hoses.

- ❏ Explain the purpose, parts, and operation of a metering valve.

- ❏ Explain the purpose, parts, and operation of a proportioning valve.

- ❏ Define the split point and slope of a proportioning valve.

- ❏ Explain the purpose, parts, and operation of a pressure differential valve.

- ❏ Describe the purpose, parts, and operation of a brake failure warning switch.

- ❏ Explain the operation of a height-sensing proportioning valve.

- ❏ Identify the different types of two- and three-function combination valves.

- ❏ Explain the purpose and describe the operation of a fluid level switch and a stoplamp switch.

Introduction

Hydraulic lines made of tubes and hoses transmit fluid under pressure from the master cylinder to each of the wheel service brakes (Figure 5-1). Several valves are used in the system to control hydraulic pressure and as safety devices. Electrical switches to operate the failure warning lamp and the stoplamps, as well as sensors to indicate low brake fluid level, also are important parts of the brake system.

Brake Lines and Hoses

Brake lines or tubing consist of steel tubes or pipes and flexible hoses connected with fittings. Rigid tubing is used everywhere except where the lines must flex. Flexing of the brake lines is necessary between the chassis and the front wheels and between the chassis and the rear axle or suspension. Brake tubing and hoses are manufactured to strict specifications developed by SAE and the International Standards Organization (ISO).

The general requirements for automotive brake fluid tubing and hoses are:

- ❏ Good corrosion resistance against chlorides (road salt)
- ❏ Strength, high burst pressure, and good fatigue or corrosion resistance
- ❏ Smooth bores allowing nonrestricted flow
- ❏ Good resistance to surface fretting and stone pecking
- ❏ Ready availability at a realistic cost
- ❏ Easy to form

Shop Manual
pages 185–203

Figure 5-1 Brake lines consist of rigid tubing or pipes and flexible hoses that carry brake fluid from the master cylinder to the service brakes.

Brake Tubes or Pipes

The hydraulic tubing used in the brake system is double-wall, welded steel tube that is coated to resist rust and other corrosion. Double-wall tubing is made in two ways: seamless and multiple ply. Each must meet the specifications of SAE Standard J1047.

> ⚠ **WARNING:** Double-wall steel brake tubing is the only type of tubing approved for brake lines. Never use 100 percent copper tubing as a replacement; it cannot withstand the high pressure or the vibrations to which a brake line is exposed. Fluid leakage and system failure will result.

There is a copper-nickel alloy brake tubing that meets SAE Standard J1047 and ISO 4038. The alloy is 10 percent nickel, 1.7 percent iron, 0.8 percent manganese, and about 90 percent copper. This tubing meets all international and U.S. requirements for brake tubing, and it has the added advantage of being more corrosion resistant. Audi, Porsche, and Aston Martin vehicles use the copper-nickel brake tubing for their hydraulic brake systems. This chapter confines the discussion to steel tubing, however, because it is the most common, and the use, treatment, and fittings for the copper-nickel alloy tubing are exactly the same as those for steel.

Seamless tubing is made by rolling a steel sheet twice around a mandrel so the edges *do not* adjoin each other to form a seam (Figure 5-2). The tubing is then run through a furnace where copper plating is applied and brazed to form the tubing into a single, seamless piece.

Multiple-ply tubing is formed as two single-wall tubes, one inside the other. The seams of each section must be at least 120 degrees apart. Then the two-ply tubing is furnace brazed just as is seamless tubing to form a single, seamless length.

All brake tubing is plated with zinc or tin for corrosion protection. In addition, all brake tubing must meet the burst specification of SAE Standard J1047, which requires that an 18-inch length of tubing must withstand an internal pressure of 8,000 psi. These requirements, by themselves, make it clear why copper tubing cannot be used for brake lines.

1. Tinned copper-steel alloy protects outer surface.
2. Long-wearing and vibration-resisting soft steel.
3. Fused copper-steel alloy unites two steel walls.

Figure 5-2 Cross-sectional view of a brake line.

Tubing Sizes. Brake tubing is made in different diameters, lengths, and shapes (Figure 5-3). The most common diameters for steel tubing or pipes in the inch system are $\frac{3}{16}$ inch, $\frac{1}{4}$ inch, and $\frac{5}{16}$ inch. Other tubing diameters from $\frac{1}{8}$ inch to $\frac{3}{8}$ inch also are available.

Some vehicles use tubing sized in metric diameters. Metric diameters are specified in millimeters by SAE Standard J1290. Common metric diameters are 4.75 mm, 6 mm, 8 mm, and 10 mm.

AUTHOR'S NOTE: It is almost impossible to get metric-sized brake tubing in a small town. Most parts vendors sell "metric" brake tubing, but in reality it is an SAE-sized tubing (usually $\frac{3}{16}$-inch inside diameter) with metric-sized and metric-threaded fasteners. A standard double flare tool will work on these lines, but the ISO flare tool will not beacuse the tubing collapses before flaring. The problem with double flaring an SAE-sized tube for connection to a vehicle's ISO fitting is the flare angles. They are a little off and will not seal properly.

Tubing Installation. Tubing installed on a car when it is manufactured is shaped properly to fit into the brake system. As are other parts made by the carmaker, tubing is referred to as an original equipment manufacturer (OEM) part. Aftermarket replacement tubes are most often available straight and in different lengths. If available, however, it is preferable to use an OEM-shaped, prefabricated tube as a replacement. Generally, a one-piece tubing is used between the

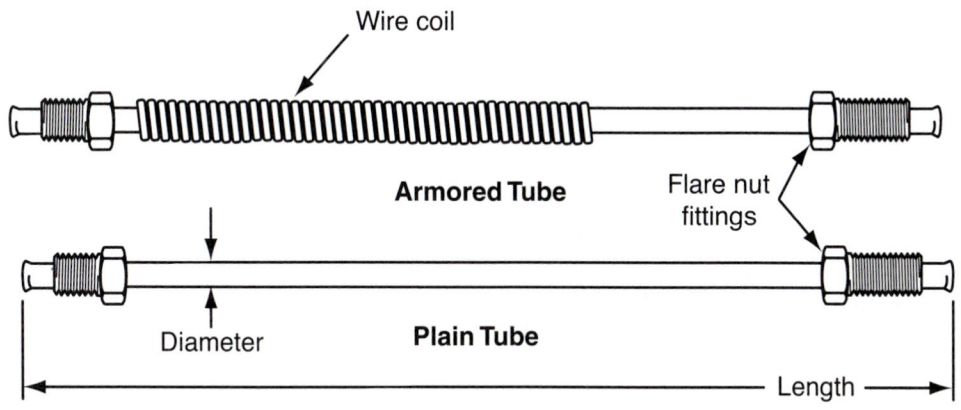

Figure 5-3 Brake tubing (line) comes in various sizes and lengths, flared with flare nuts. Armored tubing uses a wire coil for protection..

different components of the brake system. For instance, there is a brake tubing from the master cylinder to a control valve, then another tubing from the valve to either a wheel or the junction block where the tubing is branched to each wheel. If possible, do not cut the tubing to add a section. Replace the whole section of tubing instead of patching it.

Each end of the tubing has a fitting for connection into the system. The fitting, which is described later, fits over the tube or pipe and seats against a specially formed end of the tubing or pipe.

The formed end of the tubing is called a flare. There are two common types of flares formed on the end of brake tubing (Figure 5-4). The inverted double flare has the end of the tubing flared out, then it is formed back onto itself.

The ISO flare has a bubble-shaped end formed on the tubing. Each type of flare is used with a different type of fitting, and they are not interchangeable. Tools are available to form these flares when fabricating new tubing for a repair. Flares and their fittings are described in more detail later in this chapter.

The brake line tubes are routed from the master cylinder along the car frame or body toward the wheel service brakes (Figure 5-5). Clips hold the tubes in position. The clips usually have rubber isolators to cushion the tubes. Clips are important because they prevent the tubes from

ISO flares tend to seal better while using less torque on the threaded nut.

Figure 5-4 Inverted double flares and ISO flares are used on brake lines.

Figure 5-5 Typical installation of brake tubing and hoses. Brake tubing is held to the vehicle frame by clips and brackets.

Figure 5-6 Tubing may also be routed on driveline and suspension parts, such as a solid rear axle housing.

vibrating, which could cause metal fatigue and eventual rupture. Tubing also may be routed on suspension and driveline parts, such as a rigid rear axle and differential (Figure 5-6).

Brake Hoses

Brake hoses (Figure 5-7) are the flexible links between the wheels or axles and the frame or body. Hoses must withstand high fluid pressures without expansion and must be free to flex during steering and suspension movement.

Figure 5-8 shows the parts of a brake hose. The hose is made from materials that resist damage from both brake fluid and petroleum-based chemicals. Brake hoses are manufactured and tested to the specifications of SAE Standard J1401 and FMVSS 106. Among these specifications is the requirement that a brake hose must withstand 4,000 psi of pressure for 2 minutes without rupturing. The test pressure is then increased at a rate of 25,000 psi per minute until the hose bursts. The burst pressure is recorded as the final test measurement. These performance requirements make it clear why brake hoses are important safety devices.

FMVSS 106 further requires that brake hoses must be marked with raised longitudinal ribs or two $\frac{1}{16}$-inch colored stripes on opposite sides to indicate if the hose is twisted during installation. Twisting creates stress that could lead to rupture. Twisting also may cause the hose to kink, cause ply seperation, and block fluid pressure. To further prevent twisting, at least one end fitting on a hose usually can be rotated with the hose before fastening. This normally is the end at the body or frame.

Figure 5-7 Brake hoses are the flexible sections of the brake lines.

Figure 5-8 Brake hoses are made with two fabric layers alternating with two rubber layers. The outer jacket is ribbed to indicate if the hose is twisted during installation.

Brake hoses are reinforced with metal or synthetic cords to withstand high pressures. Each end has a fitting so that it can be connected to other parts of the brake system. The brake hose length is specified from the end of one fitting to the end of the other fitting. Hose diameter is specified as the inside diameter of the hose. Hoses are available in both inch and metric diameters. Brake hoses come in different sizes and lengths, with various end fittings for different vehicle requirements.

AUTHOR'S NOTE: Unlike the brake tubing, metric-sized brake hoses are, in fact, metric sized. This is primarily because the metric fittings are crimped or swagged to the hose. Using metric fittings on an SAE-sized hose could cause the same problems as a double flare fitting connected to a metric fitting.

The fittings are crimped, or swagged, to the brake hose ends at very high pressures. Clamp-on or crimped fittings used with low-pressure hydraulic hoses or oil hoses cannot be used for brake hoses.

For all of their strength and durability, brake hoses are the weakest links in the brake hydraulic system. Atmospheric ozone attacks the rubber material and, over a long period, causes the hoses to deteriorate. Slight porosity of the hose material also lets air enter the system that contaminates the brake fluid—again, over a long period of time, however.

Hoses also are subject to wear, both externally and internally. Hoses must be installed so that they do not rub against vehicle parts. Some hoses have rubber ribs around their outer circumference to protect them from rubbing on suspension and chassis parts. A bulge in a hose usually indicates that the hose may fail, due to internal wear or damage.

AUTHOR'S NOTE: There are several technical service bulletins (TSBs) on brake diagnosis and service. When starting the repair order, consult the shop's service database to determine if any of these TSBs apply to the job at hand.

Another internal hose problem can occur when the inside of the hose wears to the point that a flap of rubber loosens from the hose wall. If the flap stays secured to the wall at one end and its loose end faces the master cylinder, it can delay fluid pressure to the wheel brake and cause uneven braking or pulling. If the loose end of the flap faces the wheel brake, it can delay pressure release from the wheel brake and cause brake drag. This type of hose defect is impossible to see from the outside and is difficult to pinpoint with any kind of test. If the symptoms described above exist, hoses usually are replaced to try to eliminate the problem.

Figure 5-9 The retaining clip fits in a groove at the end of the hose fitting.

Figure 5-10 Typical brake hose mounting.

The point at which a hose connects to a rigid tube usually is secured to a bracket on the frame or body. This is the end that can be moved so the hose is not twisted. A clip fits in a groove in the end of the hose fitting (Figure 5-9), and the end of the hose is inserted through a support bracket (Figure 5-10). The steel line is then threaded into the fitting on the end of the hose. The fitting on one side of the bracket and the clip on the other side hold the hose securely in position.

Brake Fittings

The threaded parts used to connect brake hoses and tubing together and to other brake components are called **fittings** (Figure 5-11). Fittings are made from steel to withstand brake system pressures. Fittings are threaded to allow connection to other brake parts. The ends of brake tubing are formed into either an inverted double flare or an ISO flare (described later) and fitted with a flare nut. Brake hoses can have either male or female fittings. The threaded connections of master cylinders, wheel cylinders, calipers, and most valves are female. Fittings are made with both SAE inch and ISO metric threads. Threads from one system do not fit threads of the other system.

Figure 5-12 shows an assortment of different brake line fittings. Adapters can be used to connect two different sizes of fittings. Unions and tee fittings are used to connect two lines together.

A complete fitting usually consists of a nut fitted around a tube. The nut may be male (external threads) or female (internal threads) and will screw into or around its counterpart on a device or another tube. The flare is trapped and compressed by the two pieces of fitting.

Fittings is a term applied to all plumbing connections on the car or in the hose. They may be referred to as flare nut, nut, or line nut. The term *fitting* is used in this text.

Figure 5-11 Typical flare-nut fitting for an inverted flare connection.

**Step-up/step-down adapters
(double flare)**

**Brake line union
(double flare)**

Brake line flare nuts

**Brake line tee
(double flare)**

Metric line adapter

Figure 5-12 Assorted adapters and fittings are used for brake line connections.

SAE Fittings. All SAE fittings used in brake systems have a 45-degree taper on the male nut and on the inside of the flare (Figure 5-13). The tubing seat in the female fitting has a 42-degree taper. The 3-degree mismatch forms an interference fit that creates a leak-free, high-pressure seal. SAE standards also exist for 37-degree fittings, but these are not used in brake systems.

A fitting with an external taper is simply called a standard flare (Figure 5-13A). When the fitting is tightened, the tapered surfaces of the male and female fittings create the seal. A fitting with an internal taper is called an inverted or LAP flare (Figure 5-13B). These are more common than standard flares on brake tubing. The male flare fitting compresses the bell-mouthed inverted flare against the seat in the female fitting, and the tubing is sandwiched between the two halves of the fitting to form the seal.

The flared tubing end can be formed as either a single flare or a **double flare** (Figure 5-14). A single flare does not have the sealing power of a double flare and is subject to cracking. Preformed replacement brake tubing is sold with a double flare on each end and the fittings in place

Double flare refers to a tubing connection in which the end of the tubing is flared out then formed back on itself.

Figure 5-13 SAE 45-degree fittings are commonly used for brake connections.

Figure 5-14 Brake line steel tubing must have a double flare. The copper-nickel alloy tubing requires the same flaring.

on the tubing. If you cut and form a brake pipe from bulk tubing, you will have to form double flares on both ends. Flaring procedures are covered in the *Shop Manual*.

ISO Fittings. The **ISO** fitting is a metric design, originally used on imported vehicles but now common on domestic brake systems. An ISO flare is not folded back on itself as is a double flare. The unique shape of an ISO fitting causes it to be called a "bubble" flare. Figure 5-15 shows the differences between the shapes of an inverted double flare and an ISO fitting and the ways that the tubing ends are held in their fittings.

Like a standard or an inverted flare, an ISO fitting uses interference angles between the flare and its seat to form a leak-free seal. The angle of the outer surface of an ISO fitting flare is approximately 32.5 degrees. The flare seat is 30 degrees, and the angle at the end of the flare nut is 35 degrees.

ISO fitting refers to a tubing flare in which a bubble-shaped end is formed on the tubing; also called a bubble flare; it does not need a separate tubing seat as an inverted flare does.

Inverted double flare

ISO flare

Figure 5-15 ISO and inverted flares are distinguished by the shape of the flare. An ISO flare does not need a separate tubing seat as does an inverted flare.

ISO fittings have become popular with both domestic and foreign vehicle manufacturers because the outer surface of an ISO fitting will form a leak-free seal against a mating surface in a cylinder or caliper body that is simply drilled and countersunk with the right taper (see Figure 5-15). Manufacturing operations are simplified because an inverted, cone-shaped seat for an inverted flare is not required. A seat for an inverted flare requires extra machining operations or the addition of a steel insert. The seat for an ISO fittings is much simpler to machine and provides a seal that is equal to—or better than—a traditional inverted flare.

Compression Fittings. Straight compression fittings are usually found on the ends of brake hoses that attach to calipers or wheel cylinders. As the rigid fitting on the end of the hose is tightened, it compresses a soft copper washer against a flat, machined surface on the cylinder or caliper (Figure 5-16). Because the compression fitting is usually attached rigidly to the end of the hose and does not swivel, this end of the hose should be connected before the end with a swivel fitting. Compression fittings also are sometimes used on hydraulic junction blocks and valve bodies.

Repeated tightening and loosening of a compression fitting can permanently compress the copper washer. Whenever you disconnect a compression fitting, replace the washer.

Banjo Fittings. A **banjo fitting** is a circular fitting that looks like the instrument of the same name. A banjo fitting is used to attach a hose or tube to a port on a cylinder or caliper at a close right angle (Figure 5-17). Fluid passes from the brake line into the cutaway section inside the banjo fitting and then through the hollow bolt to the cylinder or caliper.

A banjo fitting is a kind of compression fitting, and both flat surfaces of the fitting are sealed with soft copper washers. The washers should be replaced anytime the banjo bolt is loosened.

Compression fittings are so named because the seal is made by compressing a washer. Change the washer if the fitting is loosened or removed.

A **banjo fitting** is a round, banjo-shaped tubing connector with a hollow bolt through its center.

Always replace the washer(s) when the banjo fitting is loosened or removed.

Figure 5-16 A copper washer is used with compressed fittings.

Figure 5-17 Banjo fittings are typically used at right-angle connections.

Brake Line and Fitting Precautions

Brake lines and fittings have important safety requirements. Service instructions in the *Shop Manual* contain specific WARNINGS and CAUTIONS where needed. The following paragraphs summarize universally important precautions and guidelines for installing these parts:

- ❑ Never use straight copper tubing in place of double-wall steel brake tubing. Copper cannot withstand the high pressure or the vibrations to which a brake line is exposed. Fluid leakage and system failure can result. The approved copper-nickel alloy tubing is accepted.
- ❑ For similar reasons, never use copper fuel line fittings as a replacement for steel brake line fittings.

- ❏ Never use spherical-sleeve compression fittings in brake lines. Spherical compression fittings are low-pressure fittings for applications such as fuel lines. They will fail and leak under the high pressures and vibrations of a brake hydraulic system.
- ❏ Do not interchange metric-sized and SAE fittings. They have different threads and cannot be mixed. Do not interchange ISO flare nuts with SAE inch-sized flare nuts. Either of these conditions can cause fluid leakage and system failure.
- ❏ Do not use low-pressure fuel or oil hoses in place of brake hoses. Hoses not made for brake system use can fail under the high system pressures and may deteriorate when exposed to polyglycol brake fluids. Fluid leakage and system failure can result.
- ❏ When installing a brake hose, be careful not to twist or kink it. Fluid leakage and system failure can result.

A BIT OF HISTORY

Dual-chamber master cylinders and split hydraulic systems became standard brake system hardware in 1967 by federal regulation. However, the first safety backup brake system was more than 30 years old by then.

In 1936, Hudson introduced *Duo-Automatic Hydraulic Brakes,* which used four-wheel mechanical parking brake linkage superimposed on the hydraulic service brake system. Override linkage on the brake pedal applied the brake shoes mechanically if the pedal dropped low enough. Hudson's competitors apparently thought the system was more monkey business than safety business, and it remained a Hudson exclusive for more than 20 years. The last *Duo-Automatic Hydraulic Brakes* were used on some 1957 models, the last of the Hudsons before the merger with Nash created American Motors.

Brake Hydraulic Valves

Shop Manual
pages 204–212

Before 1967 when most brake systems had drum brakes at all four wheels and master cylinders had only one chamber, the systems worked well enough with hydraulic pressure operating uniformly throughout the lines and cylinders. The development of combination disc and drum brakes required that the timing of pressure application had to be altered in certain ways. The requirement for dual-chamber master cylinders and split hydraulic systems also called for control valves and switches to operate warning lamps. The general kinds of control valves that may be used in late-model brake systems are:

- ❏ Metering valve
- ❏ Proportioning valve
- ❏ Pressure differential valve (warning lamp switch)

In addition, some master cylinders have a residual pressure check valve (explained in Chapter 4 in this *Classroom Manual*), and many systems have a single combination valve that incorporates several of the valves listed previously.

Metering Valve

Metering valves function only during the initial stage of braking.

Some cars with front disc and rear drum brakes have a **metering valve** in the hydraulic system to achieve balanced braking between the front and rear wheels. Figure 5-18 is a simplified brake hydraulic system diagram that shows a metering valve in relation to the front disc brakes and a proportioning valve (described later) in relation to the rear drum brakes.

Figure 5-18 Metering and proportioning valves in relation to front disc and rear brake drums.

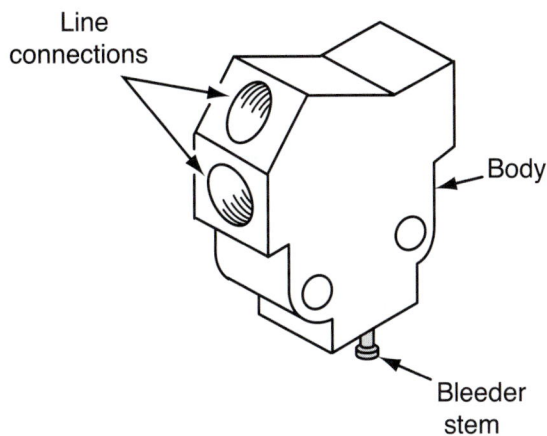

Figure 5-19 This traditional metering valve slightly delays application of front disc brakes while hydraulic pressure builds up in rear drum brake wheel cylinders.

Metering valves are used primarily on rear-wheel-drive (RWD) vehicles. A metering valve (Figure 5-19) is located in the line to the front brakes and keeps the front disc pads from operating until the rear drum brakes have started to work. The valve delays pressure application to the front disc brakes because disc brakes are fast acting, whereas drum brakes have spring tension and linkage clearance to overcome.

When the shoes contact the drums and the pressure rises in the master cylinder, the metering valve is "recentered" and pressure is allowed to the front disc brakes. This helps prevent the vehicle from nose diving on initial brake application.

Figure 5-20 is a sectional view of a typical metering valve. There is an inlet connection from the master cylinder and two outlets, one for each front wheel.

Figure 5-21A is a sectional view of a typical metering valve shown in the rest position. There is no fluid pressure applied. Fluid is returned to the master cylinder when the brake pedal is released. Fluid travels in the fluted channels of the valve stem. This allows fluid to expand and contract with temperature changes.

From master cylinder

To front wheel

Boot

To rear wheels

Metering valve stem

To front wheel

Metering valve seal

Figure 5-20 This metering valve is part of a three-part combination valve. Whether a separate component or part of a combination valve, the operation is the same.

Figure 5-21B shows the metering valve when the brakes are first applied. Fluid pressure from about 3 psi to 30 psi depresses the spring tension that held the stem valve open, item 1. As the stem valve moves, the head seals the fluted channels and blocks the lines to the front brakes.

Figure 5-21C shows the pressure increase as the driver continues to apply the brakes. This overcomes the large spring tension, item 2. The seal moves to open a passage to the front brakes. This occurs from 75 psi to 300 psi, depending on system specifications.

When the brakes are applied, fluid pressure from about 3 psi to 30 psi immediately closes the check valve at the head of the valve stem against the piston to block the lines to the front brakes. Pressure is applied to only the rear brakes until it rises enough to overcome spring tension and start to apply the brake shoes. The continuing pressure increase in the master cylinder overcomes the metering valve spring pressure against the valve piston. This occurs from 75 psi to 300 psi, depending on system specifications. Fluid pressure then is applied to the caliper pistons to operate the front disc brake pads.

When the brakes are released, the metering valve piston closes, but the valve stem passages open to allow fluid back to the master cylinder. All of this valve action takes place in a fraction of a second. The effect of the metering valve is felt during the beginning stages of all brake applications and during light brake application.

Metering valves appeared with the first disc brake systems of the mid-1960s. Vehicles of that era were almost entirely RWD vehicles with front disc and rear drum brakes. Metering valves were required for all the reasons just described. With the increased use of front-wheel drive (FWD) and diagonally split hydraulic systems used since the early 1980s, however, metering valves have been eliminated from many vehicles.

On an FWD car, 80 percent of the braking is done by the front brakes so it is desirable to apply them as quickly as possible. Until all clearance in the brake system is taken up, braking force is not great enough to overcome the torque of the front drive wheels. This driving torque and the forward weight bias of an FWD car eliminates the problem of front wheel lockup and any need for a metering valve. Furthermore, diagonally split brake systems used on most FWD cars would require two metering valves, one for the front brake on each side of the hydraulic system. Avoiding the complication of extra parts is another good reason to eliminate the metering valve.

On a vehicle with four-wheel disc brakes, the application times for the brakes at all wheels are about equal. A metering valve is therefore unnecessary.

A

Brakes
released
(rest)

To
front brake

From
master cylinder

To
front brake

B

Brakes
initial
application
(3–30 psi)

From
master cylinder

1

C

Brakes
applied
(75–300 psi)

To
front brake

From
master cylinder

1   2

To
front brake

Figure 5-21 Metering valve operation.

Proportioning Valve

The **proportioning valve** restricts fluid to the rear wheels, thereby lowering the pressure.

The amount of fluid restriction by a proportioning valve is changed proportional to the vehicle load and amount of braking force being applied.

The **proportioning valve** was introduced in 1969 to help balance front and rear pressure on cars with disc and drum brakes. A metering valve controls the *timing* of pressure application to front disc brakes. A proportioning valve controls the actual *pressure* applied to rear brakes.

Inertia and momentum cause weight to shift forward during braking. The weight shift is proportional to the braking force and the rate of deceleration. During hard braking, the weight shift unloads the rear axle and reduces traction between the tires and the road. With reduced traction, the rear brakes may lock, and the vehicle may spin. Rear brake lockup can be avoided, however, by modulating the hydraulic pressure applied to the rear brakes. The goal for the best possible braking is to maintain an equal coefficient of friction between the front and rear tires and the road.

Disc brakes require higher hydraulic pressure than do drum brakes for equal braking force at the tires. Drum brakes use mechanical servo action (explained in Chapter 8 of this *Classroom Manual*) to increase force applied to the brake shoes. Because of this servo action, drum brakes require less hydraulic pressure to maintain braking force than to establish it. Disc brakes require a constant hydraulic pressure for a given amount of braking force. Overall, disc brakes always require higher hydraulic pressure than do drum brakes. A proportioning valve does exactly what its name indicates. It proportions hydraulic pressure between disc and drum brakes to maintain equal braking force at the tires.

Proportioning valves were originally designed for use with front disc, rear drum combinations for all the reasons just explained. The earliest proportioning valves were separate components, installed in the line to the rear brakes in front-to-rear split hydraulic systems (Figure 5-22).

Some late-model vehicles with four-wheel disc brakes also have proportioning valves for the rear disc brakes. The goal of efficient braking always remains to maintain equal friction coefficient between the front and rear tires and the road. To reach this goal, the pressure to the rear disc brakes on some vehicles must be modulated to prevent lockup.

A proportioning valve has an inlet passage from the master cylinder at one end and an outlet to the rear brakes at the other end. Inside, a spring-loaded piston slides in a stepped bore. One end of the piston has a larger area than the other. The actual proportioning is done by a spring-loaded, check-valve-type stem that moves in a smaller bore through the center of the piston.

When the brakes are first applied and under light braking, the proportioning valve does nothing. Fluid enters the valve at the end with the smaller piston area (Figure 5-23), passes through the small bore around the stem, and exits to the rear brakes. The end of the valve piston at the outlet side of the valve is the end with the larger surface area. As outlet pressure rises in the valve, it exerts greater force on the piston than inlet pressure does and moves the piston toward the inlet, against spring pressure, thus closing the center valve stem and blocking pressure to the rear brakes, item 3 in Figure 5-23.

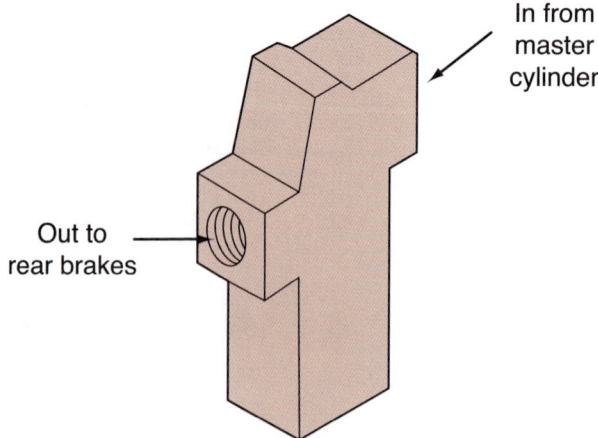

In from master cylinder

Out to rear brakes

Figure 5-22 A simple proportioning valve is installed in the line to the rear in front-to-rear split hydraulic systems.

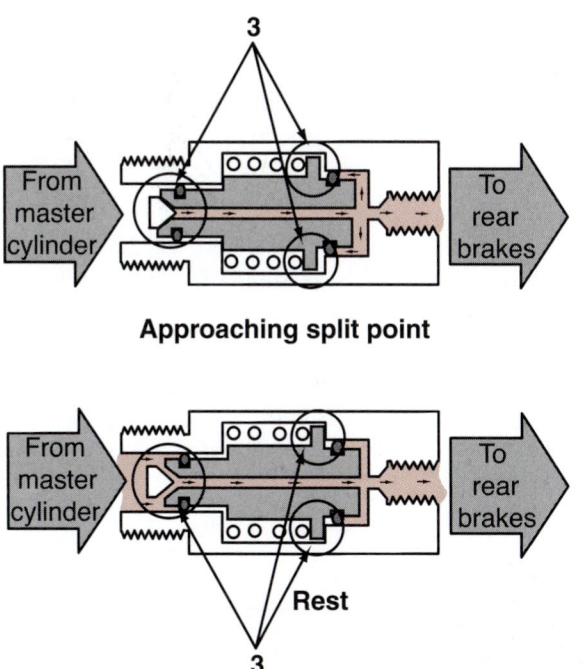

Approaching split point

Rest

Figure 5-23 This proportioning valve serves a single rear brake; a valve that serves both brakes has two outlet ports.

The pressure at which the proportioning valve closes is called the **split point** because the uniform system pressure splits at that point, with greater pressure applied to the front disc brakes and lower pressure applied to the drum brakes (Figure 5-24). As pressure continues to increase from the master cylinder, inlet pressure at the proportioning valve overcomes the pressure at the large end of the piston and reopens the valve. Fluid again flows through the center of the valve, pressure at the large end of the piston rises, and the valve closes again. The opening and closing action repeats several times per second.

Figure 5-24 Split point and slope are the operating factors of a proportioning valve.

The valve piston cycles back and forth and lets pressure to the rear drum brakes increase but at a slower rate than pressure to the front disc brakes. The pressure increase to the drum brakes above the split point is called the **slope.** The slope is the numerical ratio—or proportion—of rear drum brake pressure to full system pressure (see Figure 5-24). If half of the system pressure is applied to the rear brakes, the slope is 1:2 or 50 percent. When the brakes are released, pressure drops and the spring moves the proportioning valve piston. This opens the valve for rapid fluid return to the master cylinder.

The first proportioning valves used since the early 1970s were installed in the single line to the rear brakes. One valve controlled pressure equally to both brakes. Diagonally split brake systems separated the rear brakes from each other, however. If a proportioning valve is required, two are needed. These two valves may be housed in a single body mounted on or near the master cylinder (Figure 5-25), or the two valves may be built into the master cylinder body itself (Figure 5-26), or the valves may be separate assemblies screwed into the master cylinder outlet

Figure 5-25 These dual proportioning valves are in a separate valve body mounted away from the master cylinder.

Figure 5-26 Some master cylinders have built-in proportioning valves.

Proportioning valves

Figure 5-27 Proportioning valves are installed between the master cylinder and the output brake lines.

ports (Figure 5-27). These valves work the same as the single proportioning valve described previously.

Height-Sensing Proportioning Valve

Height-sensing proportioning valves were first used on pickup trucks in the early 1980s. The weight on the rear axle of a truck can vary greatly from an unloaded condition to fully loaded. These weight extremes affect braking balance and the amount of hydraulic pressure needed at the rear drum brakes with different loads. Because vehicle height also changes with the load in the rear of the truck, a proportioning valve that adjusts itself according to vehicle height is an effective way to provide variable pressure control to the rear brakes.

One common type of height-sensing proportioning valve has the valve mounted on a frame bracket and linkage attached to a rear spring eye (Figure 5-28). During hard braking, the weight

A **height-sensing proportioning valve** is the proportioning valve in the hydraulic system that is automatically adjusted by load and sends more fluid (pressure) to the rear brakes if the rear of the vehicle is below level.

Load sensing valve

Bracket

Bleeder

Sensor spring

Figure 5-28 This height-sensing proportioning valve is mounted on a bracket on the vehicle frame and connected to the rear spring eye.

transfer will lift the chassis in relation to the axle. When this happens, the spring eye rotates and pulls on the linkage. A lever mechanically moves the proportioning valve to reduce pressure to the rear brakes. Hydraulic pressure is adjusted automatically according to the movement of the chassis in relation to the axle.

Another height-sensing proportioning valve uses a large round ball called a G-ball (for "gravity ball") in a valve assembly called a G-valve (Figure 5-29). This assembly is mounted on the end of the proportioning valve. The ball moves back and forth in relation to the tilt of the rear of the vehicle. As the rear of the vehicle lifts during hard braking, the ball moves against the ball valve. The ball valve then controls the operation of the proportioning valve through a spring to lower pressure to the rear brakes. The advantage of this system is that there is no need for mechanical linkage to be attached to the proportioning valve.

Although height-sensing proportioning valves were first used on trucks, they also may be found on many late-model passenger cars. The weight on the rear axle of an FWD four-door sedan can vary significantly from the car with only the driver and an empty trunk to a full load of passengers and a trunk full of luggage. A height-sensing proportioning valve can modulate rear drum brake operation for the best braking action under variable load conditions. Some sedans have two height-sensing proportioning valves (Figure 5-30).

Figure 5-29 This proportioning valve uses a gravity-actuated ball, or G-ball, to correct hydraulic pressure reduction.

Figure 5-30 Some cars have two height-sensing proportioning valves.

Pressure Differential Valve (Failure Warning Lamp Switch)

A **pressure differential valve** is a hydraulically operated switch (Figure 5-31) that controls the brake failure warning lamp on the instrument panel. Each side of the pressure differential valve is connected to half of the hydraulic system (one chamber of the master cylinder). Each master cylinder piston provides pressure to a separate front/rear or diagonal hydraulic system. If one of the systems fails, the brake pedal travel will increase and more brake pedal effort will be required to stop the car. The driver might not notice a problem, however, but the lamp on the instrument panel will provide a warning in case of hydraulic failure. Figure 5-32 shows the operation of a pressure differential valve and warning lamp switch. The switch is a grounding switch, meaning it connects the light circuit to ground.

Figure 5-31 A typical individual pressure differential valve or warning lamp switch.

Figure 5-32 This warning lamp switch is part of a combination valve, but the operation is the same whether it is an individual component or combined with other valves.

AUTHOR'S NOTE: To see a simple demonstration of how the pressure differential valve works, try this. Blow up a round balloon to about half its expected size. The balloon will have a nearly perfect round shape. Wrap a thumb and one finger around the center of the balloon, dividing it into two bulges of equal sizes. Hold the thumb and finger in place while squeezing one of the bulges. That trapped air will move to the other side, making that bulge larger than the other. Unequal hydraulic pressure within the pressure differential valve does the same thing to its internal piston, except in this case an electrical switch is tripped when the piston moves.

Failure in one half of the hydraulic system causes a pressure loss on one side of the pressure differential valve. Pressure on the other side moves the valve plunger to contact the switch terminal, close the circuit, and light the lamp. All pressure differential valves work in this basic way but differ in the details of piston shape and the use of centering springs:

❏ If the center of the piston is higher than the adjacent sections, or if the switch contains two pistons joined end to end, the switch plunger drops down to close the circuit (Figure 5-33).

❏ If the center of the piston is lower than the adjacent sections, the switch plunger moves up a ramp on either side of the piston center to close the circuit (Figure 5-34).

❏ If the center of the piston is open, the piston completes the circuit when it moves and contacts the switch plunger (Figure 5-35).

Figure 5-33 Pressure differential valve (warning lamp switch) with two pistons.

Figure 5-34 Pressure differential valve (warning lamp switch) with single piston and no centering springs.

❏ If the piston does *not* have centering springs, the lamp will light the first time a pressure difference occurs and it will stay lit until the hydraulic problem is fixed. After the problem is fixed, the piston must be recentered manually as explained in the *Shop Manual*. A pressure differential valve *without* centering springs may have to be disabled when bleeding the hydraulic system.

❏ If the piston *has* centering springs, the lamp will light when a pressure difference occurs but go out when the brake pedal is released. A valve with centering springs will automatically recenter the piston after the problem is repaired and after system bleeding.

Figure 5-35 Pressure differential valve (warning lamp switch) with single piston and centering springs.

The pressure differential valve on most late-model vehicles is part of a combination valve (see Figure 5-32) or built into the master cylinder body. Some vehicles have a switch in a float in the fluid reservoir instead of a pressure differential valve. This float switch turns on the brake warning lamp when the fluid level changes to a dangerous point. This accomplishes the same goal as a pressure differential valve. Fluid level switches are described in more detail later in this chapter.

Combination Valve

Combination valves were introduced in the 1970s as a way to save money and to simplify the brake hydraulic system. A system with a combination valve has fewer fittings and short lengths of tubing. It therefore has fewer places for possible leaks and lower component costs.

As the name indicates, a **combination valve** has two or three valve functions in one valve body. The most common type is the three-function combination valve (Figure 5-36). A three-function

Figure 5-36 A typical combination brake valve with a pressure differential switch.

Figure 5-37 Cross-sectional view of a three-function combination valve.

valve has three separate sections as shown in the sectional view (Figure 5-37). One end of the combination valve houses the metering valve; the other end houses the proportioning valve. The center section has the pressure differential valve (warning lamp switch). The operation of these three sections is the same as described for the separate units.

Two-function combination valves have a pressure differential valve (warning lamp switch) with either a proportioning valve or a metering valve. A two-function combination valve with a proportioning valve can be identified by the large nut on the end that retains the proportioning parts. A two-function combination valve with a metering valve can be identified by the metering valve boot or stem on the end of the valve.

Hydraulic Pressure Control Without Valves

The installation of ABSs has increased rapidly since 1987, and antilock brakes are now standard equipment on many car lines. Antilock brakes have a single, very simple operating principle: to prevent wheel lockup by modulating hydraulic pressure to the brake at any wheel that is decelerating faster than the others and is about to lock up. An ABS accomplishes this with speed sensors at the wheels or on the driveline and a computer that processes the speed information and controls hydraulic pressure with electrically operated valves or small high-speed pumps.

AUTHOR'S NOTE: ABS uses simple function electrical solenoids that are switched on/off at a very fast rate. This switching allows instantaneous and continuous control of fluid pressures.

The pressure modulation or dynamic proportioning provided by antilock brakes is really no different than the modulation provided by metering valves and proportioning valves. In fact, ABS pressure control is more precise than any control that could be provided by a hydraulic/mechanical metering valve or a proportioning valve. As antilock brakes become standard on more vehicle models, engineers have the opportunity to eliminate these old, familiar hydraulic valves and replace them with a computer-controlled system. Chapter 10 in this *Classroom Manual* covers ABSs in detail.

Brake Electrical Warning System

Shop Manual
pages 212–223

A previous section of this chapter covers the warning lamp switch operation of a pressure differential valve. The following sections cover parking brake, fluid level, brake pad wear, stoplamp switch, and warning lamp operations.

Parking Brake Switch

The parking brake is used to hold a vehicle stationary. If the parking brake is even partially applied when driving, it will produce enough heat to glaze friction materials, expand drum dimensions, and increase pedal travel. On rear disc brake systems with integral actuators, it will distort rotors and reduce brake pad life.

A normally closed, single-pole, single-throw (SPST) switch is used to ground the circuit of the red warning lamp in the instrument cluster. This switch is located on the ratchet mechanism that locks the parking brake in place.

Vehicles with daylight running lights (DRLs) use the parking brake switch to complete a circuit that prevents the headlights from coming on if the parking brake is applied when the engine is started. When the parking brake is released, the DRLs function normally.

Brake Pad Indicators

Both domestic and import car manufacturers have built systems with an electronic wear indicator in the disc brake pads on some models. As the brake pads wear to a predetermined point, the red warning lamp notifies the driver they need attention.

In some systems a small pellet is contained in the brake pad friction material. It is wired to ground the red warning lamp circuit whenever the brakes are applied. Each set of brakes contains a pellet. Each set offers a parallel leg to ground. In other systems, the pellets are wired into a series electrical circuit. As the pellet wears, it opens the electrical circuit and turns the red warning lamp on. This system self-tests the brake pad wear detection circuit all the time.

Master Cylinder Fluid Level Switch

Because brake fluid level is important to safe braking, many vehicles have a fluid level switch, located in the reservoir, to turn on the red brake warning lamp on the instrument panel when the fluid level drops too low. This warning function is equivalent to the hydraulic failure warning provided by a pressure differential valve because fluid level in the reservoir will drop from a leak caused by hydraulic failure. Therefore, the fluid level switch has replaced the pressure differential valve on many vehicles. An added advantage of a fluid level switch is that it will alert the driver of a dangerous fluid level caused by inattention and poor maintenance practices.

Fluid level sensors are built into the reservoir body or cap. One kind of switch has a float with a pair of switch contacts on a rod that extends above the float (Figure 5-38). If the fluid level drops too low, the float will drop and the rod-mounted contacts touch a set of fixed contacts to close the lamp circuit, item 4 in Figure 5-38. Another kind of switch uses a magnet in a

Figure 5-38 Float switches such as these are used in some master cylinders to warn of low fluid level.

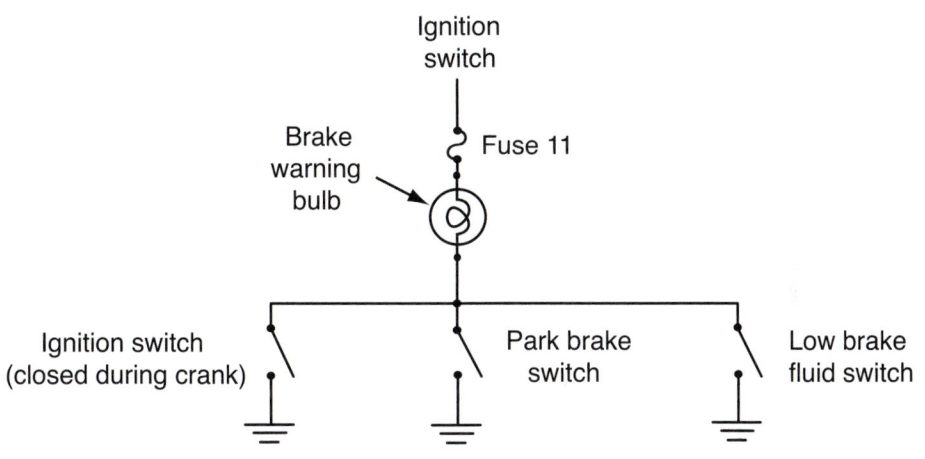

Figure 5-39 The low-fluid-level switch with the parking brake switch and the ignition switch on the ground side of the warning lamp.

movable float. If the float drops low enough, the magnet pulls a set of switch contacts together to close the lamp circuit. The contacts typically provide a ground path for the brake warning lamp (Figure 5-39).

AUTHOR'S NOTE: If the lamp is on, then the fluid level is low. Before topping off the brake fluid, however, have the brakes checked. The fluid had to go somewhere, and unless there is an external leak, the only place for the fluid to go would be to the enlarged cavities in the wheel cylinders and calipers as the brake linings wore down.

Many European vehicles have a separate indicator lamp and circuit to indicate low brake fluid level. Manufacturers such as Mercedes-Benz, BMW, Jaguar, and Porsche use an electronic module to activate a special lamp on the dash.

Figure 5-40 Mechanical stoplamp switches.

Stoplamp Switch and Circuit

Brake stoplamp switches may be operated hydraulically or mechanically. Hydraulic switches were used on older vehicles and installed in the master cylinder high-pressure chamber to be activated by system pressure. A mechanical switch is mounted on the brake pedal bracket and activated by movement of the pedal lever (Figure 5-40).

Mechanical switches have been more common for many years because they can be adjusted to light the stoplamps with the slightest pedal movement. A hydraulic switch, however, requires a certain amount of pressure before it will close the lamp circuit. Moreover, a dual-chamber master cylinder would require two hydraulic switches so a mechanical switch is simpler, more economical, and more reliable.

Stoplamp switches also may be single-function or multifunction units. Single-function switches have only one set of switch contacts that control electric current to the stoplamps at the rear of the vehicle. Multifunction switches have one set of switch contacts for the stoplamps and at least one additional set of contacts for the torque converter clutch, the cruise control, ABS, body control module (BCM), and the powertrain control module (PCM). Some multifunction switches have contacts for all of these functions. Many switches require specific adjustment procedures. Consult the service manual.

The brake lamp contacts are usually connected to the brake lamps through the turn signal and hazard flasher switch. The switch contacts for the stoplamps are always normally open contacts (Figure 5-41). Contacts for other functions may be either normally open or normally closed. In addition, some vehicles have separate brake pedal switches for the torque converter clutch, the cruise control, or the ABS. Wiring diagrams are essential for accurate identification of brake pedal switches and their functions.

Stoplamps

Stoplamps are included in the right and left taillamp assemblies. Vehicles built since 1986 also have a **center high-mounted stoplamp (CHMSL)** located on the vehicle centerline no lower than 3 inches below the rear window (6 inches on convertibles).

A three-bulb taillamp installation contains bulbs for three separate vehicle functions: the taillamps, the turn signals, and the brake stoplamps. In a three-bulb system, the stoplamps are

Stoplamp switches and lamps are more commonly referred to as brake light or stoplight switches.

A **center high-mounted stoplamp (CHMSL)** is a third stoplamp located on the vehicle centerline no lower than 3 inches below the rear window (6 inches on convertibles).

B+

Stoplamp and ABS

Torque converter clutch and cruise

Figure 5-41 Stoplamp switches control the lamp circuit on the battery (B+) side of the circuit, and the switch contacts are always open.

controlled directly by the stoplamp switch (Figure 5-42). There would be a total of six bulbs in this system.

The stoplamp switch is a normally open, momentary contact switch. The switch is attached to the brake pedal with a small clearance between the pedal arm and the switch lever or plunger (Figure 5-43). When the driver presses the pedal, the switch closes to complete the circuit and light the stoplamps. This same circuit is used to alert the ABS to monitor the wheel sensors during braking.

The stoplamp switch receives direct battery voltage through a fuse. Therefore, the lamps operate even when the ignition is off. When the normally open switch is closed, voltage is applied to the stoplamps. The lamps on both sides of the vehicle and in the CHMSL are wired in parallel. The bulbs are grounded through their mountings or by the use of remote grounds to complete the circuit.

CHMSL may be mounted in the roof line of some pickup trucks or utility vehicles.

Many vehicles have another set of contacts in the switch mounted on the brake pedal linkage. This set of contacts is normally closed and will open when the brake pedal is depressed. This set of contacts supplies battery power to the cruise control and torque converter clutch (TCC). Both the TCC and cruise control must disengage when the brakes are applied. Some vehicles use a separate switch to control these systems. On some new vehicles with computer networks that include BCM, the exterior and interior lighting control circuits are undergoing major design changes. The stoplamp circuit has undergone little change. When you service any switches attached

Figure 5-42 A three-bulb taillamp installation has separate controls for each of the three circuits.

Brakes not applied   **Brakes applied**

Figure 5-43 This mechanical stoplamp switch is mounted on the brake pedal arm and travels with it to be actuated by the end of the master cylinder pushrod.

to the brake pedal linkage, refer to a system wiring diagram to identify the type of switch, its function, and its location.

Some older vehicles use a hydraulic-actuated brake light switch. The switch is normally open (NO) and installed in a brake line, usually the front, near the master cylinder. As the brake pedal is depressed, pressure builds in the line, closing the switch and the stoplamp circuit. The switch works well, but presents some potential problems, including the possibility of leaks. In a dual system, it would be necessary to install two switches wired in parallel with each other. Very few, if any, passenger cars or light trucks use this type of stoplight switch.

Stoplamp Bulbs

Dual-Filament Bulbs. Many taillamps are two-bulb assemblies with dual-filament bulbs that perform two functions (Figure 5-44). The stoplamp circuit and the turn signal and hazard lamp circuit usually share a single dual-filament bulb, with the stoplamp circuit connected to the high-intensity filament of the bulb. The taillamps are the separate, single-filament bulbs in a two-bulb assembly.

Single-filament bulb   **Dual-filament bulb**

Figure 5-44 Single- and dual-filament bulbs cannot be interchanged.

In a two-bulb circuit, the stoplamps are wired through the turn signal and hazard switches. If neither turn signal is on, the current flows to both stoplamps (Figure 5-45). If the left turn signal is on, current for the right stoplamp is sent to the lamp through the turn signal switch and wire labeled 18 BR RD. The left stoplamp does not receive any current from the brake switch because the turn signal switch opens that circuit (Figure 5-46).

In a two-bulb circuit, the CHMSL can be wired in one of two ways. The first way is to connect the stoplamp circuit between the stoplamp switch and the turn signal switch (Figure 5-47). However, this method increases the number of conductors needed in the harness. Therefore,

Figure 5-45 Stoplamp operation with the turn signals in neutral.

Figure 5-46 Stoplamp current path through the hazard flasher with turn signal switches in left-turn mode.

most manufacturers prefer to install diodes in the wires that are connected between the left and right side bulbs (Figure 5-48). If the brakes are applied when the turn signal switch is in its neutral position, the diodes allow voltage to flow to the CHMSL. When the turn signal switch is placed in the left turn position, the left lamp must receive a pulsating voltage from the flasher. However, the steady voltage being applied to the right stoplamp would cause the left lamp to light continuously if the diode were not used. Diode 1 blocks the voltage from the right lamp, preventing it from reaching the left lamp. Diode 2 allows the voltage from the right stoplamp circuit to reach the CHMSL.

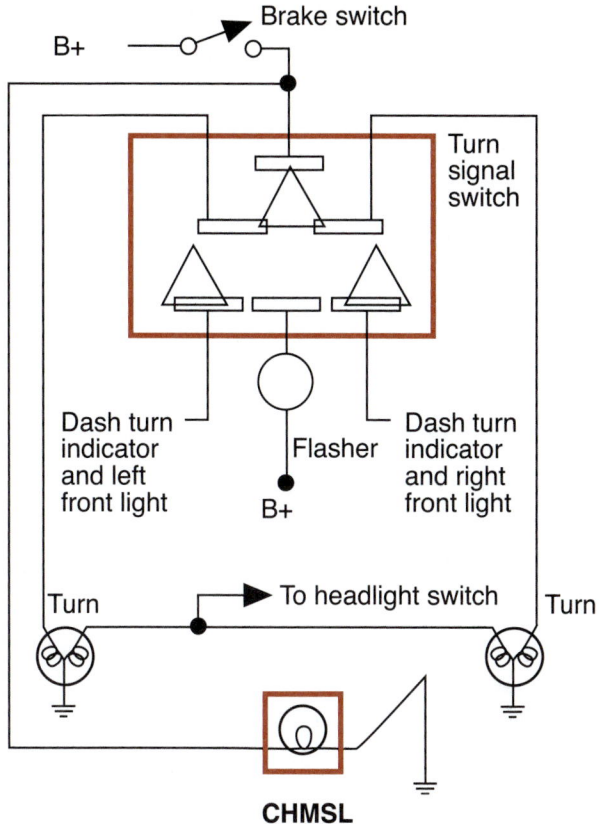

Figure 5-47 This CHMSL is connected directly to the brake switch and controlled by diodes.

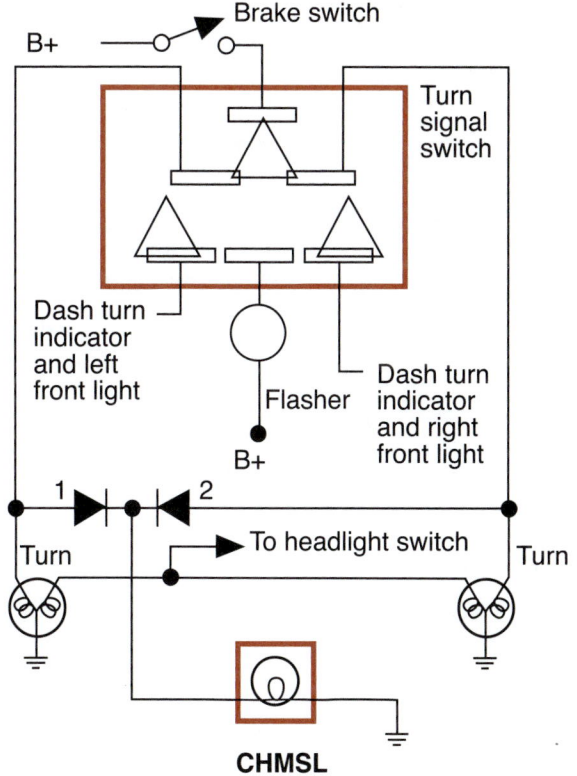

Figure 5-48 This CHMSL is wired through the turn signal switch and controlled by diodes.

Summary

Terms to Know

Banjo fitting

Center high-
mounted
stoplamp
(CHMSL)

Combination valve

Double flare

Dynamic
proportioning

Fittings

Height-sensing
proportioning
valve

ISO fitting

Metering valve

Pressure
differential valve

Proportioning valve

Slope

Split point

❏ Hydraulic lines made of tubes and hoses transmit fluid under pressure from the master cylinder to each of the wheel service brakes.

❏ Brake hoses are the flexible links between the wheels or axles and the frame or body.

❏ The threaded parts that connect brake hoses and tubing together and to other brake components are called fittings.

❏ The metering valve keeps the front disc brakes from operating until the rear drum brakes have started to work.

❏ The proportioning valve reduces the hydraulic pressure at the rear drum brakes when high pressure is required at the front disc brakes.

❏ Height-sensing proportioning valves use linkage to adjust proportioning valve operation to the height of the rear of the vehicle.

❏ A combination valve has the metering, proportioning, and pressure differential valves in one housing.

❏ A three-function combination valve has a brake failure lamp switch, proportioning valve, and metering valve combined into one housing.

❏ A two-function combination valve has a brake failure lamp switch and either a proportioning valve or metering valve combined into one housing.

❏ A brake system failure warning switch is operated by a pressure difference and lights a brake warning lamp on the instrument panel if either side of the split system fails.

❏ Brake systems also include switches for parking brakes, pad wear, stoplamps, and to warn of low brake fluid level in the master cylinder.

Review Questions

Short-Answer Essays

1. Explain why both brake tubing and flexible hoses are required in a car's brake lines.

2. Describe the difference between the double flare and ISO fitting.

3. Describe how the brake system warning lamp switch (pressure differential valve) operates when there is a hydraulic failure to one side of the split system.

4. Explain the purpose of the metering valve.

5. Describe the parts and operation of a metering valve.

6. Explain the purpose of a proportioning valve.

7. Explain why a height-sensing proportioning valve is required on some vehicles.

8. List the three sections of a three-function combination valve.

9. Discuss the routing of the current flow on a stoplamp circuit when the CHMSL is connected right behind the stoplamp switch.

10. List and explain the differences between the three-lamp, two-lamp, and one-lamp rear light system.

11. List and discuss the current flows through both filaments of a one-lamp rear light system.

Fill in the Blanks

1. Brake lines are made up of solid brake _____ and flexible brake _____.

2. Brake line flares can be _____ or _____ type.

3. The _____ valve delays pressure application to front disc brakes.

4. The _____ valve proportions the pressure between the front and rear brakes.

5. Height-sensing proportioning valves sense the height of the _____ of the vehicle.

6. Dual proportioning valve systems have a separate valve for each _____ wheel brake.

7. A pressure differential valve contains the brake warning _____ switch.

8. A three-function combination valve has a brake system failure switch, a _____ valve, and a _____ valve.

9. A two-function combination valve with a metering valve can be identified by a metering valve _____.

10. A two-function combination valve with a proportioning valve can be identified by a proportioning _____ at the rear.

Multiple Choice

1. *Technician A* says that a straight copper tube is often used as a brake line material.
   Technician B says that the lines must be made from steel.
   Who is correct?
   A. A only   **C.** Both A and B
   B. B only   **D.** Neither A nor B

2. The flare on the end of a brake line has a bubble, or ball, shape:
   Technician A says that this is a ball flare.
   Technician B says that this is an ISO fitting.
   Who is correct?
   A. A only   **C.** Both A and B
   B. B only   **D.** Neither A nor B

3. The operation of a brake warning lamp switch is being discussed:
   Technician A says that the lamp lights when there is a pressure difference between the two sides of the switch.
   Technician B says that the lamp lights when pressure is lost to the front side of the switch only.
   Who is correct?
   A. A only   **C.** Both A and B
   B. B only   **D.** Neither A nor B

4. *Technician A* says that a metering valve delays the application of the rear drum brakes.
   Technician B says that the metering valve accelerates the application of the front disc brakes.
   Who is correct?
   A. A only   **C.** Both A and B
   B. B only   **D.** Neither A nor B

5. *Technician A* says that a metering valve is necessary because drum brakes are fast acting.
   Technician B says that a metering valve is needed because disc brakes are slow acting.
   Who is correct?
   A. A only   **C.** Both A and B
   B. B only   **D.** Neither A nor B

6. *Technician A* says that a proportioning valve provides greater pressure to the rear drum brakes during normal vehicle load.
   Technician B says that a proportioning valve provides lower pressure to the front disc brakes during initial brake application.
   Who is correct?
   A. A only   **C.** Both A and B
   B. B only   **D.** Neither A nor B

7. *Technician A* says that a proportioning valve allows free flow of pressure during light braking stops.
   Technician B says that the valve restricts pressure during panic stops.
   Who is correct?
   A. A only **C.** Both A and B
   B. B only **D.** Neither A nor B

8. *Technician A* says that a diagonally split brake system may have separate proportioning valves for each rear brake.
   Technician B says that a diagonally split brake system never has a proportioning valve.
   Who is correct?
   A. A only **C.** Both A and B
   B. B only **D.** Neither A nor B

9. A two-function combination valve with a valve boot on the side is being discussed:
   Technician A says that it has a warning switch and a proportioning valve.

   Technician B says that it has a warning switch and a metering valve.
   Who is correct?
   A. A only **C.** Both A and B
   B. B only **D.** Neither A nor B

10. Brake bleeding is being discussed:
    Technician A says that two persons are required for suction bleeding.
    Technician B says that the term *bleeding* refers to the process of replacing all old fluid with new fluid.
    Who is correct?
    A. A only **C.** Both A and B
    B. B only **D.** Neither A nor B

Power Brake Systems

Upon completion and review of this chapter, you should be able to:

- ❑ Explain the relationships of atmospheric pressure and vacuum.

- ❑ Describe the relationships of atmospheric pressure and vacuum in a power brake vacuum booster.

- ❑ Describe the parts and operation of a vacuum power booster.

- ❑ Identify and describe the parts of a vacuum diaphragm assembly.

- ❑ List the three major kinds of vacuum boosters.

- ❑ Describe the parts and operation of the single diaphragm with a lever-reaction vacuum booster.

- ❑ Describe the parts and operation of the single diaphragm with a reaction-disc vacuum booster.

- ❑ Describe the parts and operation of the tandem diaphragm vacuum booster.

- ❑ Explain the parts and operation of the air and vacuum systems in a vacuum booster.

- ❑ Explain the purpose and operation of an auxiliary vacuum pump.

- ❑ Describe the parts and operation of a hydro-boost hydraulic power-assist system.

- ❑ Explain the parts and operation of the PowerMaster electro-hydraulic power-assist system.

Introduction

A power brake system is used on most cars to reduce the braking effort required from the driver. The power brake system reduces driver fatigue, increasing safety.

Increasing Brake Force Input

Four methods can be used to reduce pedal pressure requirements or to boost the force applied to the master cylinder:

1. *Pedal Force.* The simplest way to increase braking force is for the driver to step on the pedal harder. This simple approach has definite limits, however. A 120-pound driver does not have the weight and probably not the leg strength of a 220-pound driver, but the braking requirements of the car do not change to compensate for the size and weight of the driver. Driver strength then becomes the limiting factor in how the car stops.

2. *Mechanical Advantage (Leverage).* Chapter 4 of this *Classroom Manual* explains the brake pedal ratio and how it provides leverage to increase force applied to the master cylinder. Mechanical limitations allow the pedal ratio to be increased only to a given point, however. The pedal arm can be only so long to fit into the car; and the longer the pedal arm, the greater the amount of pedal travel.

3. *Hydraulic Advantage (Force Multiplication).* Just as the pedal ratio multiplies force applied to the master cylinder, hydraulic piston size can be used to multiply hydraulic pressure applied by the master cylinder to the wheel cylinders and caliper pistons. These piston size and pressure relationships also are explained in Chapter 2 of this manual. Hydraulic force multiplication has its limits too. If wheel cylinder and caliper pistons are made larger for greater force, master cylinder piston travel may increase along with brake pedal travel. All brake systems use hydraulic multiplication to increase braking force, but it has limitations just as pedal leverage does.

4. *Power Boosters.* The fourth way to increase brake application force is to install a power booster in the system, and such boosters are the subject of this chapter. There are three general kinds of power boosters. One uses intake manifold vacuum acting on a diaphragm to help the driver apply pedal force to the master cylinder. The second type uses hydraulic pressure from a hydraulic pump to operate a hydraulic booster attached to the master cylinder.

The third power brake booster is not really a brake booster, but a system that controls brake hydraulics through the use of electronics. In this system, sensors on the brake pedal mechanism combined with signals shared with the ABS and vehicle stability system apply the brakes at each wheel. This is done through various configurations that share certain properties. One property is the sharing of sensors and controlling computers. The second is the installation of electronic valves (solenoids) within the brake's hydraulic lines or at each wheel. In either case, fluid pressure can be controlled to each wheel, resulting in better braking performance and vehicle control. This electrohydraulic brake system is further discussed later in this chapter.

> **AUTHOR'S NOTE:** One of the dreams of braking engineers is a true "brake-by-wire" system. There would be no or very little hydraulics involved. Mercedes offered a vehicle in 2001 with a brake-by-wire system. One of very few recalls Mercedes has ever conducted revolved around this braking system. There were some first-generation design and manufacturing problems. It is believed by some that this ended the use off the electronic brake system on automobiles, but I believe that Mercedes, and other top-end manufacturers, will solve whatever problems were found and we will see true brake-by-wire systems offered on production vehicles by 2010.

Power boosters are add-on devices that do not alter the basic brake system. They still allow braking, even if the booster fails or loses its power supply. All boosters have a power reserve to provide at least one power-assisted stop if power is lost. FMVSS 105 contains brake performance requirements for brake systems with the power boosters disabled. Because modern brake systems are designed to include the advantage of a power booster, pedal effort increases significantly if power is lost.

Vacuum Principles

Shop Manual
pages 237–239

To understand vacuum booster systems, the relationship of **atmospheric pressure** and **vacuum** must also be understood. The air around us has weight. For example, every 1-square-inch column of air extending from the Earth's surface to the edge of the atmosphere weighs about 14.7 pounds (Figure 6-1). The weight of this air is called atmospheric pressure. Atmospheric pressure varies with altitude and temperature; but at sea level and at 68°F, it is 14.7 psi. If you were to drive up in the mountains, you would find that atmospheric pressure gets lower. As you go up in altitude, the column of air is not as high, so there is less pressure.

Vacuum is a pressure lower than atmospheric pressure. When an engine is running, the intake strokes in the cylinders create low pressure. This low pressure draws in the mixture of air and fuel. In automotive work, we commonly call this low pressure a vacuum. A true vacuum, however, is a complete absence of air and is found only in the laboratory or out in space.

Atmospheric pressure and vacuum can be used as a strong force to make things move. Figure 6-2 shows a piston that is free to move back and forth in a cylinder. One end of the cylinder is connected to a vacuum pump. The vacuum pump is used to lower the pressure on the right side of the piston to below atmospheric pressure. When atmospheric pressure is applied on the other side of the cylinder, the piston will move to the right: toward the lower pressure.

The amount of force that is created depends on the **pressure differential** or the difference between the low pressure on one side of the piston and atmospheric pressure on the other.

Figure 6-1 A 1-square-inch column of air, the height of the Earth's atmosphere, exerts 14.7 pounds of pressure on the Earth at sea level.

Figure 6-2 Vacuum (low pressure) works with atmospheric pressure to develop force.

If a perfect vacuum exists on one side of the piston and atmospheric pressure is applied to the other side, the pressure differential equals:

14.7 psi − 0 psi = 14.7 psi
(atmospheric pressure) − (vacuum) = (pressure differential)

Remember, however, that a perfect vacuum does not exist in an intake manifold, and vacuum is not usually measured in psi. To get a realistic example of a typical pressure differential in a vacuum brake booster, two factors must be accounted for.

First, pressure may be measured as either absolute (psia) or gauge (psig) and in two different measurement units: pressure in pounds per square inch (psi) or vacuum in inches of mercury (in.hg.). Atmospheric pressure at sea level may be expressed as 14.7 psia or 29.9 in.hg. Similarity, vacuum can be expressed either as 29.9 in.hg below atmospheric pressure or −14.7 psig below atmospheric pressure. However, most technicians do not normally rely on absolute pressure/vacuum measurements but use gauges in almost every instance. Gauges, vacuum and pressure, will register 0-in.hg or 0-psi at sea level. In a complete vacuum a vacuum gauge would register 29.9-in.hg. while a pressure gauge would register minus 14.7-psig. However, any math would have to use absolute measurements for accuracy.

A vacuum suspended booster has two chambers. The forward one is under vacuum and the rear can be exposed to outside air. A flexible diaphragm separates the two chapters. The amount of vacuum available in the intake manifold, and the vacuum chamber, will change as the engine load changes. To calculate the approximate booster force increase, the vacuum and pressure measurements must be in the same unit. Rounding 14.7-psia to 15-psia and 29.9-in.hg to 30-in.hg gives

Atmospheric pressure is reduced by about 1 pound per each 1,000 feet of elevation above sea level. Vehicles operated in Denver, Colorado, are tuned slightly different from vehicles on the Pacific coast.

a rough conversion factor of −2 (minus because pressure and vacuum are on opposite sides of zero). To convert from psi to in.hg multiply psi by −2 and from in.hg to psi divide in.hg. by −2.

As an example, assume that the sea-level atmospheric pressure is 15-psia, the vacuum is 20-in.hg, and the booster has 50 square inches of area.

15-psia (atmospheric pressure)

20-in.hg / -2 = −10-psig

-10-psig + 15 = 5-psia

15-psia (atmospheric pressure) -5 psia (vacuum booster pressure) = 10 pounds of differential pressure

10 pounds of pressure (psi) × 50 square inches = 500 pounds for force.

This 500 pounds from a vacuum brake booster helps the driver apply the brakes.

Vacuum and Air Systems for Power Boosters

Enough vacuum and air must be delivered to the power booster for it to work correctly. Most power boosters have the same method of delivering air, with vacuum provided by either the intake manifold or an auxiliary vacuum pump.

Air Systems

Figure 6-3 shows the air system for a typical power booster. Air enters through passages in the pedal pushrod boot. The air passes through a fine mesh material called a silencer, which slows

Figure 6-3 The air system for a vacuum brake power booster.

down the air and reduces any hissing sounds. The boot and air inlet are inside the car, so any noises could be heard by the driver. The air then passes through a filter to remove any dirt that could damage the valve. Air then flows into the power piston passages to the air valve.

Intake Manifold Systems

The vacuum for most power brake systems is supplied by the engine intake manifold. On older cars, a simple vacuum hose was attached from the intake manifold directly to the housing of the brake booster. A check valve was used to protect the booster against loss of vacuum. The problem with this system was that intake manifold vacuum decreases and increases as the engine is accelerated and decelerated. Engine vacuum also begins to drop as the engine wears.

Many late-model cars that use intake manifold vacuum as the power source for the power brakes have a vacuum reservoir (Figure 6-4). The reservoir stores vacuum so it is always available, regardless of the changing vacuum in the intake manifold.

A **check valve** is installed between the manifold and the reservoir to prevent air from entering the reservoir during wide-open throttle. The check valve also is a safety device, protecting the system from losing vacuum in case of a leaking supply line or other failure in the vacuum supply. The check valve is typically mounted on the front of the booster where the vacuum hose is connected.

When vacuum is supplied by an auxiliary vacuum pump, vacuum action on the booster diaphragm is the same as it is for manifold vacuum. Only the vacuum source is different. Auxiliary vacuum pumps are described later in this chapter.

Vacuum Check Valves

Whether a power brake system has just a vacuum booster or a vacuum booster and a vacuum reservoir, it must include one or more vacuum check valves. The check valve uses a spring-loaded ball or disc to close the vacuum line if pressure in the line gets higher (closer to atmospheric pressure) than the vacuum in the booster or reservoir.

Figure 6-5 is a sectional view of a typical vacuum check valve. The valve has a small disc backed up by a spring. Manifold vacuum pulls and holds the disc off its seat. Vacuum is allowed into the booster. When vacuum drops below the calibration of the spring, the spring moves the

Figure 6-4 Many power brake systems have vacuum reservoirs to ensure a steady vacuum supply to the booster.

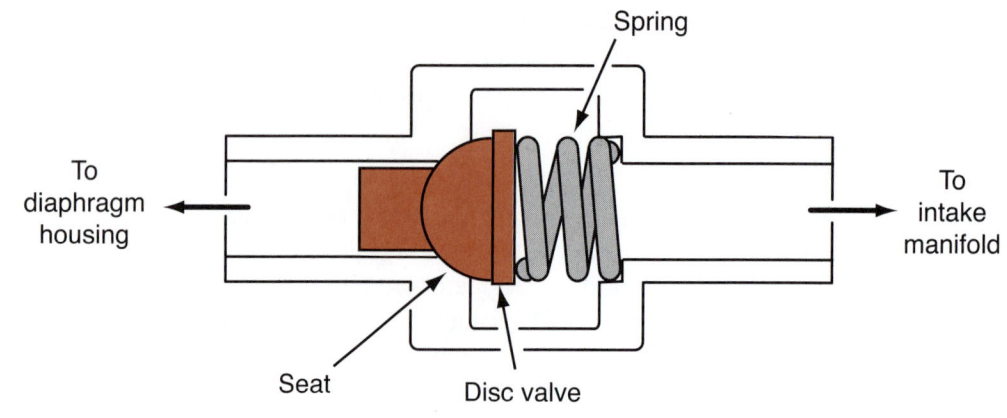

Figure 6-5 This cutaway view shows the parts of a typical vacuum check valve.

When a vehicle is powered with a diesel engine and uses a vacuum power booster, the vehicle must be equipped with a vacuum pump. Diesel engines do not create a vacuum in their intake manifold.

disc against its seat. When the disc is seated, the valve closes and does not allow vacuum to leak out of the system. The vacuum check valve has a secondary function of keeping fuel vapors out of the vacuum booster. Without a check valve, vacuum in the booster could draw part of the air-fuel mixture into the booster when the engine is at wide-open throttle with little or no manifold vacuum. As an extra safety precaution, some systems have a charcoal filter between the manifold and the vacuum check valve to trap fuel vapors before they can get near the booster.

Most systems have a single check valve as part of the inlet fitting of the vacuum booster (see Figure 6-3 and Figure 6-6). Some vehicles, however, have an in-line check valve in the vacuum line. If the power brake system has a vacuum reservoir, as well as the booster, it will have two check valves. One valve is part of the booster inlet fitting; the other is between the intake manifold and the reservoir (see Figure 6-4 and Figure 6-7).

AUTHOR'S NOTE: A worn or mistuned engine can cause damage to the check valve by pushing carbon and combustions gases into the check valve. The customer may not even notice that the boost is not working right after engine start up. If the other vacuum hoses show carbon or heat damage, inspect the vacuum check valve.

Figure 6-6 A typical vacuum brake booster.

Figure 6-7 Booster vacuum connection to engine.

Vacuum Power Boosters

Vacuum-operated power boosters have different shapes, but they all work the same way. The booster is mounted between the master cylinder and the engine compartment bulkhead or fire wall. The booster is between the brake pedal pushrod and the master cylinder (Figure 6-8). A vacuum hose is connected from the engine intake manifold to the booster (Figure 6-9). The following sections describe the operation and construction of common vacuum boosters.

Shop Manual
pages 244–250

Figure 6-8 A vacuum booster mounted on the vehicle. Note the area close around the booster.

Figure 6-9 A vacuum hose from the intake manifold supplies vacuum to the booster.

Vacuum Booster Construction

Most vacuum boosters have essentially the same parts. Figure 6-10 is a disassembled view of a typical vacuum booster. The parts are contained in a steel housing or shell that is divided into front and rear halves, held together with interlocking tabs. The rear housing has mounting studs to mount the unit on the fire wall.

The **diaphragm** is a large, hemispherically shaped part made from rubber. The flexible diaphragm moves back and forth as it is acted on by atmospheric pressure and vacuum. The center of the diaphragm is supported by a metal or plastic diaphragm support.

The brake pedal is connected to the brake pedal pushrod, which is contained in the booster. One end of the pushrod sticks out of the center of the rear housing when the booster is assembled. A rubber boot seals the area between the pushrod and the housing. The other end of the pushrod is attached to the power piston.

AUTHOR'S NOTE: One thing that should be clarified here is the front and back of a brake booster. The rear or back of the booster is that part nearest the driver and that abuts the bulkhead. The front or forward part of the booster is where the master cylinder is installed.

Figure 6-10 Disassembled vacuum booster. This is not a job for inexperienced technicians and is shown only to illustrate internal components.

The power piston is attached to the center of the diaphragm. The piston is often described as being suspended in the diaphragm. The power piston contains and operates the vacuum and air (atmospheric) valves that control the diaphragm. The power piston also transmits the force from the diaphragm through a piston rod to the master cylinder. Because the pedal pushrod is mechanically connected through the booster, any vacuum failure will not cause loss of braking action.

The piston rod also may be attached to another rod called a reaction retainer. A large coil spring is mounted between the front housing and the power piston and diaphragm assembly. The spring returns the diaphragm and piston to the unapplied position when the driver releases the brake pedal.

Two valves are located on the pedal side of the diaphragm. One, called the air valve, controls the flow of air at atmospheric pressure into one side of the booster. The other, called the vacuum valve, controls the buildup of vacuum in both sides of the booster. Both valves are connected to, and operated by, the pushrod.

Diaphragm Suspension

The operation and construction of a vacuum booster and the booster diaphragm often are described as **vacuum suspended** or **atmospheric suspended.** The brake boosters on almost all vehicles built since the mid-1970s have vacuum-suspended diaphragms. This means that when the brakes are released and the engine is running, vacuum is present on both sides of the diaphragm. When the pedal is pressed, atmospheric pressure is admitted to the rear of the diaphragm to develop booster force.

On some older vehicles, the vacuum boosters had atmospheric-suspended diaphragms. When the brakes are released, atmospheric pressure is present on both sides of the diaphragm. When the pedal is pressed, vacuum is admitted to the front of the diaphragm to develop booster force. Atmospheric-suspended diaphragms have several weaknesses that make them less practical than vacuum-suspended diaphragms, however.

By itself, an atmospheric-suspended booster has no vacuum reserve for power braking with the engine off. A vacuum-suspended booster holds enough residual vacuum for two or three moderate power-assisted stops or at least one hard stop with the engine off. Also, if an

Diaphragms can be and are used in thousands of applications. The booster diaphragm separates the pressure chamber from the vacuum chamber.

atmospheric-suspended booster is applied with low vacuum (braking immediately after acceleration, for example), power assist may be less than normal. For these reasons, a power brake system with an atmospheric-suspended booster usually has a vacuum reservoir.

Vacuum reservoirs also are used with some vacuum-suspended boosters to ensure adequate vacuum with a lot of vacuum-operated devices such as emission controls and air conditioning. A reservoir is not as essential with a vacuum-suspended booster, however.

Types of Vacuum Boosters

Three general types of vacuum power boosters are (Figure 6-11):

1. Single diaphragm with a reaction disc

Figure 6-11 Three types of vacuum boosters.

2. Single diaphragm with a reaction lever

3. Tandem booster (dual diaphragm) with a reaction disc

The single diaphragm with a reaction disc provides force (or reacts back) to the brake pedal through a rubber reaction disc. The single diaphragm with a reaction lever reacts back to the pedal through a lever arrangement. The tandem diaphragm has two diaphragms that work together to provide force. The following sections explain how a single diaphragm booster with a reaction disc works. The other two booster types operate similarly.

Vacuum booster operation can be divided into the following five stages:

1. Brakes not applied (released)

2. Moderate brake application

3. Brakes holding

4. Full brake application

5. Brakes being released

Brakes Not Applied. When the brakes are off, the return spring for the input pushrod holds the pushrod and the air control valve rearward in the power piston (Figure 6-12). The rear of the air control valve seats to close the atmospheric port. The plunger also compresses the air valve against its spring to open the vacuum port. This valve action closes the booster to the atmosphere and opens a passage between the front and rear of the booster chamber. Equal vacuum is present on both sides of the diaphragm. The power piston return spring holds the diaphragm rearward so no force is applied to the output pushrod and master cylinder.

Moderate Brake Application. When the driver applies the brakes, pedal pressure overcomes the input pushrod return spring to move the pushrod and air control valve forward (Figure 6-13). Spring pressure then moves the air valve to close the vacuum port to the rear chamber. As the air control valve continues to move forward, it opens the atmospheric port to the rear chamber. Vacuum (low pressure) still exists in the front chamber of the booster, and higher pressure (atmospheric) exists in the rear. The resulting pressure differential moves the diaphragm and power piston forward against the return spring to apply force to the master cylinder pushrod.

Figure 6-12 When brakes are released, vacuum is present on both sides of the diaphragm, and the air valve is closed to exclude atmospheric pressure.

Air valve

Floating control valve

Filter

AIR

Silencer

Air passage

Vacuum

Atmospheric pressure

Figure 6-13 Air fills the rear chamber when brake application opens the air valve and closes the vacuum valve. The front chamber is kept in vacuum.

Brakes Holding. As long as the driver maintains unchanging foot pressure on the pedal, the input pushrod does not move. The diaphragm and power piston continue to move forward until the air control valve on the piston seats against the rear of the vacuum valve plunger to close the atmospheric port (Figure 6-14). This all happens very quickly, and as long as the atmospheric port is closed, the diaphragm does not move. It is suspended by a fixed pressure differential between the front and rear chambers of the booster. The booster always seeks the holding position when pedal force is constant or unchanging.

Full Brake Application. If the driver increases pressure on the pedal, she or he may push hard enough to force the air control valve plunger fully forward against the power piston. This action closes the vacuum port to the rear chamber and fully opens the atmospheric port. Maximum atmospheric pressure then exists in the rear chamber, and the booster supplies its maximum power assist. This condition sometimes is called the booster vacuum runout point. At this point, additional braking force can be applied to the master cylinder, but it must come entirely from foot pressure on the pedal. Because the booster is supplying its maximum force at this point, the driver will feel the pedal become harder to press with additional foot pressure.

Brakes Being Released. When the driver releases the pedal, the input pushrod spring moves the pushrod and the air control valve rearward in the power piston. The rear of the air control valve

Figure 6-14 When the brakes are held at midpoint, both valves are closed to trap vacuum and atmospheric pressure and keep the diaphragm in a balanced position.

closes the atmospheric port, and opens the vacuum port. Vacuum then is applied to the rear chamber, and pressure equalizes on both sides of the diaphragm.

The large return spring moves the diaphragm, the power piston, and the output pushrod rearward. The booster returns to the released position as was shown in Figure 6-12.

Brake Pedal Feel

A power brake booster must provide some kind of physical feedback to the driver. This feedback is called brake pedal feel. Without pedal feel, the booster would apply the power assist in sudden steps and cause very poor braking control. Also without pedal feel or feedback, when the booster reached the holding position, the driver would feel only the foot pressure applied to the pedal, not the actual force applied to the master cylinder. In fact, many early power brake systems had very poor pedal feel.

Pedal feel is provided in modern vacuum boosters by a **reaction disc** or a **reaction plate and levers.** Reaction simply means that as the driver's foot, the pedal, and the booster apply force to the master cylinder, an equal force develops in the opposite direction. The driver normally feels this equal and opposite reaction force as the resistance to application force. Simply put, when the brake pedal is applied as hard as possible, it stops moving. This means that the reaction force has equaled the maximum force that is capable of being applied to the pedal.

AUTHOR'S NOTE: A simplistic means of understanding pedal feel is to press on a small coil spring with either your hand or foot. The more you press, the more reaction force is generated to force your foot or hand back (feel).

Using reaction force to provide pedal feel requires some interesting engineering. If the hundreds of pounds of force developed by the booster were applied as reaction, or feedback, to the driver's foot, the brakes could never be applied past the booster vacuum runout point. Theoretically, the booster could even force the pedal back to the unapplied position. To get around these problems, power boosters use the reaction disc or the reaction plate and levers to feed back only 20 percent to 40 percent of the booster force. The feedback force applied as pedal feel is always less than booster output force, but it also is always proportional. Whether the booster is applying 100 pounds or 400 pounds of force on the master cylinder, the feedback will be a constant percentage of whatever the output force is.

Reaction-Disc Booster

In a booster with a reaction disc, the input pushrod and vacuum valve plunger bear on a rubber disc (Figure 6-15). This reaction disc is located in the power piston and compresses under the force of the pedal. Its ability to compress lets it absorb reaction force back from the master cylinder when the brakes are applied. As the disc compresses and feeds back reaction force to the pedal pushrod, it also modulates the action of the vacuum and air control valves to adjust pressure on the diaphragm. The harder the brake pedal is pressed, the more the disc compresses and the greater the feedback feel applied to the pedal.

Plate-and-Lever Booster

A plate-and-lever booster (Figure 6-16) uses a lever mechanism to react force back to the brake pedal. The connection between the pedal pushrod is through the reaction plate and levers in the power piston.

The earliest power brake felt as if the driver was pushing his or her foot into a sponge without any brake feel.

Figure 6-15 Reaction-disc vacuum booster.

Figure 6-16 Reaction plate-and-lever vacuum booster.

When the brakes are first applied, the fixed ends of each lever are in contact with the power piston. The other ends are spring loaded and free to move. The force back to the driver's foot on the brake pedal is kept low. As brake application continues, the springs deflect enough to allow contact between the movable ends of the levers and the vacuum and air valves. The operation of the air valve and vacuum valve begins to bleed out vacuum and adds atmospheric pressure to the brake pedal side of the diaphragm. The reaction force back to the brake pedal increases. The lever and reaction plate provide a resistance and feel to the pedal similar to a nonpower brake system.

Tandem Boosters

Automotive engineers are constantly working to save car weight and space. Lighter cars have been made possible by making components out of lighter materials and making components smaller. Master cylinders are one example. They are now much smaller than they were in the past and are made from lighter materials. Brake power boosters have not escaped this trend.

The amount of braking power from a vacuum booster is directly related to the area of the diaphragm. A booster with a smaller diaphragm would provide less power assist. Engineers solved this problem by designing a smaller diaphragm housing and using two smaller diaphragms in **tandem.** The amount of force is proportional to the total area of both diaphragms. Two 10-square-inch diaphragms can provide the same power as one 20-square-inch diaphragm.

Figure 6-17 is a disassembled view of a **tandem booster;** Figure 6-18 is a sectional view. The front and rear housings are the same as those of a single-diaphragm booster. The booster contains two separate diaphragms, however, both with support plates. Some units have a housing divider between the diaphragms and others do not. Both diaphragms are connected to the power piston. A reaction disc or lever assembly is used, just as in a single-diaphragm booster.

Figure 6-17 Disassembled view of a tandem or dual diaphragm booster.

Figure 6-18 Sectional view of a tandem booster.

The two diaphragms work the same way as a single diaphragm except that the air and vacuum valves must control vacuum and atmospheric pressure on both diaphragms at the same time. When the brakes are released, vacuum is on both sides of each diaphragm. During brake application, the air valve and vacuum valve operate to admit air on the brake pedal side of both diaphragms. With a vacuum on the master cylinder side of both diaphragms and atmospheric pressure on the opposite sides, power assist is developed.

Auxiliary Vacuum Pumps

On late-model cars, vacuum is used to power air conditioning and heating control systems, cruise control, and several emission control systems. The more vacuum-operated devices a car has, the less vacuum that is available to power the brakes. In addition, diesel engines do not develop intake manifold vacuum, so diesel-powered cars and light trucks require another vacuum source for vacuum-assisted power brakes. All of these conditions led to the introduction of the auxiliary vacuum pump.

An auxiliary vacuum pump can be driven by an engine drive belt, by a gear or cam from the engine, or by an electric motor. For most belt-driven installations, the pump is mounted on a bracket at the front of the engine, and a pulley is attached to the pump drive shaft. The pulley is driven by an engine drive belt. Some belt systems do not drive the pump directly but mount the pump on the back of the alternator (Figure 6-19). The alternator is driven by the belt, and the pump is driven by an extension of the alternator shaft.

Alternator   Vacuum pump

Figure 6-19 This auxiliary vacuum pump is mounted on the rear of the alternator and driven by the alternator shaft.

A vacuum pump also can be driven directly by the engine. The pump shown in Figure 6-20 is installed in place of an ignition distributor on a diesel engine and driven by a gear on the camshaft.

An electric vacuum pump can be located any place in the engine compartment. The pump controller is connected to the wiring harness and gets electrical inputs from a low-vacuum switch in the booster or from the vehicle onboard computer. The controller has an on–off switch and a timer relay that control the motor. The electric motor gets its power from the controller. The electric motor drives either a vane pump or a diaphragm pump. The diaphragm pump moves a diaphragm to create vacuum. The vane pump has a set of vanes that rotate in an eccentric cavity.

Figure 6-20 A common mounting for an auxiliary vacuum pump used on diesel-fueled engines.

The vanes pump air out of the cavity and create vacuum. Hoses are connected to the inlet and outlet of the pump.

Hydraulically Assisted Power Brakes

Diesel engines do not produce intake manifold vacuum, and many gasoline engines with several kinds of vacuum-operated equipment also have very low intake manifold vacuum. One way to handle these vacuum conditions is to add an auxiliary vacuum pump, described previously. Another way is to eliminate vacuum as a power source and use hydraulic power instead. The three kinds of hydraulic boosters are:

Shop Manual
pages 257-265

1. A mechanical hydraulic power-assist system operated with pressure from the power steering pump. This unit is a Bendix design called hydro-boost.

2. An electrohydraulic power-assist system with an independent hydraulic power source driven by an electric motor. This unit is a General Motors design called PowerMaster.

AUTHOR'S NOTE: PowerMaster systems are used on very few vehicles now. They have been replaced for the most part by hydro-boost-type systems. Therefore, PowerMaster information contained in this chapter and the *Shop Manual* has been reduced accordingly.

3. An electrohydraulic system in which the brake fluid flow and pressure are controlled by solenoids.

The following sections explain the parts and operation of these two systems.

The fluid used for boosting with the hydro-boost system is not the same as the fluid within the brake system proper.

Hydro-Boost Principles

Bendix developed the **hydro-boost** system to be used with the light-duty diesel engine and engines with reduced manifold vacuum due to emission controls.

The hydro-boost power booster (Figure 6-21) fits in the same place as a vacuum booster, between the brake pedal and the master cylinder. Similar to a vacuum booster, the hydro-boost unit multiplies the force of the driver's foot on the brake pedal.

The hydraulic systems for power steering and power brakes are completely separate from each other in a hydro-boost system. The brake hydraulic system uses DOT 3 or DOT 4 brake fluid as specified by the carmaker. The power steering system uses the manufacturer's specified power steering fluid. The fluids cannot be mixed or substituted one for the other.

Figure 6-22 is a simplified block diagram of the main parts of a hydro-boost system. The power steering pump develops pressure that is routed to the hydraulic booster assembly. The hydraulic booster helps the driver apply force to the master cylinder pushrod. The booster has a large **spool valve** that controls fluid flow through the unit. When the brakes are applied, the spool valve directs pressure to a power chamber. The boost piston in the power chamber reacts to this pressure and moves forward to provide force to the master cylinder primary piston. The master cylinder operates the same as a conventional master cylinder. The boost piston in the hydro-boost does the same job as the vacuum diaphragm and power piston in a vacuum unit.

An **accumulator** is a pressurized storage reservoir. Loss of the power steering drive belt on a hydro-boost system means total loss of brake boost. The accumulator provides sufficient boost for one to three controlled stops.

The accumulator can be used as a fluid shock absorber or as an alternate pressure source. A spring or compressed gas behind a sealed diaphragm provides the accumulator pressure.

Hydro-boost refers to the hydraulic power brake system that uses the power steering hydraulic system to provide boost for the brake system.

A **spool valve** is a cylindrical sliding valve that uses lands and valleys around its circumference to control the flow of hydraulic fluid through the valve body.

Figure 6-21 A hydro-boost unit attached to its master cylinder.

Figure 6-22 The hydro-boost system shares hydraulic power with the power steering system.

Hydro-Boost Operation

Hydro-boost operation can be divided into the following five stages:

1. Brakes not applied (released)
2. Moderate brake application
3. Brakes holding
4. Brakes being released
5. Reserve brake application

Brakes Not Applied. Figure 6-23 shows the booster in the unapplied position (brake pedal released). Fluid under pressure from the power steering pump enters the booster through the pump port in the housing. The fluid is directed through internal passages to the spool valve. The spool valve is held rearward by its spring so that the **lands** and **valleys** direct the fluid from the pump directly through the valve to the steering gear. The spool valve also opens a passage to vent the power cavity back to the reservoir. The power cavity has no pressure, so no force is applied to the output pushrod.

Moderate Brake Application. Moderate brake application causes the input pushrod to press on the reaction rod in the end of the power piston. The reaction rod moves forward and causes the lever to pivot on the power piston and to move the spool valve forward in its bore (Figure 6-24). Spool valve movement opens the pump inlet port to the power chamber and closes the vent port. Hydraulic pressure now increases in the power chamber and moves the power piston forward to operate the master cylinder.

As the spool valve moves forward, it also restricts fluid flow out to the steering gear. Closing the vent port and restricting flow to the steering gear causes pressure to rise in the brake booster. The farther the valve moves, the more the flow to the steering gear is restricted and the more the booster pressure increases. At maximum boost, hydraulic pressure in the brake booster can exceed 1,400 psi.

Brakes Holding. As long as the driver maintains unchanging foot pressure on the pedal, the input pushrod does not move. As pressure rises to this holding point, the power piston moves forward and makes the lever pivot on the lever pin, which moves the spool valve back toward the rear of its bore. This closes the fluid inlet port and reopens the bypass port to the steering gear. The vent port stays closed, however, so pressure in the power chamber reaches a steady state. The booster always seeks the holding position when pedal force is constant or is unchanging.

This all happens very quickly. As booster pressure increases, fluid flows through a small port to a space behind the reaction rod. This creates a counterforce that moves the rod and lever rearward, which also moves the spool valve rearward and moderates application force. The reaction rod provides the same kind of proportional feedback brake feel that the reaction disc or levers in a vacuum booster provide.

Brakes Being Released. When the brake pedal is released, the spool valve moves fully rearward to the unapplied position and vents pressure from the pressure chamber. Return springs on the power piston and lever quickly return the piston to the released position. Spool valve movement also blocks the fluid inlet port to the power chamber and lets fluid bypass the booster and flow to the steering gear. The hydro-boost returns to the released position as was shown in Figure 6-23.

Reserve Brake Application. A failure in the power steering system, such as a broken power steering hose, a broken power steering drive belt, pump failure, or a stalled engine could cause a loss

Figure 6-23 When the brakes are released, power steering fluid flows past the spool and continues to the steering gear.

Figure 6-24 During brake apply, fluid at pump pressure is directed around the spool valve into the power cavity and charges the accumulator. Pressure is applied to the power valve. Fluid flow to the steering is restricted.

of pressure to the hydro-boost system. The hydro-boost has a backup system, powered by an accumulator, that allows two or three power brake applications. When the pressure from the accumulator is needed, the gas or spring pressure inside the accumulator moves the pressurized hydraulic fluid to the brake booster.

Figure 6-25 shows the hydro-boost accumulator. During normal operation, the pump pressure fills the accumulator cavity in front of the piston. The piston compresses the accumulator spring. A check ball and plunger in the passage to the accumulator allow pressure in but prevent it from escaping. This keeps the spring compressed or charged. A loss of pressure from the pump opens a passage for the pressure stored in the accumulator to be routed to the power chamber to help the driver apply the brakes. The accumulator stores enough energy for two or three brake applications.

Loss of hydraulic power from the hydro-boost only means that the driver loses power assist. It does not mean there is a loss of braking. A mechanical connection exists from the brake pedal

Figure 6-25 In case of pump failure, fluid under accumulator pressure is directed through the check valve to the power cavity. Fluid is prevented from entering the steering, pump, and return ports.

through the pedal rod, through the input rod, through the power piston to the output pushrod. The driver's pedal effort will increase, but the brakes will still work.

Early hydro-boost units (primarily those built before 1978) have a spring-loaded accumulator. Most later models have an accumulator charged with pressurized nitrogen. In 1981, a system called hydro-boost II was introduced in which the gas-charged accumulator is built into the power piston inside the assembly. Regardless of accumulator design and location, operating principles remain the same.

Vacuum Booster with Brake Assist (BA)

The vacuum booster has recently been fitted with electronic sensors and actuators that will apply the brakes more quickly during panic stops (Figure 6-26). A **brake assist** unit is added to the vacuum booster. This unit detects the brake pedal application speed; in other words, how fast the driver is pushing on the pedal. During a panic stop, the driver naturally tries to move the pedal as fast as possible so the brakes will react quickly. When this condition is detected, the BA will more quickly activate the brake booster or signal the **electronic brake system (EBS)** hydraulic modulator. The brake fluid pressure is increased much more rapidly, and the brakes, in turn, react almost instantaneously. Although the time difference between non-BA units and assisted units is measured in fractions of a second, it may make the difference between vehicle control and a possible accident.

PowerMaster® General Operation

The potential problems of having the power brake booster share a hydraulic system with the power steering led to the development of an independent hydraulic brake booster called the **PowerMaster** system. PowerMaster has a self-contained booster built into the master cylinder. This system does not use the power steering pump for hydraulic power. Instead, it has its own pump driven by an electric motor. These systems require DOT 3 brake fluid, which provides adequate lubrication for the pump. Fluid is supplied by the master cylinder reservoir both for the normal operation and for the PowerMaster booster.

Figure 6-27 shows the basic parts of the PowerMaster system. Hydraulic flow and pressure for the system come from a pump driven by an electric motor. The pump and motor are mounted directly below the master cylinder.

Figure 6-26 The brake assist unit measures pedal application speed and either actuates the vacuum booster or sends a signal to the EHB control module.

Figure 6-27 The major parts of a PowerMaster power brake assembly.

PowerMaster master cylinders are not considered "stand-alone" components because the master cylinder and booster are in one assembly. Hydro-boost and vacuum boost systems usually have a master cylinder or booster that can be replaced individually.

The PowerMaster master cylinder has three fluid partitions instead of the two found in other dual master cylinders (Figure 6-28). Two reservoir chambers serve the wheel brakes as in a common dual-chamber master cylinder. The largest reservoir chamber contains fluid for the PowerMaster booster. The bottom of this chamber has two ports: one supplies fluid to the pump, and the other is a return port from the booster. The fluid level in the PowerMaster booster chamber or the reservoir always looks as if it were low except when the accumulator is completely discharged. Do not add fluid to this chamber unless the accumulator is discharged; otherwise the fluid may overflow when the accumulator does discharge.

Figure 6-28 The PowerMaster reservoir has three chambers. Two supply the service brakes and one supplies the booster.

Figure 6-29 A PowerMaster valve position and fluid flow with brakes applied.

Although the PowerMaster booster uses the same DOT brake fluid as the wheel brakes, fluid should never be interchanged between the booster chamber of the reservoir and the chambers for the wheel brakes. The rubber diaphragm in the reservoir cover isolates only the wheel brake chambers from the atmosphere. Fluid flows to and from the booster chamber much faster and in greater volume than it does from the wheel brake reservoir chambers. Therefore, the booster reservoir chamber is vented to the atmosphere to prevent pressure or vacuum locking in the reservoir. Moisture absorbed from the air is not critical to booster operation because fluid for the booster is not subjected to high temperatures that could cause it to boil.

When the brakes are applied on a PowerMaster system, the discharge valve closes, blocking return fluid to the reservoir. As the pushrod travels farther, the apply valve opens allowing pressurized fluid into the booster cavity, which increases the driver's brake input (Figure 6-29). Further movement of the pushrod allows more fluid in, and pressure increases. When the brakes are released, the discharge valve opens and the apply valve closes. This releases the fluid for return to the reservoir, and boost is lost.

Electro-Hydraulic Brake (EHB)

As discussed before, an **electro-hydraulic brake (EHB)** system is not a direct power booster, but the end result amounts to the same. EHB and BA units share information with ABS, TCS, and electronic suspension. The most common sensors are the yaw, wheel speed, and steering wheel angle. The most common electronic actuator is the hydraulic modulator, which is shared with ABS and TCS. In truth, the EHB will replace the ABS and TCS as used today. In Figure 6-30, the EHB is compared to the BA unit discussed earlier. If the BA unit is removed (right side of figure), a conventional brake system with the ABS would be displayed. In this case, the BA unit is part of the EHB. The left side of Figure 6-30 shows a complete setup for an EHB. Note that the brake components at the wheels are the same as the conventional brakes found in 2006 vehicles. The main components of an EHB are the electronic control unit and the electronic pedal module. A pedal

CONVENTIONAL BRAKE SYSTEM

FUTURE EHB BRAKE SYSTEM

Figure 6-30 Note that the primary visual difference between the two layouts is the absence of the brake booster on the future system. The EHB control module is more complex than the hydraulic module.

module/master cylinder is shown in Figure 6-31. The main advantage of an EHB is shorter stopping distances and better vehicle control.

When the brake pedal is depressed on an EHB system, pedal movement and speed are detected by a sensor module and the signal is sent to the EHB control module mounted on or near the hydraulic module. This unit is very similar in appearance and operation to the hydraulic modulator used with current ABS technology. The pedal sensor module also has components that will create a reaction or feedback to the driver's input.

The EHB control module collects data from the yaw and wheel speed sensor and commands the individual hydraulic valves within the hydraulic module to apply fluid to the brakes. There is one valve per wheel. In this manner, the EHB can control pressure to each wheel individually, preventing wheel lockup while maintaining maximum stopping power. In the opposite direction, if a wheel loses traction on acceleration, the EHB with the proper programming can act as a TCS.

Figure 6-31 An EHB electronic pedal module with a master cylinder.

The EHB can be an active brake system once it is married completely into other braking, steering, and ride control systems.

During normal operation of the EHB, the master cylinder supplies pressurized fluid to the modulator. Its function as the primary brake pressure-generator stops at that point, but it still provides the backup or redundant measure in event of electrical component failure. If the EHB should fail in part or completely, the driver will still have the ability to brake the vehicle using the brake's conventional operational hydraulics. As with the loss of a vacuum booster, the driver will have to apply more force, but the vehicle will be controllable and can be stopped.

Summary

❏ Vacuum is simply air pressure below atmospheric pressure.

❏ Vacuum power brakes are by far the most common kind of power brakes. They operate on power provided by the pressure differential between atmospheric pressure and vacuum.

❏ The basic vacuum power booster consists of a diaphragm in a housing along with vacuum and air valves.

❏ The air valve is the vacuum booster valve that controls atmospheric pressure on the vacuum diaphragm.

❏ The vacuum valve is the valve in the power booster that controls vacuum on the vacuum diaphragm.

❏ The power piston is a part of the power booster that controls the air and vacuum valves and transmits force to the master cylinder.

❏ The diaphragm return spring returns the diaphragm and power piston to the unapplied position when brakes are released.

❏ The three types of vacuum power boosters are the single diaphragm with lever reaction, the single diaphragm with disc reaction, and the tandem diaphragm with disc reaction.

❏ The air system for a vacuum booster consists of a silencer and air filter.

❏ A check valve and a reservoir prevent vacuum loss. Vacuum for some systems is provided by an auxiliary vacuum pump.

❏ The two major hydrauically assisted power brake systems are hydro-boost and PowerMaster.

❏ The brake assist (BA) unit measures brake pedal application speed and activates the booster quicker or signals the electronic brake module.

❏ The hydro-boost system uses a spool valve to control the flow of pressurized fluid to a power chamber.

Terms to Know

Accumulator

Atmospheric pressure

Atmospheric suspended

Brake assist

Check valve

Diaphragm

Electro-hydraulic brake (EHB)

Electronic brake system (EBS)

Hydro-boost

Lands

PowerMaster

Pressure differential

Reaction disc (or reaction plate and levers)

Terms to Know (continued)

Spool valve

Tandem

Tandem booster

Vacuum

Vacuum suspended

Valleys

❏ An accumulator stores energy for several brake applications if the hydraulic system fails.

❏ The PowerMaster system has an electric motor and vane pump that supply brake fluid under pressure from the master cylinder reservoir for hydraulic assist.

❏ An electro-hydraulic brake (EHB) combines signals from sensors shared with ABS, TCS, steering, and suspension systems to control braking forces at individual wheels.

Review Questions

Short-Answer Essays

1. Explain how vacuum is used to provide a power assist.
2. What is the purpose of the air and vacuum valves in a vacuum power system?
3. What is the purpose of the power piston in a vacuum power unit?
4. List and describe three different types of vacuum boosters.
5. List and describe the different types of drive systems for auxiliary vacuum pumps.
6. Why have hydraulic power units replaced vacuum units on some cars?
7. Describe the main parts of a hydro-boost power brake system.
8. Explain the purpose of an accumulator in a hydraulic power system.
9. Explain the general operation of the BA system used on a vacuum booster.
10. Explain the purpose of the brake pedal module on an EHB system.

Fill in the Blanks

1. The _____ valve controls airflow to the pedal side of the vacuum booster diaphragm.

2. The _____ valve controls vacuum to the pedal side of the vacuum booster diaphragm.

3. There is vacuum on both sides of the power booster diaphragm when the brakes are _____.

4. There is vacuum on one side of the power booster diaphragm when the brakes are _____.

5. There is atmospheric pressure on one side of the booster diaphragm when the brakes are _____.

6. Vacuum for a power brake system can be supplied by the _____ or _____.

7. Hydraulic pressure for power brakes can be provided by the _____ _____ or _____ pump.

8. A spool valve is used to direct flow to the power chamber in the _____ hydraulic power brake system.

9. Vacuum is air pressure less than _____ pressure.

10. A hydraulic _____ can store energy for brake application if the system has a hydraulic failure.

Multiple Choice

1. *Technician A* says that sometimes vacuum is on both sides of the diaphragm in a vacuum booster. *Technician B* says that sometimes vacuum is on one side of the diaphragm in a vacuum booster.
   Who is correct?
   - **A.** A only
   - **B.** B only
   - **C.** Both A and B
   - **D.** Neither A nor B

2. *Technician A* says that during brake application, atmospheric pressure is on both sides of the booster diaphragm. *Technician B* says that during brake application, atmospheric pressure is on one side of the booster diaphragm.
   Who is correct?
   - **A.** A only
   - **B.** B only
   - **C.** Both A and B
   - **D.** Neither A nor B

3. *Technician A* says that when the brakes are released, vacuum is on both sides of the booster diaphragm. *Technician B* says that when the brakes are released, atmospheric pressure may be on one side of the booster diaphragm.
   Who is correct?
   - **A.** A only
   - **B.** B only
   - **C.** Both A and B
   - **D.** Neither A nor B

4. *Technician A* says that vacuum is exhausted through the check valve from the booster as soon as the engine is turned off. *Technician B* says that vacuum is on both sides of the booster diaphragm when the engine is turned off.
   Who is correct?
   - **A.** A only
   - **B.** B only
   - **C.** Both A and B
   - **D.** Neither A nor B

5. The use of auxiliary vacuum pumps for power brakes is being discussed:
   Technician A says that the pumps can be driven by an accessory drive belt. *Technician B* says that the pumps can be driven by an electric motor.
   Who is correct?
   - **A.** A only
   - **B.** B only
   - **C.** Both A and B
   - **D.** Neither A nor B

6. An intake manifold vacuum check valve for vacuum power brakes is being discussed:
   Technician A says that the valve closes to retain vacuum in the booster. *Technician B* says that the valve closes to keep atmospheric pressure from reaching the booster.
   Who is correct?
   - **A.** A only
   - **B.** B only
   - **C.** Both A and B
   - **D.** Neither A nor B

7. *Technician A* says that a hydro-boost system uses brake fluid for the booster. *Technician B* says that a PowerMaster system uses brake fluid for the booster.
   Who is correct?
   - **A.** A only
   - **B.** B only
   - **C.** Both A and B
   - **D.** Neither A nor B

8. The hydraulic systems for the hydro-boost and PowerMaster systems are being discussed:
   Technician A says that the hydro-boost system is powered by the power steering pump. *Technician B* says that the PowerMaster system is powered by a pump driven by an electric motor.
   Who is correct?
   - **A.** A only
   - **B.** B only
   - **C.** Both A and B
   - **D.** Neither A nor B

9. An EHB is being discussed:
   Technician A says that a BA sensor is part of the pedal module. *Technician B* says that the hydraulic modulator is basically the same as the current ABS hydraulic modulator.
   Who is correct?
   - **A.** A only
   - **B.** B only
   - **C.** Both A and B
   - **D.** Neither A nor B

10. *Technician A* says that energy can be stored by a spring in an accumulator. *Technician B* says that energy can be stored by gas under pressure in an accumulator.
    Who is correct?
    - **A.** A only
    - **B.** B only
    - **C.** Both A and B
    - **D.** Neither A nor B

Disc Brakes

Upon completion and review of this chapter, you should be able to:

❏ Explain the advantages and disadvantages of disc brakes.

❏ Describe the basic parts of a disc brake assembly.

❏ Describe the different kinds of rotors and the types of hub-and-rotor assemblies.

❏ Describe how a caliper works to stop a vehicle.

❏ Name the kinds of friction material used on disc brake pads and explain the advantages and disadvantages of each.

❏ Identify the parts of a typical caliper.

❏ Describe the two principal kinds of caliper designs and the variations of each.

❏ Describe how a low-drag caliper works compared to other kinds of calipers.

❏ Describe the different kinds of parking brakes used with rear disc brakes.

❏ Explain how brake pad wear indicators operate.

Introduction

Disc brakes are used on the front wheels of almost all cars and light trucks built since the early 1970s and on all four wheels of many vehicles.

Although no law or regulation requires disc brakes on the front wheels of cars and light trucks, the brake performance requirements of FMVSS 105 make front disc brakes virtually mandatory. Disc brakes provide greater stopping power than do drum brakes, with less brake fade and fewer problems such as grabbing and pulling. Disc brake efficiency became even more important with the development of FWD cars, which require 60 percent to 80 percent of the stopping power at the front wheels.

This chapter explains the construction and operation of disc brake systems and begins with a summary of their advantages and disadvantages.

Disc Brake Advantages and Disadvantages

Disc brakes have several very important strong points and a few relatively small disadvantages. The principal advantages of disc brakes are strong fade resistance, self-adjustment, and reduced pulling and grabbing. Disc brake disadvantages are generally minor, but they include the lack of self-energizing (servo) action, noise, and poorer parking brake operation with complicated linkage.

Fade Resistance

Remember that brake operation is a process of changing kinetic energy (motion) into thermal energy (heat) through the application of friction (Figure 7-1). When any brake installation reaches its limit of heat dissipation, **brake fade** sets in. Brake fade is simply the loss of braking power due to excessive heat that reduces friction between brake linings and the rotors or drums.

One major factor that makes disc brakes more fade resistant than drum brakes is that the friction surfaces are exposed to more airflow. Many brake rotors also have cooling passages or fins to dissipate heat even more (Figure 7-2).

Another factor that contributes directly to heat dissipation and fade resistance is the **swept area** of a disc brake rotor. With drum brakes, almost the entire swept area is in direct contact with

Shop Manual
pages 277–281

The **swept area** is the total area of the brake drum or rotor that contacts the friction surface of the brake lining.

Figure 7-1 The clamping action of friction material in the rotating disc or rotor creates heat.

Figure 7-2 Cooling fins increase heat dissipation in many rotors, and centrifugal force helps to keep the rotor clean by throwing off dirt and water.

the friction materials during the entire braking operation. With disc brakes most of the swept area is cooling while only a small area directly contacts the friction materials at any given time (Figure 7-3). The greater the swept area, the greater the heat dissipation ability. Although a brake drum can have a relatively large swept area, the entire area is on the inner drum surface. The swept area of a disc brake comprises both sides of the rotor. For any given wheel size, the swept area of a disc brake will always be larger than the swept area of a drum brake. For example, a 10-inch-diameter rotor will have almost 50 percent more swept area than a 10-inch drum.

Freedom from mechanical fade is another disc brake advantage. Mechanical fade is a problem that can occur with drum brakes when the drum becomes very hot and expands outward, away from the brake shoes. The shoes then must travel farther to contact the drum surface with normal braking force. As a result, the pedal drops lower as the brakes are applied. The increased heat at the braking friction surfaces also reduces the coefficient of friction. The combined result is brake fade.

Disc brakes do not suffer from mechanical fade because the rotor does not expand away from the pads. If anything, the rotor expands very slightly toward the pads if it becomes very hot.

Disc brake design also reduces the effects of lining fade, water fade, and gas fade. Lining fade occurs when the linings are overheated and the coefficient of friction drops off severely. Some heat is needed to bring brake linings to their most efficient working temperature. The coefficient of friction rises, in fact, as brakes warm up. If temperature rises too high, however, the coefficient of friction decreases rapidly. Because disc brakes are exposed to more cooling airflow than drum brakes, lining fade due to overheating is reduced.

Water fade occurs when water is trapped between the brake linings and the drum or rotor and reduces the coefficient of friction. Severe water exposure can eliminate friction almost entirely and prevent braking until the friction surfaces dry themselves.

60 Square inches

Swept area,
one surface

100 Square inches

Swept area,
two surfaces

Figure 7-3 For any given size wheel, a disc brake always has 35 percent to 50 percent more swept area to dissipate heat.

Gas fade is a condition that occurs under hard braking when hot gases and dust particles are trapped between the brake linings and the drum or rotor surfaces. These gases build up pressure that acts against brake force on the drum or rotor surface. More important, these hot gases actually lubricate the friction surfaces and reduce the coefficient of friction.

The basic design differences between disc and drum brakes make disc brakes much more self-cleaning to greatly reduce or eliminate water and gas fade. The rotor friction surfaces are perpendicular to the axis of rotation, and centrifugal force works to remove water and gas from the braking surfaces. In addition, the leading edges of the brake pads help to wipe the rotor surfaces clean, and many pad linings have grooves to drain water and help prevent gas buildup.

Disc Brake Self-Adjustment

Brake adjustment is the process of compensating for lining wear and maintaining correct clearance between brake linings and the drum or rotor surfaces. Disc and drum brakes both require adjustment; but unlike drum brakes, disc brakes do not need cables, levers, screws, struts, and other mechanical linkages to maintain proper pad-to-rotor clearance.

When disc brakes are applied, the caliper pistons move out far enough to apply braking force from the pads to the rotors. When the brakes are released, the caliper pistons retract only far enough to release pressure (Figure 7-4). The pads always have only a few thousandths of an inch clearance from the rotor, regardless of lining wear. In later sections of this chapter, brake caliper operation and disc brake self-adjustment are explained in detail.

Drivers must beware when driving around construction or road repair sites on clear days. Sufficient water may be on the road to cause water fade and cause a rear-end collision in the stop-and-go traffic at these sites.

Figure 7-4 When the brakes are released, the caliper piston seal holds the piston in a retracted position (A). When the brakes are applied, the seal extends (B) and then returns to the relaxed position to retract the piston and provide clearance between the pads and the rotor.

Brake Servo Action, Pulling, and Grabbing

Drum brakes have an operating feature by which they develop mechanical leverage as the lining contacts the drum. One end of the lining contacts the drum before the other does and becomes a pivot point as friction increases quickly. The brake shoe becomes a self-energizing lever and adds its own mechanical leverage to hydraulic force to help apply the brakes (Figure 7-5).

This self-energizing operation may be confined to one shoe of a drum brake installation, which is the leading shoe in relation to the direction of wheel rotation. When the self-energizing operation of one shoe applies mechanical force to the other shoe to assist its application, it is called servo action (Figure 7-6). Servo action can be both an operational advantage and a disadvantage; but in either case, the same kind of servo action does *not* exist in disc brakes.

Servo action increases braking force at the wheels, but it must develop smoothly as the brakes are applied. If servo action develops too quickly or unevenly, the result can be brake grabbing or lockup. This is usually experienced as severe pull to one side or the other or reduced vehicle control during braking. Because disc brakes do not develop mechanical servo action, they are free from severe grabbing or pulling. Slight pull may occur in a front disc brake installation due to a sticking caliper piston or problems in hoses or valves. Disc brake pull, however, is usually never as severe as drum brake pull can be. Because equal hydraulic pressures are applied to both surfaces of the rotor during braking, no distortion of the rotor occurs regardless of the severity or duration of application. This is another reason that grabbing and pulling are reduced with disc brakes.

> A servo action is using one component to increase the input force so the second component of the assembly is applied with greater force.

AUTHOR'S NOTE: Many times the customer will define a slight pulling to one side as a drift; for example, "the vehicle tends to drift toward one side as I apply the brakes slowly." Such a slight pull or drift may be caused by a brake or alignment problem on either side of the vehicle.

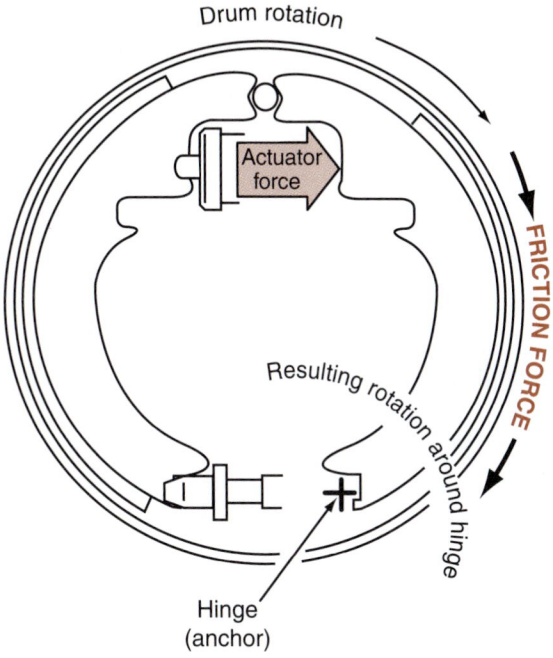

Figure 7-5 Drum rotation adds leverage to the brake shoe as it contacts the drum. This is called self-energizing action.

Figure 7-6 Self-energizing action of the primary shoe applies force to the secondary shoe. This is called servo action.

Lack of servo action in disc brakes may eliminate some operational problems, but it also means that disc brakes require higher application force from the hydraulic system than drum brakes require. The greater size ratio between master cylinder pistons and caliper pistons, along with universal use of power boosters with disc brakes, provides the necessary braking force without requiring undue pedal effort from the driver.

Disc Brake Noise

Many disc brake silencers do not eliminate the squeaking noise. They cause the frequency to change so the human ear will not hear it.

A common complaint about disc brakes is that they are noisy. The metallic brake pad used by many manufacturers is the loudest of all. Noise usually occurs when the brakes are applied and released and most commonly results from slight high-speed rattling of the pads. Modern disc brakes built for sale in North America use various devices and techniques to minimize noise. Antirattle springs and clips hold pads securely in calipers and help to dampen noise. Adhesive to hold pads to pistons and caliper mounting points further dampens squeaks and rattles. On some installations, metal or plastic shims between pads and calipers and pistons help to reduce vibration and noise.

Simple squeaks and rattles may be annoying to the driver, but they are generally not a problem for brake operation. When the noise becomes a continuous grinding or loud scraping sound as the brakes are applied, it often indicates that the linings have worn to the metal surfaces of the pads. These sounds mean that it is time for brake service before vehicle safety is jeopardized further.

Many disc brake pads have audible wear indicators, which are small steel pins or clips that rub on the rotor to create a constant squeal when the pads have worn to their minimum thickness. However, with the radio on and traffic noise the sound created by the wear indicators is masked out. As a result the driver may not be aware of the problem until the rotor and possibly the caliper piston are ruined. In an effort to combat this problem, many disc brake pads are equipped with an electronic wear indicator (Figure 7-7). A ground wire within a warning light circuit is installed onto the inboard pad. When the pad wears down, the wire touches the rotor, grounding the circuit and lighting a warning light in the instrument panel (Figure 7-8).

Disc Brake Parking Brake Disadvantage

Drum brakes provide a better static coefficient of friction than do disc brakes. The brake linings grab and hold the drums more tightly than disc pads can hold a rotor. The servo action of drum

Figure 7-7 The brake caliper fits in a U-shape over the rotor.

Figure 7-8 Pads, lined with friction material, are forced against the rotating disc to stop the car.

brakes contributes to this feature, as does the larger area of brake shoe linings compared to disc brake pads.

Because most brake installations have discs at the front and drums at the rear, the parking brake weaknesses of disc brakes are not a problem. Four-wheel disc brake installations, however, must have some way to mechanically apply the rear brakes. With movable (floating or sliding) calipers, this is usually done with a cam-and-lever arrangement in the caliper that mechanically moves the piston to develop clamping force. Disc brakes with fixed calipers usually have small, cable-operated brake shoes that grip a small drum surface toward the center of the rotor. Some late-model rear disc brakes with sliding calipers also have small drum-type parking brakes built into the rotors.

A BIT OF HISTORY

The kind of disc brakes that we know today were originally called spot brakes because caliper action clamped a friction pad against a rotor. Braking action takes place at the "spot" of the brake pad. The earliest automotive disc brakes, however, actually used a pair of circular discs lined with friction material. These systems consisted of large finned housings mounted on the hubs or axles and in turn held the wheels and tires. Two lined pressure plates inside each housing were forced outward hydraulically during brake application. The plates—or discs—did not rotate but clamped against the inside of the rotating housing to provide braking force. A ramp-and-ball mechanism on each disc provided servo action to aid brake application.

These early disc brake systems appeared on the subcompact Crosley and heavyweight Chrysler Imperial models of 1949. Although they provided improved braking action, they were complicated, costly, and prone to trouble. They also added too much unsprung weight to each wheel. The Crosley version lasted one model year. Chrysler discontinued its system in 1955. Ten years later, modern disc (spot) brakes were taking over the industry.

Disc Brake Construction

The principal parts of a disc brake are a rotor, a hub, and a caliper assembly (Figure 7-7). The rotor provides the friction surface for stopping the wheel. The wheel is mounted to the rotor hub by wheel nuts and studs. The hub houses wheel bearings that allow the wheel to rotate. The rotor has a machined braking surface on each side.

The hydraulic and friction parts are housed in a caliper that straddles the outside diameter of the rotor. When the brakes are applied, pistons inside the caliper are forced outward by hydraulic pressure. The pressure of the pistons is exerted through the pads or shoes in a clamping action on the rotor. A splash shield on most installations helps to keep water and dirt away from the rotor and caliper and directs airflow to the rotor for improved cooling. Figure 7-8 illustrates basic disc brake operation.

Rotors, Hubs, and Bearings

The disc brake **rotor** has two main parts: the hub and the braking surface (Figure 7-9). The hub is where the wheel is mounted and contains the wheel bearings. The braking surface is the machined surface on both sides of the rotor. It is machined carefully to provide a friction surface

Shop Manual
pages 281–294

The **rotor** is the rotating part of a disc brake that is mounted on the wheel hub and contacted by the pads to develop friction to stop the car. It is also called a disc.

Figure 7-9 Major parts of a typical rotor and hub assembly.

for the brake pads. The entire rotor is usually made of cast iron, which provides an excellent friction surface. The rotor side where the wheel is mounted is the outboard side. The other side, toward the center of the car, is the inboard side.

The size of the rotor braking surface is determined by the diameter of the rotor. Large cars, which require more braking energy, have rotors measuring 12 inches in diameter and larger. Smaller, lighter cars can use smaller rotors. Generally, manufacturers want to keep parts as small and light as possible, while maintaining efficient braking ability.

The rotor is protected from road splash on the inboard side by a sheet metal splash shield that is bolted to the steering knuckle (Figure 7-10). The outboard side is shielded by the vehicle wheel. The splash shield and wheel also are important in directing air over the rotor to aid cooling.

Fixed and Floating Rotors

Rotors can be classified by the hub design as fixed (with an integral hub) or floating (with a separate hub). A **fixed rotor** has the hub and the rotor cast as a single part (Figure 7-11) and is commonly known as a one-piece rotor.

Floating rotors and their hubs are made as two separate parts (Figure 7-12), and may be known as a two-piece rotor or just rotor. The hub is a conventional casting and is mounted on wheel bearings or on the axle. The wheel studs are mounted in the hub and pass through the rotor center section. This kind of rotor is called a hubless or floating rotor. One advantage of this design is that the rotor is less expensive and can be replaced easily and economically when the braking surface is worn beyond machining limits.

Composite Rotors

Traditionally, brake rotors, as were drums, were manufactured as a single iron casting. The development of floating, two-piece rotors and the need to reduce vehicle weight led to the development of **composite rotors.** Composite rotors are made of different materials, usually cast iron and

Figure 7-10 The splash shield is mounted on the steering knuckle inboard of the rotor.

Figure 7-11 The wheel hub and the rotor are cast as a single part in a fixed or integral rotor.

Figure 7-12 A rotor that is cast as a separate part and fastened to the hub is called a floating rotor.

steel, to reduce weight. The friction surfaces and the hubs are cast iron, but supporting parts of the rotor are made of lighter steel stampings. The steel and iron sections are bonded to each other under heat and high pressure to form a one-piece finished assembly (Figure 7-13). Composite rotors may be fixed components with integral hubs, or they may be floating rotors mounted on a

Figure 7-13 The cast-iron friction surfaces are cast onto the steel web during manufacture to form a composite rotor. The steel web of this rotor also contains the parking brake drum.

separate hub. Because the friction surfaces of composite rotors are cast iron, the wear standards and refinishing methods are generally the same as they are for other rotors.

Solid and Ventilated Rotors

A solid rotor is usually a floating rotor. Some solid rotors are "throwaways," meaning they cannot be machined. They are very cheap and are installed on very small cars.

A rotor may be solid or it may be ventilated (Figure 7-14). A **solid rotor** is simply a solid piece of metal with a friction surface on each side. A solid rotor is light, simple, cheap, and easy to manufacture. Because solid rotors do not have the cooling capacity of ventilated rotors, they usually are used on small cars of moderate performance or on rear disc brake installations.

Figure 7-14 A solid rotor, left, and a large ventilated rotor, right.

A **ventilated rotor** has cooling fins cast between the braking surfaces to increase the cooling area of the rotor. When the wheel is in motion, the rotation of these fins in the rotor also increases air circulation and brake cooling.

Although ventilated rotors are larger and heavier than solid rotors, these disadvantages are more than offset by their better cooling ability and heat dissipation. Some heavy ventilated rotors may have weights between the fins, or in an area near the fins, ground away to balance the rotor.

Some ventilated rotors have cooling fins that are curved or formed at an angle to the hub center. These fins increase centrifugal force on rotor airflow and increase the air volume that removes heat. Such rotors are called **unidirectional rotors** because the fins work properly only when the rotor rotates in one direction. Therefore, unidirectional rotors cannot be interchanged from right to left on the car and the fins of the installed rotor must point forward when viewed from the top.

A few high-performance sports cars have solid rotors with holes drilled through the friction surfaces. These drilled rotors are not made to increase cooling so much as they are to release water and hot gases from the rotor surface that can cause water or gas fading. Drilled rotors are typically very light and can have very short service lives. Therefore, they are used mostly on race cars and dual-purpose, high-performance cars.

Rotor Hubs and Wheel Bearings

Tapered roller bearings, installed in the wheel hubs, are the most common bearings used on the front wheels of RWD vehicles and the rear wheels of FWD cars. The tapered roller bearing has two main parts: the inner bearing cone and the outer bearing cup. A tapered roller bearing has the rollers set at an angle to the centerline of the bearing assembly. The rollers are held in place by the bearing cage mounted to the inner race. The race is the precision machined area where the rollers run. The bearing cup acts as the outer race and can usually be removed from the bearing assembly. The bearing fits into the outer cup or race, which is pressed into the hub. This provides two surfaces, an inner cone and outer cup, for the rollers to ride on. See Chapter 3 for details on tapered roller bearings and their service.

Almost every FWD vehicle has a sealed double-row ball bearing pressed into the front hubs (Figure 7-15). The bearing usually requires no periodic service and is replaced when necessary.

Figure 7-15 Most FWD vehicles have a pressed-in, double-roll sealed ball bearing.

The bearing is pressed into the hub from the inboard side and is retained with a snapring. The FWD axle stub slides through the inner bearing race. A typical sealed bearing will last 100,000 miles, and some will last until the vehicle rusts away.

Brake Pads

Shop Manual
pages 284–294

Each brake caliper contains two **brake pads.** Pads are the disc system braking friction surfaces. They perform the same function as the shoes in a drum brake system. The brake pads are positioned in the caliper on the inboard and outboard sides of the rotor. The caliper piston, or pistons, forces the brake pad linings against the rotor surfaces to stop the car.

Fundamentally, a brake pad is a steel plate with a friction material lining bonded or riveted to its surface (Figure 7-16). The simple appearance of a brake pad, however, hides some sophisticated engineering that goes into its design and manufacturing. Chapter 2 of this *Classroom Manual* introduced the subject of friction materials and summarized the common requirements of both drum brake linings and disc brake pads. Friction materials are classified as organic (nonmetallic), semimetallic, fully metallic, and—in some advanced systems—synthetic.

Brake Friction Materials (Drum and Disc)

The friction material used for disc brake pads is generally harder than that used on drum brake linings because the friction surface is smaller and higher pressures are used to push the pads into contact with the rotor. The friction material used for drum brake lining is generally softer than that used on disc brake pads because the friction surface is larger.

Asbestos has excellent friction qualities and long service life. Therefore, it was the most common brake lining material for years. The health hazards of asbestos have led to its drastic reduction or removal from most brake friction materials, however. Today's basic types of disc pad lining materials are organic, semimetallic, metallic, and synthetic.

Organic Linings. Organic linings usually wear faster than do semimetallic linings, but they have the benefit of breaking in faster. **Organic linings** are made from nonmetallic fibers bonded together to form a composite material. For decades, asbestos was the main ingredient in organic linings. Asbestos is actually an inorganic mineral but not truly a metal. Regardless of its chemistry, the organic name stuck to friction material based on asbestos.

Figure 7-16 Parts of a disc brake pad.

Health regulations have caused asbestos content in organic linings to drop from as high as 75 percent to below 25 percent. The goal is to eliminate asbestos from all brake linings, and it has been achieved for friction materials made in North America. Today's organic brake linings contain the following kinds of materials:

- ❏ Friction materials and **friction modifiers,** which are the materials that provide the frictional stopping power. Some examples are graphite, powdered metals, and even nut shells.
- ❏ **Fillers** are secondary materials added for noise reduction, heat transfer, and other purposes.
- ❏ **Binders** are glues that hold the other materials together.
- ❏ **Curing agents** accelerate the chemical reaction of the binders and other materials.

Organic linings have a high coefficient of friction for normal braking, they are economical, they are quiet, they wear slowly, and they are only mildly abrasive to rotors. Organic linings fade more quickly than other materials, however, and they do not operate well at high temperatures. High-temperature organic linings are available for high-performance cars, but they do not work as well at low temperatures and they wear faster than common organic linings.

Semimetallic Linings. Semimetallic materials are made from a mixture of organic or synthetic fibers and certain metals molded together; they do not contain asbestos. **Semimetallic linings** are harder and more fade resistant than organic materials but require higher brake pedal effort.

Most semimetallic linings contain about 50 percent iron and steel fibers. Copper also has been used in some semimetallic linings and, in smaller amounts, in organic linings. Concerns about copper contamination of the nation's water systems has led to its reduced use in brake linings, however.

> **AUTHOR'S NOTE:** A few vehicles, and not necessarily high-end ones, will work quietly only with the manufacturer's brand of brake pads. Aftermarket pads, particularly semimetallic pads, make a lot of noise even when first installed. This is usually caused by the composition of the rotor's friction area, the friction area finish, and the makeup of the pads themselves. At times the only possible solution is to buy pads from the local dealership. Luckily this problem applies to only a few models of a few brands.

Semimetallic linings operate best above 200°F to 250°F (90°C to 120°C) and actually must be warmed up to bring them to full operating efficiency. Consequently, semimetallic linings often have poorer operating characteristics than organic linings at low temperatures.

Semimetallic linings often are blamed for increased rotor wear, but this is not entirely true. Early semimetallic linings were more abrasive than current materials, which may cause no more wear with the properly matched rotors than organic materials. Also, the better heat transfer characteristics of semimetallic linings can reduce rotor temperatures and help to counteract abrasiveness. Many small, FWD cars built since the early 1980s have smaller front brakes that require the better high-temperature friction characteristics and heat transfer abilities of semimetallic linings.

Currently, semimetallic linings are used only on front disc brakes of passenger cars and light trucks. The lighter braking loads on rear brakes, particularly on FWD cars, may never heat semimetallic linings to their required operating efficiency. Semimetallic linings also have a lower static coefficient of friction than organic linings, which makes them inferior for parking brake use.

Metallic Linings. Fully metallic materials were used for many years in racing, particularly in the heyday of drum brakes. **Metallic lining** is made from powdered metal that is formed into blocks by heat and pressure. These materials provide excellent resistance to brake fade but require high brake pedal pressure and create the most wear on rotors and drums. Metallic linings work very poorly until they are fully warmed. Improved high-temperature organic linings and semimetallic

A **friction modifier** is a class of materials used in brake linings to modify the final coefficient of friction of the linings.

Curing agent refers to a class of materials used in brake linings to accelerate the chemical reaction of the binders and other materials.

Semimetallic friction material can seriously damage or destroy some types of rotor.

materials have made metallic linings almost obsolete for late-model automotive use. Metallic linings are extremely noisy, which must be considered when talking choices to a customer.

Synthetic Linings. The goals of improved braking performance and elimination of the disadvantages of current lining materials has led to a new generation of **synthetic lining** friction materials. They are called synthetic because that term generally describes nonorganic, nonmetallic, and nonasbestos materials. Two principal kinds of synthetic materials show promise as brake linings: fiberglass and **aramid fibers.** Synthetic lining may be used on disc or drum brakes.

Fiberglass was introduced as a brake lining material to help eliminate asbestos. As does asbestos, it has good heat resistance, good coefficient of friction, and excellent structural strength. The disadvantages of fiberglass are its higher cost and its reduced friction at very high temperatures. Overall, fiberglass linings perform similarly to organic linings, but the higher costs have confined their use primarily to rear drum brake linings.

Aramid fibers are a family of synthetic materials that are five times stronger than steel, pound for pound, but weigh little more than half what an equal volume of fiberglass weighs. Friction materials made with aramid fibers are manufactured similarly to organic and fiberglass linings. Aramid fibers have a coefficient of friction similar to semimetallic linings when cold and close to that of organic linings when hot. Overall, the performance of aramid linings is somewhere between organic and semimetallic materials but with much better wear resistance and longevity than organic materials.

Ceramic Brake Pads. Many automotive manufacturers are now supplying brake pads of **ceramic** materials as the pad of choice. Akebono is supplying many vehicle manufacturers with their OEM ceramic brake pads. In addition, some aftermarket vendors such as Raybestos Brakes' "Quiet Stop" and NAPA's "Ceramix" are two brands of ceramic pads used to replace semimetallic types. Combining ceramic material and copper fibers is the typical formula for ceramic pad manufacture. Many of the drawbacks of semimetallic pads have been reduced or eliminated entirely.

The steel in semimetallic pads tends to make noise, wear rotors, and create lots of brake dust, which affects the appearance of the wheel and rims. Ceramic and copper composite pads make much less noise, are not as damaging to the rotors, and create virtually no dust. These benefits are much more satisfactory to the owner/driver. Ceramic pads are usually slotted vertically, horizontally, or diagonally to change the noise frequency to a range beyond human hearing. The slots also provide an escape route for gases and dust, thereby reducing brake fades at higher temperatures. Some ceramic pads have chamfered leading and trailing edges, which also help to reduce noise. Porsche offers ceramic pads and rotors on one of its top-of-the-line vehicles.

Ceramic brake pads do have a few disadvantages over semimetallic. The main disadvantage is primarily the result of improper pad/rotor configuration. Some operators may wish to add "high-performance" ceramic pads in lieu of the specified semimetallic pads. Many times, this configuration would warp or wear the rotor(s) very quickly, even during normal driving and braking. The high-performance ceramic pad does not dissipate heat as quickly as semimetallic pads and the heat is trapped in the rotor, caliper, and brake fluid. As mentioned earlier, there are aftermarket ceramic pads for many cars, but they are made to directly replace those specified by the vehicle manufacturer. The technician or service writer needs to caution the customer on the use of ceramic brake pads. Most of the time the customer only wishes to keep the wheels and rims looking neat.

Another item the technician and customer must consider is the cost of ceramic pads. Typically, a set of ceramic brake pads cost three to four times the cost of standard semimetallic or organic pads. The labor cost should be the same. In most cases, ceramic brake pads are overkill braking for the typical driver and vehicle. However, the appearance of the vehicle may be of more concern to the customer then the price of the parts.

Friction Material Selection

You can identify the friction material of a new brake pad from a code printed on the edge of the pad or shoe (Figure 7-17). This code is called the **Automotive Friction Material Edge Code.**

Friction material

Metal backing

Figure 7-17 Automotive Friction Material Edge Code.

Letters and numbers in the code identify the manufacturer, the material, and, of most importance, the cold and hot coefficients of friction. In the example shown in Figure 7-17, the first letters "NRSS" identify the manufacturer. The numbers "12041" identify the lining material, and the last two letters "FF" identify the cold and hot coefficients of friction, respectively.

From a service standpoint, the friction codes are probably the most important. The technician may be able to compare them to a decoding chart in brake parts catalogs to determine the friction material and whether it is recommended for a specific vehicle (Table 7-1). The following chart gives definitions of different DOT Automotive Friction Edge Codes. Note that EE coded pads have roughly the same coefficient of friction of steel on steel or bare caliper piston against steel rotor. A pad with the edge HH would probably not work well on a conventional vehicle when noise, brake wear, and cost are considered. The two different temperatures given are considered as cold and hot.

It is important to use the recommended friction material when replacing brake pads or drum brake shoes. The incorrect type of friction material can affect the stopping characteristics of the car. These codes, however, indicate only the coefficient of friction. They do not address lining quality or its hardness.

Hard and *soft* are terms applied to linings within a general category of material. Thus, any particular organic lining may be considered as a hard or a soft organic material. Overall, organic

The codes on some inexpensive brake pads often cannot be read because of the printing and paint used.

Table 7-1

DOT Edge Code	Coefficient of Friction (C.F.) @ 250°F and @ 600°F	Fade probability
EE	0.25 to 0.35 both temps	0–25% @ 600°F
FE	0.25 to 0.35 @ 250°F temp 0.35 to 0.45 @ 600°F temp	2% to 44% fade at 600°F
FF	0.35 to 0.45 @ both temps	0–22% fade at 600°F
GG	2.45 to 0.55	Very rare
HH	0.55 to 0.65	Carbon/Carbon only. Glow at about 3000°F

linings have been considered softer than semimetallic linings, and semimetallic linings are considered softer than fully metallic linings. Typically, a hard lining has a low coefficient of friction but resists fade better and lasts longer than a soft lining. A soft lining has a higher coefficient of friction but fades sooner and wears faster than a hard lining. Additionally, a soft lining is less abrasive on rotor and drum surfaces and operates more quietly than a hard lining.

It also is important to know that carmakers sometimes specify different friction materials for the inboard and outboard disc pads. You may find original equipment installations that have an organic disc pad on one side of the caliper and a semimetallic disc pad on the other. It also is common to use linings with a lower coefficient of friction on the rear brakes than on the front brakes to minimize rear brake lockup.

Although the coefficient of friction and hardness of linings can vary quite a bit for the same kind of material, these general rules for the coefficient of friction will help you compare different materials:

Organic—cold: 0.44; warm: 0.48

Semimetallic—cold: 0.38; warm: 0.40

Metallic—cold: 0.25; warm: 0.35

Synthetic—cold: 0.38; warm: 0.45

When troubleshooting a brake performance problem, a noise problem, or a problem with premature rotor wear, it often is a good idea to install new brake pads that match the original equipment exactly. This establishes a baseline of what should be original brake performance for further diagnosis.

Friction Material Attachment

Both disc brake and drum brake linings are attached by rivets or adhesive bonding or a combination of riveting and bonding to the metal backing (Figure 7-18).

Bonded lining

Riveted lining

Figure 7-18 The linings can be bonded (top) or riveted (bottom) to the disc brake pads.

Riveted Linings. In a riveted brake pad or shoe assembly, the **riveted lining** is attached to the steel pad or shoe by copper or aluminum rivets. Riveting allows a small amount of flexing between the pad and lining to absorb vibration and reduce noise. Riveting also is very reliable, and rivets maintain a secure attachment at high temperature and high mileage.

Rivets require that about one-third to one-quarter of the lining thickness remain below the rivet for secure attachment. This places the rivet head closer to the lining friction surface and reduces the service life or mileage of the pad. In some cases, the holes above the rivet heads can trap abrasive particles that can score the rotor.

Historically, riveting was the most common method for lining attachment. Since bonding was perfected, however, rivets are used more with semimetallic linings than with organic linings.

Bonded Linings. Bonding is a method of attaching the friction material to the pad or shoe with high-strength, high-temperature adhesive.

Bonded linings can provide longer service life because more material is available for lining wear before the steel pad or shoe contacts the rotor or drum. This feature can be misleading and provides false comfort to many motorists, however. If a driver neglects lining wear until a disc pad hits the rotor, it often severely scores the rotor and destroys it. Minor scoring from rivet heads often can be cleaned up by resurfacing, however.

Bonded linings do not have the flexibility between the pad and shoe lining that riveted linings have. Therefore, they can be more noisy than riveted linings. A noise complaint often can be fixed by replacing bonded linings with riveted linings that have the recommended friction characteristics.

Mold-Bonded Linings. In a disc brake pad assembly with **mold-bonded lining,** an adhesive is applied to the pad and the uncured lining material is poured onto the pad in a mold. The assembly is then cured at high temperature to fuse the lining and adhesive to the pad. Holes drilled in the pad are countersunk from the rear so that the lining material flows through the holes and rivets itself to the pad as it cures. Many high-performance pads are made this way to avoid the stress-cracking problems associated with rivets while providing very secure pad attachment.

Pad-to-Caliper Attachment

Most brake pads are held in the caliper by locating tabs formed on the end of the metal backing plate (Figure 7-19). The pads also may be retained by retaining pins that go through holes in the metal pad backing.

Many pads have antirattle or support clips, which are spring steel clips that hold the pads in position to keep them from rattling when the pads are out of contact with the rotor. These small parts are called **pad hardware** and should be replaced when the pads are replaced. Figure 7-20 shows a selection of typical pad hardware.

Brake Pad Wear Indicators

Many current brake systems have **pad wear indicators** or sensors to warn the driver that the lining material has worn to its minimum thickness and that the pads require replacement. The most common types are audible contact sensors and electronic sensors.

The audible sensor is the oldest type. The audible system uses small spring clips on the brake pads and lining that make a noise when the linings are worn enough to be replaced. The spring clips are attached to the edge of the brake shoe or into the shoe from the pad side. They are shaped or positioned to contact the rotor when the lining wears to where it should be

Pad hardware refers to miscellaneous small parts, such as antirattle clips and support clips, that hold brake pads in place and keep them from rattling.

Figure 7-19 Tabs on the metal backing plate locate the pad in the caliper.

Figure 7-20 Typical antirattle and retaining hardware for brake pads.

New pad	**Worn pad**

Figure 7-21 As the linings wear, the wear indicator eventually hits the rotor and creates noise to warn the driver that new pads are needed.

replaced. When the linings wear far enough, the sensor contacts the rotor (Figure 7-21) and makes a high-pitched squeal to warn the driver that the system needs servicing. This squeal usually goes away when the brakes are applied. Unfortunately, many drivers are deaf to the sound of these wear indicators when the radio is on or there is a lot of traffic noise, or they just ignore noise.

AUTHOR'S NOTE: In defense of some drivers, we experienced an incident where the brake pads were completely gone, the pad metal backing was rubbing on the rotors, and on one side the backing was gone to the point where the caliper piston was almost out of the bore. I questioned the owner, a fellow faculty member, as to why the car was not brought in when it started to make a lot of noise during braking. The owner stated that there was never any noise, and that the car just got to the point where it did not want to stop. As soon as the owner left and before we put the car on the lift, I drove it around the parking lot several times and applied the brakes both in easy and in panic modes just to prove a point. But the owner was right. The car did not stop well, and there was absolutely no noise that would be associated with extremely worn pads. So sometimes when the owners say there was no noise, they may be correct.

Electronic sensors provide a warning lamp or message on the instrument panel to inform the driver of brake pad wear or circuit problems. The electronic sensor uses pellets embedded into the friction material for circuit completion. Early systems used a grounding logic to complete a parallel electrical circuit and turn on the warning lamp (Figure 7-22, Item A). If a circuit wire became grounded, this too would turn on the warning lamp. An open circuit, however, could not be detected and may disable the detection circuit. Later systems were designed to use an open logic to do the same thing. Open logic allows the system to detect both opens and grounds in the circuit (Figure 7-22, Item B).

A

Parallel circuit

B

Series circuit

Figure 7-22 Electronic wear sensors are used in both parallel (A) and series (B) circuits.

A   BIT   OF   HISTORY

Disc brakes were developed for racing, and the first practical disc brake system was used in 1953 on a C-type Jaguar that won the 24-hour endurance race at Le Mans. The effectiveness of the brakes in this event led automotive engineers to develop disc brakes for passenger cars by the 1960s and 1970s.

Caliper Construction and Operation

Shop Manual
pages 298–313

The disc **brake caliper** converts hydraulic pressure from the master cylinder to mechanical force that pushes the brake pads against the rotor. The caliper is mounted over the rotor (Figure 7-23). Although there are many design differences among calipers (described in a later section of this chapter), all calipers contain these major parts:

- ❏ Caliper body or housing
- ❏ Internal hydraulic passages
- ❏ One or more pistons
- ❏ Piston seals
- ❏ Dust boots
- ❏ Bleeder screw

Figure 7-24 is a sectional view of a caliper. A brake line from the master cylinder is attached to the caliper body. When the brakes are applied, fluid under pressure from the master cylinder enters the caliper. The caliper has at least one large hydraulic piston located in a piston bore. During braking, fluid pressure behind the piston increases. Pressure is exerted equally against the bottom of the piston and the bottom of the cylinder bore. The pressure applied to the piston is transmitted to the inboard brake pad to force the lining against the inboard rotor surface. Depending on caliper design, fluid pressure may be routed to matching pistons on the outboard side of

Figure 7-23 The caliper is bolted to the caliper support, which is bolted to the steering knuckle.

Figure 7-24 Cross section of a floating caliper. Sliding calipers are similar.

the caliper, or the pressure applied to the bottom of the cylinder can force the caliper to move inboard on its mount (Figure 7-25). In either case, the caliper applies mechanical force equally to the pads on both sides of the rotor to stop the car.

Caliper Body

The caliper body is a U-shaped casting that wraps around both sides of the rotor (Figure 7-26). Almost all caliper bodies are made of cast iron. Single-piston caliper bodies are usually cast in one piece, but caliper bodies with pistons on the inboard and outboard sides of the rotor are usually cast in two pieces and bolted together.

Front calipers are mounted on **caliper supports** or adapters that may be an integral part of the **steering knuckles.** The steering knuckle provides a spindle for the wheel to rotate on, steering linkage attachments, and connections to the vehicle's suspension systems. Although one-piece steering knuckle and caliper support forgings are the simplest and most economical

A **steering knuckle** is the outboard part of the front suspension that pivots on the ball joints and lets the wheels turn for steering control.

Figure 7-25 Hydraulic pressure in a sliding or floating caliper forces the piston and one pad in one direction and the caliper body and the other pad in the opposite direction. This is an application of one of the Laws of Motion: For every action there is an equal and opposite reaction.

Figure 7-26 The brake caliper straddles the rotor.

way to manufacture these components, most front caliper supports are separate pieces that bolt to the steering knuckle. Rear calipers are mounted on caliper supports that bolt to the rear suspension or the axle housing.

Most caliper bodies have one or two large openings in the top of the casting through which the lining thickness can be inspected. Although these openings are handy inspection points, that is not their primary purpose. Caliper bodies are cast with these openings to reduce weight and, of more importance, to minimize large masses of iron that can cause uneven thermal expansion of the assembly. A few caliper bodies are without these openings because they are not needed structurally.

Caliper Hydraulic Passages and Lines

The fitting for the brake line is located on the inboard side of the caliper body, as is the bleeder screw. The bleeder screw is almost always at the top of the caliper housing. A movable caliper with one or two pistons on the inboard side has hydraulic passages cast into the inboard side of the body. A fixed caliper with pistons on both the inboard and the outboard sides requires crossover hydraulic lines to the outboard side. These usually are cast into the caliper halves and sealed with O-rings when the caliper body is bolted together (Figure 7-27). Some fixed calipers, however, have external steel tubing to carry fluid to the outboard pistons.

Caliper Pistons

Depending on its design, a caliper may have one, two, three, or four pistons, with single-piston calipers being the most common by far. The piston operates in a bore that is cast and machined into the caliper body (Figure 7-28). The piston is the part that actually converts hydraulic pressure to mechanical force on the brake pads. To do its job, the piston must be strong enough to operate with several thousand pounds of pressure, and it must resist corrosion and high temperatures.

The inner side of the piston contacts the brake fluid in the caliper bore, and the outer side bears against the steel brake pad. The brake pads operate at temperatures above the boiling point of brake fluid, and the piston must help to insulate the fluid from this extreme heat. The piston surface that contacts the pad is hollow or cup shaped, which reduces weight and reduces the area available to absorb heat from the pad. The ability to absorb heat and not transfer it to the fluid also is an important design consideration for the piston materials. Caliper pistons typically are made of steel, cast iron, **phenolic plastic,** and aluminum.

Chrome-plated cast-iron and steel pistons are the most common. Iron pistons were used in many early disc brakes, but steel is more common in late-model designs. Steel pistons are strong and thermally stable, but they are heavier than desired in late-model brake assemblies, and they can conduct excessive heat to the brake fluid.

To reduce both weight and heat transfer, manufacturers turned to phenolic plastic pistons in the mid-1970s. Today, phenolic plastic pistons are original equipment on about half the light-duty vehicles sold in North America. Phenolic plastic pistons are strong, light, excellent insulators, immune from corrosion, and economical to make. In addition, the plastic surface is not as slippery as chrome plating and grips the piston seals better than a steel piston does.

Phenolic pistons do have some disadvantages; among them are a tendency to wear faster and score more easily than steel pistons. Early designs also tended to stick in their bores due to

Phenolic plastic is plastic made primarily from phenol, a compound derived from benzene; phenol is also called carbolic acid.

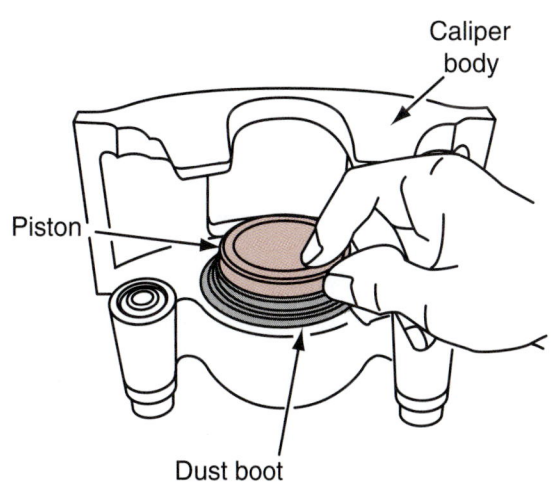

Figure 7-27 This cross section shows the fluid crossover passages in a fixed caliper.

Figure 7-28 The piston operates in a machined bore in the caliper body.

caliper bore varnish and corrosion. Improved dust boots and seals have reduced that problem, however. Phenolic plastic pistons also may be damaged if mishandled.

Aluminum would seem to be a good material for caliper pistons because of its light weight. In fact, aluminum has been used for pistons in some high-performance brakes where weight is critical, but it has serious drawbacks that limit its use in passenger car and light truck brakes.

Aluminum expands faster that iron or steel, and aluminum pistons must be made with more clearance in their caliper bores. This additional clearance, in turn, increases the possibility of leakage, as does aluminum's greater tendency toward scoring and corrosion. The major problem with aluminum caliper pistons, however, is their ability to transfer heat. Aluminum is a very poor thermal insulator and increases the danger of boiling the brake fluid.

Caliper Piston Seals

Brake calipers require seals to keep fluid from escaping between the pistons and their bores, but caliper piston seals also perform other functions. Disc brakes do not have return springs to move the pads out of contact with the rotor. This is accomplished by the caliper piston seal. Movable calipers use **fixed seals,** also called **square-cut piston seals** or **lathe-cut seals.** Older, fixed calipers used a type of lip seal called a stroking seal.

Fixed Seals. A fixed seal, used with a movable caliper, is installed in a groove in the inner circumference of the caliper bore. The piston fits through the inside of the seal and is free to move in the seal. The outer circumference of the seal remains in a fixed position in the caliper bore. Many seals are square or rectangular in cross section, but others have different cross-sectional shapes. Because all seals are not identical, it is important to be sure that replacement seals match the shape of the originals.

During braking, the piston seal is deflected or bent by the hydraulic pressure. When the pressure is released, the seals relax or retract, pulling the pistons back from the rotors. Figure 7-29 is a cross section of a piston and seal that shows the seal action from relaxed (before application) to an applied position and back to a relaxed position. The seal flexing releases pressure from the rotor but maintains only very slight clearance of 0.001 inch or less between the pad and the rotor. The seal fits closely around the piston and holds it in position. Because the seal is installed at the outer end of the caliper bore, it keeps dirt and moisture out of most of the bore and away from the piston. This minimizes corrosion and damage to the caliper bores and pistons.

As the brake linings wear, the piston can move out toward the rotor to compensate for wear. The seal continues to retract the piston by the same amount, however. Thus the piston can travel outward, but its inward movement is restricted by the flexibility of the seal. This action provides the inherent self-adjusting ability of disc brakes, and pedal height and travel remain constant throughout the life of the brake linings.

Steel pistons used with fixed seals have very close clearances in their bores. Manufacturers typically specify 0.002 inch to 0.005 inch of clearance. Phenolic plastic pistons require slightly more clearance to allow for more expansion. They are typically installed with 0.005 inch to 0.010 inch of clearance. The close fit of the pistons in their bores and the close running clearances between pads and rotors keep pistons from cocking in the calipers and minimize piston knock back due to rotor runout or warping. Close piston-to-bore clearances also keep the seals from flexing too much and rolling in their grooves.

Low-Drag Caliper Seals. Low-drag calipers increase the clearance between the brake pads and rotors when the brakes are released. This increased clearance reduces friction and improves fuel mileage. In a low-drag caliper, the groove for the fixed seal has a tapered outer edge. This lets the seal flex farther as the brakes are applied. This increased flexing outward as the brakes are applied is matched by equal inward flexing as the brakes are released. The result is more clearance between the pads and rotors.

Before application

Seal deflection

During application

After application

Figure 7-29 These cross sections show how a seal deflects during brake application and then returns to the relaxed position to retract the piston.

Low-drag calipers require a quick take-up master cylinder as described in Chapter 4 of this *Classroom Manual.* The quick take-up master cylinder provides more fluid volume with the initial pedal movement to take up the greater pad-to-rotor clearance. The combination of low-drag calipers and quick take-up master cylinder maintain normal brake pedal height and travel.

Stroking Seals. Older, fixed-caliper disc brakes have piston seals called **stroking seals.** A stroking seal is similar to a lip seal and is installed at the rear of the piston (Figure 7-30). Unlike a fixed seal that is installed in a groove in the caliper bore, a stroking seal is installed on the piston and moves with it. Sealing is provided by the seal lip, and the seal flexes to retract the piston similar to the way a fixed seal works. Pistons used with stroking seals are installed with slightly more clearance than pistons with fixed seals.

Because stroking seals are installed at the rear of the pistons, the caliper bore can be exposed to moisture and dirt if the dust boot fails. This can lead to scored bore walls and leakage as the piston moves out in its bore. Stroking seals also provide less support for the piston in the bore and can let the piston cock or tilt slightly. If a piston tilts in its bore, it is more likely to be knocked back in the bore as the rotor turns. Piston travel will then increase on the next brake application. Piston knock back can cause excessive and inconsistent pedal travel. Because of these reasons—caliper corrosion and leakage and excessive pedal travel—stroking seals are not used in late-model brake designs.

Caliper Dust Boots

A rubber boot fits around every caliper piston to keep dirt and moisture out of the caliper bore (Figure 7-31). The opening in the center of the boot fits tightly around the outer end of the piston.

Figure 7-30 A stroking seal is like a lip seal and is installed toward the rear of the piston. The seal moves with the piston but exerts the same kind of force as a square-cut seal to retract the piston.

Figure 7-31 A rubber boot keeps dirt and moisture out of the caliper bore.

The outer circumference of the boot may be attached to the caliper body by a retaining ring in the caliper or by tucking it into a groove inside the bore.

Types of Disc Brakes

Although all disc brakes have the same kinds of common parts—calipers, rotors, pistons, pads, and so on—they are commonly classified into two groups in terms of caliper operation: fixed and movable. Fixed calipers are bolted rigidly to the caliper support on the steering knuckle or on the rear axle or suspension. Fixed calipers have pistons and cylinders (bores) on the inboard and outboard sides. Hydraulic pressure is applied equally to the inboard and outboard pistons to force the pads against the rotor. Movable calipers slide or float on the caliper support. Sliding or floating calipers have a piston only on the inboard side, and hydraulic pressure is applied to the piston to force the inboard pad against the rotor. At the same time, hydraulic pressure on the bottom of the caliper bore forces the caliper to move inboard and clamp the outboard pad against the rotor with equal force.

Fixed Calipers

A **fixed caliper brake** has a brake caliper that has piston(s) on both the inboard and the outboard sides and that is bolted to its support and does not move when the brakes are applied.

Fixed caliper disc brakes are the oldest type of disc brake. Their use on production cars in high volume dates to the mid-1960s. Probably the best known **fixed caliper brake** is the four-piston Delco Moraine brake used on Chevrolet Corvettes from 1965 through the early 1980s. Fixed caliper brakes from other suppliers such as Bendix, Budd, and Kelsey–Hayes were used by other domestic carmakers through the mid-1970s, and some late-model imported vehicles continue to have fixed caliper brakes.

AUTHOR'S NOTE: Ford and other light- and medium-duty truck manufacturers (F-350 and above) were using fixed, four-piston calipers up through the mid-1990s. This was primarily due to the length of the pads used on these trucks. Compared to lighter trucks and passenger cars, the pads were about twice as long and could not be evenly applied along their full length by one piston and a sliding caliper without making the single piston very large.

Figure 7-32 A fixed caliper is mounted rigidly over the rotor and has one or two pistons on the inboard and outboard sides to apply the brake pad.

A fixed caliper is bolted to its support and does not move when the brakes are applied. A fixed caliper must have pistons on both the inboard and the outboard sides. The most common are four-piston brakes, with two pistons on each side of the caliper (Figure 7-32). Two-piston fixed caliper brakes were common on some lighter imported vehicles, and a few three-piston designs also were manufactured (with one large inboard and two small outboard pistons).

The pads for a fixed caliper brake are held in the caliper body by locating pins. They are mounted more loosely than the pads of a movable caliper brake because the outboard pad cannot be attached solidly to the caliper body. Antirattle clips and springs reduce vibration and noise, however.

Fixed calipers must have equal piston areas on each side so that equal hydraulic pressure will apply equal force to the pads (Figure 7-33). When the brakes are released, the pistons are retracted by their seals as explained earlier in this chapter.

Because the body of a fixed caliper brake is mounted rigidly to the vehicle, it must be aligned precisely over the brake rotor. Piston travel must be perpendicular to the rotor surface, and the pads must be exactly parallel to the rotor (Figure 7-34). If the caliper is misaligned or cocked on its support, the pads will wear unevenly and the pistons may stick in their bores. Just as important, the caliper must be centered over the rotor so that the inboard and outboard pistons move equal distances to apply the pads. If the caliper is offset to one side or the other, unequal piston travel can create a spongy brake pedal feel and uneven pad wear. Moreover, it may be more difficult to bleed all air from the caliper during service. Shims often are used to align fixed calipers on their supports.

Fixed calipers are large and heavy, which provides good heat dissipation and strength to resist high hydraulic pressures. Because a fixed caliper is mounted rigidly to the vehicle, it does not tend to flex under high temperature and pressure from repeated use. Fixed calipers can provide consistent braking feel under repeated, hard use.

The use of fixed calipers has declined over the past 30 years, and they are not found on many late-model, lightweight cars. The greater weight of a fixed caliper is its greatest disadvantage, and that caliper weight increases the percentage of unsprung weight on the steering knuckle. Fixed calipers also are more expensive to make than floating or sliding calipers, as well as harder

Figure 7-33 Equal hydraulic pressure on both sides of a fixed caliper applies equal force to the inboard and outboard pistons.

Figure 7-34 A fixed caliper must be parallel to the rotor surface (Dimension D) and spaced equally on each side of the rotor (Dimension C).

to service. Because fixed calipers are built with a two-piece body, they require more time to service and leaks have more opportunities to develop around O-rings and crossover line fittings. In summary, cost, weight, and complexity have made fixed caliper disc brakes less common on late-model vehicles.

Floating Calipers

By far, most late-model cars and light trucks have disc brakes with movable calipers. Movable caliper brakes are further subdivided into **floating calipers** and sliding calipers, which identify the ways in which the calipers are mounted on their supports.

Floating calipers began to appear on domestic and some imported vehicles in the late 1960s. A floating caliper brake has a one-piece caliper body and usually one large piston on the inboard side. Some medium-duty trucks, particularly Fords, and some late-model GM FWD cars have floating caliper brakes with two pistons on the inboard side (Figure 7-35).

The caliper is mounted to its support on two locating bolts or guide pins that are threaded into the caliper support (Figure 7-36). The caliper slides on the pin in a sleeve or bushing. The bushing may be lined with Teflon or have a highly polished surface for low friction. The pins let the caliper move in and out and provide some flexibility for lateral movement to help the caliper

Figure 7-35 Some late-model floating calipers have two pistons.

Figure 7-36 A floating caliper rides on two bolts or guide pins in the caliper support.

stay aligned with the rotor. The pads are attached to the piston on the inboard side and to the caliper housing on the outboard side.

When the brakes are applied, the fluid pressure behind the piston pushes the piston outward in its bore. The pressure is transmitted directly to the inboard pad, which is forced against the inboard rotor surface. The pressure applied to the bottom of the cylinder bore forces the caliper to slide or move on the guide pins toward the inboard side. This movement causes the outboard section of the caliper to apply equal pressure against the back of the outboard pad, forcing it against the outboard rotor surface.

In a floating caliper, the piston and the bottom of the caliper bore are both hydraulic pressure surfaces of equal size, and equal pressure is applied to both the piston and the caliper. Hydraulic pressure against the bottom of the caliper bore creates a reaction force that moves the caliper body inward as the piston moves outward. As a result, the rotor is clamped between the piston on one side and the caliper body on the other.

When the driver releases the brakes, the pressure behind the piston drops. The seal relaxes, or springs back, and moves the piston back. As the piston moves back, the caliper relaxes and moves in the opposite direction on the guide pins to the unapplied position. The piston seals provide the self-adjusting action and required pad-to-rotor clearance described earlier in this chapter.

Floating calipers (and sliding calipers, described in the next section) are lighter, easier to service, and cheaper to build than fixed calipers. Because of their simpler design, they also are less likely to develop leaks. The flexible mounting of a floating caliper provides some beneficial self-alignment of the caliper with the rotor. If the flexibility becomes excessive, however, the pads can wear at an angle or become tapered, which decreases pad life.

Sliding Calipers

Sliding calipers operate on exactly the same principles as floating calipers, but their mounting method is different. The caliper support has two V-shaped surfaces that are called abutments or ways (Figure 7-37). The caliper housing has two matching machined surfaces. The caliper slides

Figure 7-37 A sliding caliper moves on machined ways on the caliper support.

onto the caliper support, where the two parts are held together with a caliper support spring, a key, and key retaining screw. An antirattle spring is used to prevent noise from vibration.

When a sliding caliper is replaced, the caliper ways on the caliper support should be inspected closely and caliper movement should be checked to be sure that the replacement caliper slides correctly. It may be necessary to polish the caliper ways with a fine file or emery cloth for proper clearance.

The pins on a floating caliper should be lubricated according to the carmaker's instructions, but lubrication is even more important on a sliding caliper. Both the caliper ways and the mating surfaces on the caliper should be cleaned and then lubricated with high-temperature brake grease, which ensures smooth operation, reduces wear, and prevents corrosion on the sliding surfaces.

Rear Wheel Disc Brakes

Rear wheel disc brake calipers may be fixed, floating, or sliding, all of which work just as do the front brake assemblies described previously. The only difference between a front and rear disc brake caliper is the need for a parking brake in the rear.

Some rear disc brakes have a small brake drum built into the center of the rotor (Figure 7-38). This is commonly known as a **drum-in-hat** design. Two brake shoes are expanded into the brake drum when the parking brake is applied (Figure 7-39). The adjustment of this system is covered in Chapter 9 of the *Shop Manual*. Most rear disc brakes rely on a cable that moves the caliper piston to apply the brake (Figure 7-40). The piston is connected to a lever system that can be operated either mechanically by the parking brake lever or pedal or by hydraulic force from the master cylinder. Chapter 9 covers parking brakes in detail.

Shop Manual
pages 337–342

Performance Disc Brakes

The overall operation of disc brake systems for racing or high-performance vehicles is exactly the same as that for a small subcompact passenger car. The components are made of different materials, however, and are usually a little larger. Many calipers manufactured for racing vehicles are made of aluminum or an aluminum alloy. Pads and, possibly, rotors could be made of ceramic. Ceramic's main disadvantage is that it does not dissipate heat well. The heat is trapped in the caliper, rotor, pads, and eventually the brake fluid. Carbon fiber may be another material that could be used in racing and eventually on production vehicles. Carbon fiber is light and fairly resistant to heat, but it is expensive. One system used on racing vehicles to dissipate heat that generally has not been installed on production vehicles is the air duct system that moves air from the front bumper or grille

Parking brake drum

Rotor surface

Figure 7-38 This parking brake system has a drum internal to the disc. It is known as a drum-in-hat design.

Figure 7-39 Shown is the shoe-type parking brake used with some rear disc brake systems. The actual size is not much larger than the size shown.

Figure 7-40 Most rear disc brakes have a lever-and-screw mechanism to apply the service brakes for parking.

and directs that air over the braking mechanism. In general terms, this ductwork is not required on production vehicles.

Some aspects of racing rotors are making it to the production line. The slotted and/or drilled rotors are being installed on many production vehicles. In this case, the industry is finally catching up to motorcycles, which have been using drilled rotors for years. Drilled rotors offer two advantages that take nothing away from braking performance. The first is weight. Drilling a hole removes metal, hence producing a weight reduction. The second advantage is heat dissipation. Passing air can flow through the holes and remove more heat from the rotor. This airflow does nothing to decrease the stability of the vehicle during braking action. The main advantage of slotted rotors is the removal of heat and dispersion of any gas that may form between the pad and rotor. The material used in rotor construction is of a different makeup than the material used for production rotors, so they are more resistant to warping during the hard braking required of a race vehicle; this material is usually cast iron or steel.

Racing calipers may be made of cast aluminum because it allows for weight decrease. However, aluminum does not dissipate heat as well as steel or other heavier metals. In most races, this does not cause a lot of concern because the calipers are supposed to last only for a specific time or distance. Selecting a racing caliper requires the vehicle crew to estimate the amount of braking and how hard the braking will be during an entire race. The caliper manufacturers and the vehicle crews have gotten this estimate down to a near-perfect science, although there were instances when the rotor blew apart during a race, thereby damaging most of the adjacent components. At times the fluid heats up so much in the caliper and attaching hose that vapors increase rapidly and the brakes "go away." Like most other components on a race vehicle, short tracks that require a lot of hard braking require the vehicle to have a larger rotor, pads, and caliper. The long, high-speed-track vehicle has smaller braking components because in a perfect race they need to work only during a pit stop.

The final consideration when selecting a racing caliper and pads is drag. As mentioned earlier, a low-drag caliper where the pads actually separate from the rotors reduces that slight amount of drag inherent in standard disc brakes. Because brake drag reduces the horsepower available to drive the vehicle and increases fuel consumption, low-drag calipers are the best choice for racers. As a reminder, if low-drag calipers are used, then some version of a quick takeup master cylinder must be used or the brakes may engage too late.

Racing pads may be carbon fiber, ceramic, or semimetallic. Selecting the pads involves several considerations: weight, rotor wear, and durability. As mentioned earlier, the pads are selected based on the use they will endure during a race. The pads are no good, however, if they tend to wear the rotor excessively. The pads and rotors have to match for good braking effect throughout the entire race.

Like most racing equipment, the racing brake components may lead the way to better brakes for production. The manufacturer has two problems adapting racing equipment to production, however: durability and cost. Professional racers with some well-paying sponsors can afford to buy calipers, rotors, and pads that are thrown away after the race is completed. The average consumer is not willing to pay this additional cost without some really great improvements in durability, however.

Summary

❑ The major parts of a disc brake are the rotor, the hub, the caliper, and the pads.

❑ The hub contains the wheel bearings and is where the wheel is mounted.

❑ Rotors come in two types: solid or ventilated.

❑ The four general kinds of friction pad materials are asbestos, organic, semimetallic, and metallic.

❑ The coefficient of friction of the pad materials is indicated by letters and numbers stamped on the edge of the friction material. These letters and numbers are known as the Automotive Friction Material Edge Code.

Terms to Know

Aramid fibers

Automotive Friction Material Edge Code

Binder

Terms to Know (continued)

Bonded lining

Brake caliper

Brake fade

Brake pad

Caliper support

Ceramic pad

Composite rotor

Curing agent

Drum-in-hat

Filler

Fixed caliper brake

Fixed rotor

Fixed seal

Floating caliper

Floating rotor

Friction modifier

Gas fade

Lathe-cut seal

Metallic lining

Mold-bonded lining

Organic lining

Pad hardware

Pad wear indicators

Phenolic plastic

Riveted lining

Rotor

Semimetallic lining

Sliding caliper

Solid rotor

Square-cut piston seal

Steering knuckle

Stroking seal

Swept area

Synthetic lining

Unidirectional rotor

Ventilated rotor

Water fade

❏ Antirattle clips and springs are used to keep the pads from making noise when the pads are not in contact with the rotor.

❏ Brake pad wear indicators alert the driver when the brake linings need to be serviced.

❏ Piston seals are used to seal the hydraulic brake fluid and retract the piston and pad from the rotor.

❏ The two types of calipers are fixed and movable. Movable calipers can be either the floating style or the sliding style.

❏ Low-drag calipers are movable calipers that move the pads away from the rotor, leaving no contact or drag.

❏ Rear-wheel disc brakes are similar to front-wheel disc brakes but must operate from hydraulic pressure and from the parking brake cables.

Review Questions

Short-Answer Essays

1. List the basic parts of a disc brake.

2. List the components of a typical caliper.

3. Why do manufacturers use two-piece rotors on some vehicles?

4. In addition to keeping road splash off the brake assembly, why are splash shields used on disc brakes?

5. Why is asbestos no longer used for brake pad and shoe linings?

6. What are common brake pad and shoe friction materials?

7. What is the difference between fixed and movable calipers?

8. How does a low-drag caliper operate?

9. How is a rear disc brake activated with the emergency brake lever or pedal?

10. What is the difference between an audible wear sensor and a electronic feedback sensor?

Fill in the Blank

1. In a disc brake assembly, the wheel is attached to the _____.

2. The two different kinds of rotors are the _____ rotor and the _____ rotor.

3. In a two-piece rotor/hub assembly, the rotor is separate from the _____.

4. As the friction pad material is pushed against the rotor, _____ is generated and must be dissipated.

5. The common types of pad friction material are _____, _____, _____, and _____.

6. The two types of caliper mountings are _____ and _____.

7. The caliper may contain a steel or phenolic _____.

8. The caliper makes use of a square-cut _____.

9. Rear disc brakes with expanding brake shoes in a drum use _____ action to set the parking brake.

10. The high squeal noise heard from worn-out disc brake pads may be created by the _____.

Multiple Choice

1. The advantages and disadvantages of disc brake systems are being discussed.
   Technician A says that disc brakes are more likely to develop brake fade and water fade.
   Technician B says that reduction of mechanical fade and gas fade is an advantage of discbrakes.
   Who is correct?
   A. A only   **C.** Both A and B
   B. B only   **D.** Neither A nor B

2. The operation of disc brake systems is being discussed.
   Technician A says that composite rotors were developed to save weight.
   Technician B says that a fixed rotor assembly is part of the hub.
   Who is correct?
   A. A only   **C.** Both A and B
   B. B only   **D.** Neither A nor B

3. Disc brake calipers are being discussed.
   Technician A says that a single caliper may have one, two, or more pistons.
   Technician B says that the piston seal is known as a square-cut seal.
   Who is correct?
   A. A only   **C.** Both A and B
   B. B only   **D.** Neither A nor B

4. *Technician A* says that solid rotors can usually be machined several times before their minimum thickness limits are reached.
   Technician B says that composite rotors have cast-iron hubs with steel friction surfaces.
   Who is correct?
   A. A only   **C.** Both A and B
   B. B only   **D.** Neither A nor B

5. When discussing disc brake work:
   Technician A says that organic linings are a composite material made from bonding nonmetallic fibers.
   Technician B says that semimetallic linings require higher brake pedal effort but are more fade resistant compared to organic linings.
   Who is correct?
   A. A only   **C.** Both A and B
   B. B only   **D.** Neither A nor B

6. Two technicians are discussing a disc brake caliper:
   Technician A says that fixed calipers use one or two pistons on each side of the rotor to apply the brakes.
   Technician B says that sliding calipers typically use one piston on each side of the rotor.
   Who is correct?
   A. A only   **C.** Both A and B
   B. B only   **D.** Neither A nor B

7. *Technician A* says that a return spring is used to retract a caliper piston.
   Technician B says that the piston seal retracts the caliper piston when hydraulic pressure is released.
   Who is correct?
   A. A only   **C.** Both A and B
   B. B only   **D.** Neither A nor B

8. *Technician A* says that inner and outer pad friction materials may have different coefficients.
   Technician B says that identical inboard and outboard pad linings should always be used when replacing brake pads.
   Who is correct?
   A. A only   **C.** Both A and B
   B. B only   **D.** Neither A nor B

9. *Technician A* says that when the driver hears the audible brake pad wear indicator, he should immediately park the car and not use it until the brakes are replaced.
   Technician B says that the wear pad indicators alert the driver to worn pads that should be replaced soon.
   Who is correct?
   A. A only   **C.** Both A and B
   B. B only   **D.** Neither A nor B

10. *Technician A* says that synthetic brake linings are used for many types of brakes.
    Technician B says that synthetic brake linings are made of aramid or fiberglass.
    Who is correct?
    A. A only   **C.** Both A and B
    B. B only   **D.** Neither A nor B

Drum Brakes

Upon completion and review of this chapter, you should be able to:

❏ Describe the basic parts of a drum brake assembly.

❏ Describe how a drum brake stops a vehicle.

❏ Describe different types of brake shoes and explain how they operate.

❏ Describe the three types of friction linings and how they are attached to the shoes.

❏ Describe the components that make up a wheel cylinder and the purpose of each one.

❏ Describe the two major drum brake designs and how they differ in operation.

❏ Describe the different types of self-adjusters used on duo-servo and leading-trailing shoe brake systems and how they work.

Introduction

For more than 80 years, drum brakes at all four wheels were the standard of the automobile industry. Although disc brakes have become standard equipment at the front of most cars and light trucks, drum brakes continue in use as the most common choice for rear-wheel brakes. No law or regulation requires disc brakes on the front wheels, but the brake performance requirements of FMVSS 105 make front disc brakes virtually mandatory. Nevertheless, drum brakes have certain advantages that will make them an important part of vehicle engineering and service for many more years.

This chapter explains the construction and operation of modern drum brake systems and begins with a summary of their advantages and disadvantages.

Drum Brake Advantages and Disadvantages

The major drum brake advantages are self-energizing and servo action, lack of noise, and efficient parking brake operation without complicated linkage. Drum brake disadvantages include poorer heat dissipation and less fade resistance than disc brakes, lack of self-adjustment without special linkage, and a greater tendency than disc brakes to pull and grab.

Drum Brake Self-Energizing and Servo Action

One end of the lining on one shoe of a drum brake contacts the drum before the other end does and becomes a pivot point as friction increases quickly. The brake shoe becomes a self-energizing lever and adds its own mechanical leverage to hydraulic force to help apply the brakes (Figure 8-1).

When a component is operated or applied and that component uses the initial input force to increase that force and transfer it to a matching component, the entire operation is said to be **self-energizing.** This self-energizing operation may be confined to one shoe of a drum brake installation, which is the leading shoe in relation to the direction of wheel rotation. When the self-energizing operation of one shoe applies mechanical force to the other shoe to assist its application, it is called servo action (Figure 8-2). Self-energizing operation and servo action can have both operational advantages and disadvantages. Self-energizing drum brakes that use servo operation can always apply more stopping power for a given amount of pedal force than disc brakes can.

Drum Brake Pulling and Grabbing

Servo action increases braking force at the wheels, but it must develop smoothly as the brakes are applied. If servo action develops too quickly or unevenly, the result can be brake grabbing or

Self-energizing operation is the action of a drum brake shoe when drum rotation increases the application force of the shoe by wedging it tightly against the drum.

Figure 8-1 Drum rotation adds leverage to the brake shoes as they contact the drum. This process is called self-energizing action.

lockup. This is usually experienced as severe pull to one side or the other or reduced vehicle control during braking.

Lack of Noise

Drum brakes are almost noise free. Heavy return springs and hold-down springs hold the brake shoes against the wheel cylinders, anchor pins, and backing plates. Linings are securely bonded or riveted to the brake shoes, and the entire assembly is encased in the brake drum. About the only time that noise is a concern with drum brakes is when it is the sound of steel brake shoes grinding against the brake drum after the linings have worn away.

Drum Brake Parking Brake Operation

Drum brakes provide a better static coefficient of friction than do disc brakes. The brake linings grab and hold the drums more tightly than disc pads can hold a rotor. In addition, the self-energizing and servo actions of drum brakes contribute to this feature, as does the larger area of brake shoe linings compared to disc brake pads.

Because most brake installations have discs at the front and drums at the rear, the parking brake weaknesses of disc brakes are not a problem. Compared to the complicated mechanisms used to mechanically apply parking brakes on rear disc brakes, parking brake linkage for drum brakes is quite simple. In almost all cases, the driver operates a pedal or a lever that pulls on cables attached to the rear brakes. The cables operate levers that mechanically apply the brake shoes. All mechanical motion is in a straight longitudinal line from the front to the rear of the vehicle.

Drum Brake Self-Adjustment

Prior to the installation of self-adjusters, periodically the vehicle brakes would have to be adjusted by a technician or an experienced owner.

Brake adjustment is the process of compensating for lining wear and maintaining correct clearance between brake linings and the drum or rotor surfaces. Disc and drum brakes both require adjustment,

Figure 8-2 Self-energizing action of the primary shoe applies force to the secondary shoe. This process is called servo action.

but self-adjustment is a basic feature of disc brake design. Drum brakes need extra cables, levers, screws, struts, and other mechanical linkage just to provide self-adjustment and proper lining-to-drum clearance.

Fade Resistance

Brake fade is the loss of braking power due to excessive heat that reduces friction between brake linings and the rotors or drums. One factor that contributes to heat dissipation and fade resistance is the swept area of brake drum or rotor, which is the total area that contacts the friction surface of the brake lining. The greater the swept area, the greater the surface available to absorb heat. Although a brake drum can have a relatively large swept area, the entire area is on the inner drum surface. The swept area of a disc brake comprises both sides of the rotor. For any given wheel size, the swept area of a disc brake will always be larger than the swept area of a drum brake (Figure 8-3). For example, a 10-inch-diameter rotor will have almost 50 percent more swept area than a 10-inch drum.

Mechanical fade is a problem that occurs with drum brakes when the drum becomes very hot and expands outward, away from the brake shoes. As a result, the shoes then must travel farther to contact the drum surface with normal braking force, and the pedal drops lower as the brakes are applied. The increased heat at the braking friction surfaces also reduces the coefficient of friction. The combined result is brake fade.

60 Square inches

Swept area,
one surface

100 Square inches

Swept area,
two surfaces

Figure 8-3 For any given wheel size, a disc brake always has 35 percent to 50 percent more swept area to dissipate heat.

Lining fade occurs when the linings are overheated and the coefficient of friction drops off severely (Figure 8-4). Some heat is needed to bring brake linings to their most efficient working temperature. The coefficient of friction rises, in fact, as brakes warm up. If the temperature rises too high, however, the coefficient of friction decreases rapidly. Because disc brakes are exposed to more cooling airflow than drum brakes, lining fade due to overheating is reduced.

Water fade occurs when water is trapped between the brake linings and the drum or rotor and reduces the coefficient of friction. Gas fade is a condition that occurs under hard braking when hot gases and dust particles are trapped between the brake linings and the drum or rotor and build up pressure that acts against brake force. These hot gases actually lubricate the friction surfaces and reduce the coefficient of friction.

AUTHOR'S NOTE: You can use a piece of sandpaper to demonstrate the effect of heat on friction. Slide the sandpaper slowly over a hard surface and you can feel the drag or friction. Speed up the sliding action and shortly you will feel the heat and the reduction of the drag.

The basic design differences between disc and drum brakes make disc brakes much more self-cleaning to greatly reduce or eliminate water and gas fade. The rotor friction surfaces are perpendicular to the axis of rotation, and centrifugal force works to remove water and gas from the braking surfaces. Additionally, the leading edges of the brake pads help to wipe the rotor surfaces clean, and many pad linings have grooves to drain water and help prevent gas buildup. Due to

Figure 8-4 Initially, the coefficient of friction increases with heat, but very high temperatures cause it to drop off and cause the brakes to fade.

the design of the drum brake, it will trap heat, water, and gases inside the drum. This makes the drum brake more susceptible to brake fade and pull.

Drum Brake Construction and Operation

The basic parts of a drum brake are: a drum and hub assembly, brake shoes, a backing plate, a hydraulic wheel cylinder, shoe return springs, hold-down springs, and an adjusting mechanism (Figure 8-5). Rear brakes also include a parking brake. The drum and shoes provide the friction surfaces for stopping the wheel. The drum has a machined braking surface on its inside circumference.

Shop Manual
pages 359–363

Figure 8-5 Basic parts of all drum brakes. A leading-trailing system is shown.

Figure 8-6 The hydraulic and friction components are mounted on the backing plate.

The wheel is mounted to the drum hub by nuts and studs. The hub and drum may be a one-piece casting, or the drum may be separate from the hub or axle flange and fit over the wheel studs for installation. Separate or floating drums are sometimes held to the hub or axle flange by thin push-on nuts commonly referred to as speed nuts or Tinnerman nuts. Screws may also be used to hold drums on. The hub houses the wheel bearings that allow the wheel to rotate.

The hydraulic and friction components are attached to the **backing plate** (Figure 8-6), which is mounted on the axle housing or suspension. The backing plate provides the mounting platform for the brake assembly and closes the rear of the mounted drum. The drum encloses and rotates over these components. When the brakes are applied, hydraulic pressure forces the pistons in the wheel cylinder outward. The pressure on the pistons is transmitted to the shoes as the shoes are forced against a pivot or anchor pin and into the rotating drum. Through this action, the shoes tightly wrap up against the drum to provide the stopping action.

The frictional energy on the drum surface creates heat. This heat is dissipated to the surrounding air as the drum rotates with the wheel. Some drums are finned to help them dissipate heat more easily.

Brake Drums and Hubs

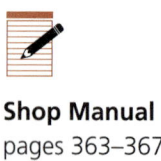

Shop Manual
pages 363–367

The brake drum mounts on the wheel hub or axle and encloses the rest of the brake assembly except the outside of the backing plate. Brake drums are made of cast iron, steel and cast iron, or aluminum with an iron liner. In all of these variations, iron provides the friction surface because of its excellent combination of wear, friction, and heat-dissipation characteristics.

The drum is a bowl-shaped part with a rough cast or stamped exterior and a machined friction surface on the interior. The open side of the drum fits over the brake shoes and other parts mounted on the backing plate (Figure 8-7). It is the side of the drum that is visible when the wheel assembly is removed. The closed side of the drum is called the **drum web** and contains the wheel hub and bearings or has mounting holes through which the drum is secured to the axle flange or separate hub.

Tapered roller bearings, installed in the hubs, are the most common bearings used on the rear wheels of FWD cars. The tapered roller bearing has two main parts: the bearing cone and the outer

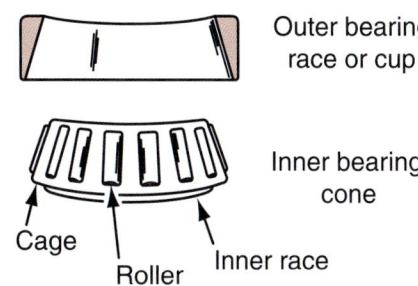

Figure 8-7 The drum fits over the brake shoes and other parts mounted on the backing plate.

Figure 8-8 The outer race or cup is a separate part of a tapered roller bearing. The remaining parts are combined in one assembly.

cup (Figure 8-8). The bearing cone contains steel tapered rollers that ride on an inner cone and are held together by a cage. The bearing fits into the outer cup, which is pressed into the hub to provide two surfaces, an inner cone and outer cup, for the rollers to ride on. Chapter 3 in this *Classroom Manual* contains more detailed information on different kinds of wheel bearings.

On RWD vehicles the brake assembly is mounted to the backing plate, which is fastened to the rear axle housing (Figure 8-9). The axle extends outward past the brake assembly and has the wheel studs installed into the axle flange. The axle bearing fits into the end of the axle housing.

Figure 8-9 Drum removed with the axle partially withdrawn. Shown is the general layout of RWD axle and brake drum.

The drum is fitted over the studs and held in place by the placement of the wheel assembly. The outer end of the axle, and the brake assembly, are enclosed within the drum. Information and service on axle bearings may be found in *Today's Technician Manual Transmissions and Transaxles*.

Some iron and aluminum drums have fins cast into the outer circumference to aid in cooling (Figure 8-10). Some drums in older vehicles also may have a coil spring wrapped around the outer circumference. The spring is a vibration and noise damper that quiets high-pitched noise as the linings contact the drum. If the noise was not subdued, it would be heard as annoying brake squeal.

The braking surface area of a drum is determined by the diameter and the depth (width) of the drum. Large cars and trucks, which require more braking energy, have drums that measure 12 inches in diameter or larger. Smaller vehicles use smaller drums. Generally, manufacturers try to keep parts as small and light as possible, while still providing efficient braking. Brake drums can be categorized as either solid or composite and by the different ways in which the drums are made.

Solid Cast-Iron Drums. A solid drum is a one-piece iron casting (Figure 8-11). Cast iron has excellent wear characteristics and a coefficient of friction that make it ideal as a braking friction surface. Iron also dissipates heat very well, and it is easy to machine when refinishing is necessary.

Along with these advantages, cast-iron drums have some disadvantages. A large iron casting can become brittle and may crack if overstressed or overheated. Small cracks may be almost invisible to the naked eye, but they can lead to drum failure or heat checking and glazing.

A one-piece cast-iron drum is the heaviest of all drum types, which is both an advantage and disadvantage. The weight and mass of a one-piece drum make it very good at absorbing and dissipating heat. The drum weight, however, adds a lot of weight to the vehicle, and it is all unsprung weight at the wheels. Engineers consider brake drum weight more a disadvantage than an advantage for late-model cars, so more and more vehicles are being built with composite drums of the following kinds.

Steel and Iron Drums. Steel and iron **composite drums** are made in two ways. The most common type has a stamped steel web mated with the edge of a cast-iron drum (Figure 8-12). The other type of composite drum has a centrifugally cast-iron liner inside a stamped steel drum. To make this kind of drum, a stamped steel drum is rotated at high speed while molten iron is poured into it. Centrifugal force causes the molten iron to flow outward and bond tightly to the inner circumference of the steel drum.

Cooling fins

Cast iron

Figure 8-10 This drum has cooling fins cast into its outer circumference.

Figure 8-11 Cross section of a cast-iron drum.

Figure 8-12 Cross section of a composite drum.

Figure 8-13 Cross section of a bimetallic drum.

Steel and iron composite drums are lighter and cheaper to make than one-piece cast-iron drums, but they are less able to absorb and dissipate heat and resist fade. They work well on the rear drum brakes of compact cars, however.

Bimetallic Aluminum Drums. Bimetallic drums consist of an aluminum outer drum cast around an iron liner (Figure 8-13). They are almost three times lighter than a one-piece cast-iron drum, and they cool much better. Bimetallic drums are much more expensive than one-piece cast-iron drums, however. Aluminum-iron bimetallic drums were more common in the 1950s when cars had four-wheel drum brakes. They were commonly used on luxury or high-performance cars whose price supported the high cost of these drums.

Brake Shoes and Linings

Brake shoes are the components that mount the friction material or linings that contact the drums. The shoes also provide the area to mount the shoe to the backing plate. Brake shoes are usually made from welded steel, although some drum brakes use aluminum shoes. The outer part is called the **table** (or sometimes the rim) and is curved to match the curvature of the drum (Figure 8-14). The brake lining is riveted or bonded to the brake shoe table (Figure 8-15). Many brake shoes have small notches or nibs along the edge of the table that bear against the backing plate and help to keep the shoe aligned in the drum. The **web** is the inner part of the shoe that is perpendicular to the table and to which all of the springs and other linkage parts attach.

Duo-servo and leading-trailing nonservo brakes are described in detail later in this chapter, but all drum brakes have a pair of shoes at each wheel. The front shoe on duo-servo brakes is called the primary shoe, and the rear shoe is called the secondary shoe. The front and rear shoes on leading-trailing brakes are called just that: the leading shoe and the trailing shoe.

On leading-trailing brakes, the lining on the leading and trailing shoes is usually the same length and positioned in the same location on each shoe. On a duo-servo brake, however, the primary shoe lining is shorter than the secondary lining and may even have a different coefficient of friction. As explained later in this chapter, the primary shoe of a duo-servo brake is self-energized by drum rotation and then applies servo action to the secondary shoe. The secondary shoe applies most of the stopping force and thus requires a larger lining. The coefficient of friction for the secondary lining is often higher than for the primary lining to provide good stopping power. However, the lower coefficient of friction for the primary lining helps to keep the brakes from applying too harshly and locking.

Lining for the secondary shoe is almost always centered from top to bottom on the shoe table. Primary lining may be centered from top to bottom on the table, or it may be offset toward either the top or the bottom (Figure 8-16), depending on the operating characteristics of a specific brake installation.

Table refers to the outer surface of a brake shoe to which the lining is attached.

Some aftermarket brake shoe sets are fitted with equal amounts of lining on each shoe. This allows one brake shoe set to replace several types of original manufacturer shoe.

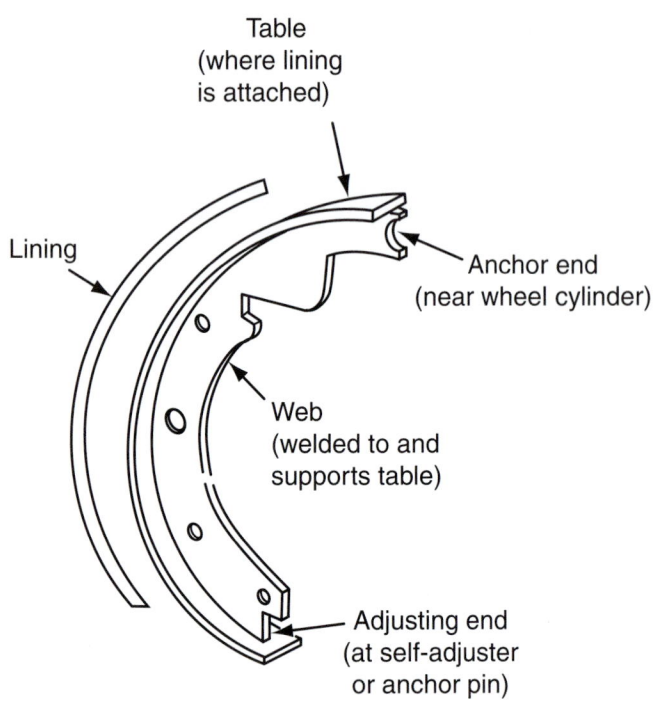

Figure 8-14 A typical brake shoe with the lining detached.

Bonded lining

Riveted lining

Figure 8-15 Linings may be bonded or riveted to the shoe.

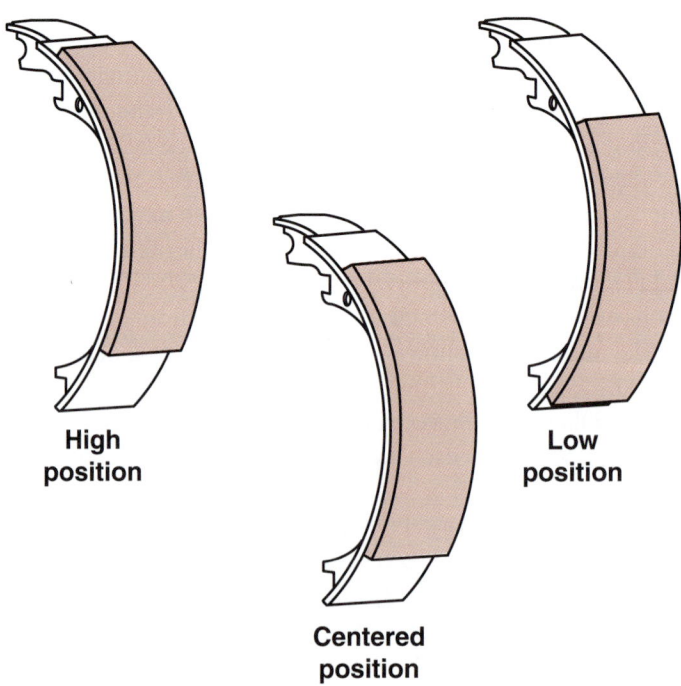

High position

Centered position

Low position

Figure 8-16 Lining can be attached to the shoe in different positions, depending on the desired stopping characteristics.

Brake Friction Materials. Both drum and disc brake friction materials are covered in Chapter 7 of this *Classroom Manual.*

Lining-to-Drum Fit. When disc brakes are applied, a flat pad contacts a flat rotor, so the surface fit of the pad to the rotor is usually not a concern. With drum brakes, however, a semicircular lining contacts a circular drum, and the lining does not move in a straight line toward the drum. The complete lining surface does not contact the drum all at the same time, so the surface fit of the lining to the drum is an important operating consideration.

The leverage that operates in drum brakes and the motion of brake shoes as they are applied work against full lining contact with the drum. Therefore, lining shape must be adjusted to overcome these natural conditions and work toward providing full lining-to-drum contact.

The hydraulic force of the wheel cylinder pistons and the rigid positions of the shoe anchors tend to force the ends of the shoes into contact with the drum before the center of the shoes. Most drum brake noise complaints are caused by binding at the ends of the shoes as they contact the drum before the center does. More important, if the full surface of new brake linings does not contact the drum as the linings are broken in, the linings can overheat and become glazed, which reduces braking effectiveness.

To compensate for these natural characteristics of drum brakes, linings are fitted to shoes to provide more clearance at the ends than at the center. This process is known as **arcing** the shoes, and it forms either cam-ground or undersized linings. Arcing or cam-grounding of the lining is sometimes very visible on some brake shoes. On others the arc or ground is not noticeable without close inspection. A **cam-ground lining** is thinner at the ends than at the center (Figure 8-17), and the lining surface is not a portion of a circle with a constant radius. An undersized lining has a uniform thickness and a constant radius (Figure 8-18), but it has a smaller outside diameter than the inside diameter of the drum. A fixed-anchor arc is a variation of an undersized lining in which the overall arc is offset slightly so that one end of the lining is thicker than the other (Figure 8-19).

Arc grinding, or arcing, of new brake linings used to be a standard part of brake service. Today, however, brake shoe linings are seldom arced as part of the brake installation job. In fact, most modern brake lathes no longer have the attachments for shoe arcing.

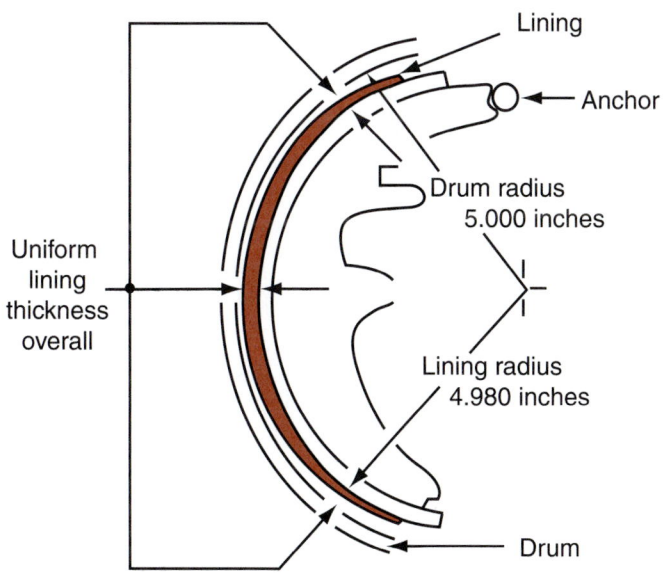

Figure 8-17 A cam-ground lining is thinner at the ends than in the center (clearances exaggerated).

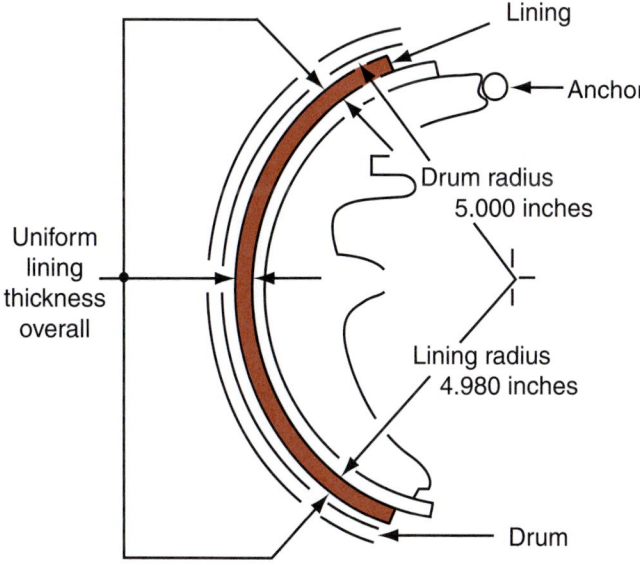

Figure 8-18 An undersize lining has a shorter radius than the drum radius (clearances are exaggerated).

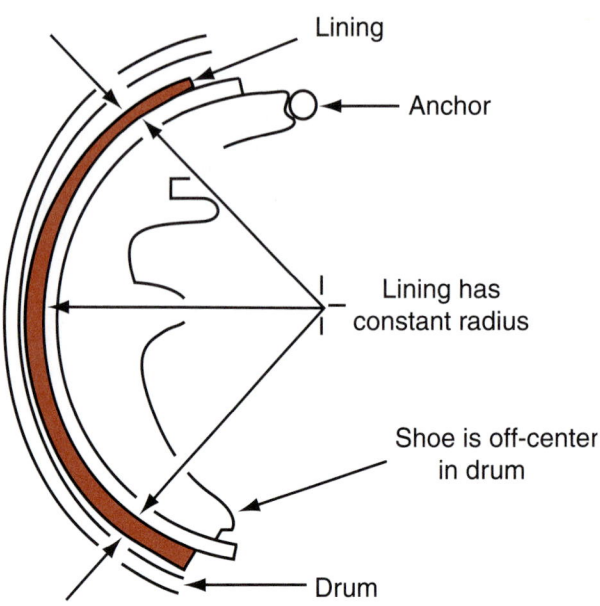

Figure 8-19 The overall arc of a fixed-anchor lining is offset slightly, and one end of the lining is thicker than the other (clearances exaggerated).

Lined shoes are sold with the lining surface already contoured for proper drum fit. Several factors led to the decline of arc grinding, and concerns about airborne asbestos and brake dust in general were the leading cause. In addition, rear drum brakes used with front disc brakes provide a smaller percentage of overall braking for late-model FWD cars. Brake linings and drums are less prone to overheating, and lining wear is not as severe as it was in the past. Therefore, arcing requirements are not as great, and precontoured linings are more easily provided by manufacturers.

Backing Plate

The backing plate is bolted either to the steering knuckle on the front suspension or to the axle flange or hub at the rear (Figure 8-20). The backing plate is the mounting surface for all other brake parts except the drum. The circumference of the backing plate is curved to form a lip that fits inside the drum circumference and helps to keep dirt, water, and road debris out of the brake assembly.

Brake **shoe anchors** are attached to the backing plate to support the shoes and keep them from rotating with the drum. Most modern brakes have a single anchor that is either a round post or a wedge-shaped block. Some older drum brakes and a few late-model versions have separate anchors for each shoe.

All backing plates have some form of shoe-support pads stamped into their surfaces. The edges of the shoes slide against these pads as the brakes are applied and released. The pads thus keep the shoes aligned with the drum and other parts of the assembly. A light coat of brake grease should be applied to the pads to aid shoe movement and reduce noise.

Many backing plates have **piston stops,** which are steel tabs at the ends of the wheel cylinder. These stops keep the pistons from coming out of the cylinder during service. The wheel cylinder must be removed from a backing plate with piston stops for disassembly.

Wheel Cylinders

Wheel cylinders convert hydraulic pressure from the master cylinder to mechanical force that applies the brake shoes. Historically, many kinds of wheel cylinders have been used on different kinds of drum brakes. These included single-piston cylinders and stepped-bore cylinders with pistons of different diameters. Some wheel cylinders were installed to slide on the backing plate as they applied the brake shoes. Wheel cylinders for late-model drum brakes, however, are almost

A **shoe anchor** is a large pin (post) or block against which a drum brake shoe pivots or develops leverage.

Shop Manual
pages 373–374, 382–383

Anchor pin

Support pads

Support pads

Backing plate

Figure 8-20 The backing plate holds the complete drum brake assembly. The shoe slides across the lubricated shoe contact pads as they are applied and released.

all two-piston, straight-bore cylinders that are mounted rigidly to the backing plate. The basic parts of all cylinders are the body, the pistons, the cups (seals), the spring and cup expanders, and two dust boots (Figure 8-21). Some cylinders include two shoe links, or pushrods, to transfer piston movement to the shoes. In other designs, the shoe webs bear directly against the cylinder pistons.

The wheel cylinder body is usually cast iron, but a few aluminum cylinder bodies have been used. The cylinder bore is finished to provide a long-wearing, corrosion-resistant surface. A fitting for a hydraulic line is provided at the center of the cylinder, between the two pistons. The bleeder screw also is tapped into the cylinder body at its highest point.

Some imported vehicles have been built with rear wheel cylinders plumbed in series. In this design, a brake line runs from the master cylinder to one rear wheel that does not have a bleeder screw. Instead, a second line runs from the first wheel cylinder to the other rear wheel cylinder that does have a bleeder. Both rear cylinders must be bled from the single bleeder in one cylinder.

Some wheel cylinders act as the anchor on one end and the shoes are linked together at the other end. As a result there are no metal anchors.

Dust boot

Cup

Cup expander

Piston

Shoe link (pushrod)

Inlet port

Spring

Bleeder screw near top

Figure 8-21 Almost all modern drum brakes use straight-bore, two-piston wheel cylinders such as the one shown in this cross section.

Bleeder screw

Link Boot Piston Cup Return spring with cup expanders Cup Piston Boot Link

Figure 8-22 An exploded view of a typical wheel assembly.

A piston is installed in each end of the wheel cylinder, and the inside end of each piston is sealed with a cup seal. Hydraulic pressure against the seal forces the seal lip to expand against the cylinder bore and form a leak-free seal. Most wheel cylinders also have spring-loaded metal **cup expanders** that bear against the inner sides of the cups (Figure 8-22). Piston cup expanders were made to keep the lips of the seal in place when the brakes are released. This allows the master cylinder's brake residual check valve specifications to be lowered or the valve to be removed entirely. The spring also takes up any slack between the pistons and their pushrods and helps to center the pistons in the cylinder bore.

Each end of the wheel cylinder is sealed with a dust boot to keep dirt and moisture out of the cylinder. The dust boots also prevent minor fluid seepage from getting onto the brake linings.

Return and Hold-Down Springs

Shop Manual
pages 360–362, 374–382

Strong **return springs** retract the shoes when the brakes are released and hold them against their anchors and the wheel cylinder pushrod. Most return springs are tightly wound coil springs in which the coils touch each other when retracted (Figure 8-23). Return springs often are color coded to indicate different tension values among springs of the same size and shape.

###	Orange	3-7/16"	35 lb.
###	White	3-3/16"	50 lb.
###	Purple	3-3/8"	45 lb.
###	Bronze	2-5/8"	55 lb.
###	Black	4-1/4"	50 lb.
###	Blue	3-3/16"	65 lb.

Figure 8-23 Return springs are identified by their part number, free length, and tension. Most are color coded to identify tension differences among similar-looking springs.

Figure 8-24 Typical return spring installation on a duo-servo system.

Figure 8-25 Various kinds of hold-down springs are used on different drum brakes.

The type, location, and number of springs vary, but return springs are installed either from shoe to shoe or from each shoe to an anchor post (Figure 8-24).

Hold-down springs, clips, and pins also come in various shapes and sizes, but all have the same purpose of holding the shoes in alignment with the backing plate. Hold-downs must hold the shoes in position while providing flexibility for their application and release. Figure 8-25 shows some of the varieties of hold-down springs and clips.

Some early 1990s GM cars and some imported vehicles use a single large horseshoe-shaped spring that acts as both a shoe return spring and a hold-down spring (Figure 8-26).

Figure 8-26 The big horseshoe-shaped spring serves as both a shoe return spring and shoe hold-down on some late-model GM rear brakes.

Shop Manual
pages 374–389

Self-adjusters refer to cable, lever, screw, strut, or other linkage parts that provide automatic shoe adjustment and proper lining to drum clearance as a drum brake lining wears.

Self-Adjusters

As the brakes are applied, the lining wears and becomes thinner. As the distance between the shoe and the drum becomes greater, the shoes must move farther before the lining contacts the drum. Therefore, the pistons of the wheel cylinder also must move farther, and more brake fluid must come from the master cylinder to apply the brakes. To provide more brake fluid, brake pedal travel will increase.

Brake adjustment compensates for lining wear and maintains correct clearance between brake linings and the drum. Disc and drum brakes both require adjustment, but self-adjustment is a basic feature of disc brake design. Drum brakes, however, need extra cables, levers, screws, struts, and other linkage to provide self-adjustment and proper lining-to-drum clearance. Many different kinds of manual and automatic adjustment devices have been used on drum brakes over the years, but today automatic **self-adjusters** that operate star wheel or ratchet adjustment mechanisms are the most common. Automatic self-adjusters use the movement of the brake shoes to maintain proper shoe/drum clearance.

Duo-Servo Star Wheel Adjusters. The bottoms of the shoes in duo-servo brakes are connected to each other by a link, and the shoes are held against the link by a strong spring (Figure 8-27). The link keeps the shoes aligned with each other and transfers the servo action of one shoe to the other during braking. It is not attached to the backing plate, but it moves back and forth with the shoes.

Figure 8-27 The link between the shoes of a duo-servo brake contains the adjustment star wheel.

One half of the link is internally threaded, and the other is externally threaded. The externally threaded part has a **star wheel** near one end. When the two halves of the link are assembled, rotating the star wheel screws them together or apart. In this way, the adjustment link is lengthened or shortened to adjust the lining-to-drum clearance.

On manually adjusted duo-servo brakes, an adjusting tool is inserted through a hole in the backing plate to engage the star wheel and rotate it to adjust the link. Even with self-adjusters, the initial adjustment of a duo-servo brake is made this way after new shoes are installed and the drum is resurfaced. Further adjustment is done automatically, however, by the self-adjustment mechanism for the life of the linings.

Duo-servo brakes commonly have self-adjusters operated by the secondary shoe. Either a cable, a heavy wire link, or a lever is attached to the secondary shoe. The cable, link, or lever is attached to a smaller lever or **pawl** that engages the star wheel (Figure 8-28). During braking in reverse, the secondary shoe moves away from the anchor post. If the lining is worn far enough and the shoe can move far enough, it will pull the cable, link, or lever to move the pawl to the next notch on the star wheel. When the brakes are released, the pawl spring moves the pawl to rotate the star wheel (Figure 8-29). The rotation of the star wheel then expands the link to take up the clearance between the linings and the drum.

A **pawl** is a hinged or pivoted component that engages a toothed wheel or rod to provide rotation or movement in one direction while preventing it in the opposite direction.

Figure 8-28 A cable-operated self-adjuster on a duo-servo brake moves the lever (pawl) upward when the secondary shoe moves off its anchor during reverse braking.

Figure 8-29 As the adjuster spring pulls the lever (pawl) downword, it turns the star wheel one notch to expand the adjuster link.

Figure 8-30 Some duo-servo self-adjusters are operated by a link and a lever instead of a cable.

Quite a bit of variety exists in the specific linkage used on different duo-servo brakes. Cable- and link-operated self-adjusters usually move the pawl to engage the next notch in the star wheel as the brakes are applied in reverse. Lever-operated self-adjusters (Figure 8-30) usually move the pawl to rotate the star wheel as the brakes are applied. Some cable-operated self-adjusters move a pawl mounted under the star wheel to adjust the brakes during application, not release. These cable-operated adjusters usually have an **overload spring** in the end of the cable that lets the cablemove without breaking if the pawl or star wheel is jammed.

The overload spring also prevents overadjustment during very hard braking when the drum may distort and let the shoes move farther out than normal. To prevent this, the overload spring stretches with the cable and does not let the pawl actuate. Overadjustment will not occur during hard braking when the brakes are in normal adjustment.

Left-hand and right-hand threaded star wheel adjusters are used on the opposite sides of the car. Therefore, parts must be kept separated and not intermixed. If a star wheel adjuster is installed on the wrong side, the adjuster will not adjust at all and may unadjust, causing excessive shoe/ drum clearance.

Leading-Trailing-Shoe Star Wheel Adjusters. Even more variety exists in the linkage used for star wheel adjusters on leading-trailing-shoe brakes. The star wheel self-adjusters can be operated by either the leading or the trailing shoe and can work whenever the brakes are operated in either forward or reverse.

The star wheel is usually part of the parking brake strut that is mounted between the two shoes. A pawl can be operated by either the leading or the trailing shoe to engage and rotate the star wheel as the brakes are applied or released.

Leading-Trailing-Shoe Ratchet Adjusters. A lever-latch ratchet adjuster has a large lever and small latch attached to the leading shoe. Teeth on the lever and the latch form a ratchet mechanism that moves the shoes outward to take up excess clearance as the brakes are applied.

Another kind of ratchet adjuster consists of a pair of large and small toothed ratchets and a spacer strut mounted between the shoes (Figure 8-31). The strut is connected to the leading shoe through the hand brake lever and the inner edge of a large ratchet. As the gap between the shoes and the drums becomes greater, the strut and the leading shoe move together to close the gap.

Figure 8-31 The ratchet-type self-adjuster for leading-trailing brakes is operated by the parking brake lever.

More movement will cause the large ratchet on the trailing shoe to rotate inward against a small spring-loaded ratchet and reach a new adjustment position.

Other similar systems are actuated by the parking brake and adjust when the parking brake is applied and released. An adjustment strut is attached to the parking brake lever (Figure 8-32). As the lining wears, application of the parking brake will restore the proper lining clearance.

The strut-quadrant self-adjuster is a variation of the strut-rod adjuster shown in Figure 8-32. The strut-quadrant adjuster has a rotating semicircular quadrant as half of the engaging ratchet mechanism.

Leading-Trailing-Shoe Cam Adjusters. Cam-type adjusters use cams with an adjuster pin that fits in a slot on the shoes (Figure 8-33). As the brake shoes move outward, the pin in the slot moves the cam to a new position if adjustment is needed. Shoe retraction and proper lining clearance are always

Many drivers do not regularly use the parking brake. The result is excessive wear and replacement on the front brakes and almost none on the rear brakes.

Figure 8-32 Semiautomatic self-adjuster for leading-trailing brakes.

Figure 8-33 Cam-type self-adjuster for leading-trailing brakes.

maintained because the pin diameter is smaller than the width of the slot on the shoe. These brakes can be adjusted even while the vehicle is at a standstill because brake pedal application is all that is needed to move the cams into proper adjustment. Because the brakes can be adjusted completely with one application of the brake pedal, this adjuster is sometimes called the one-shot adjuster.

> **AUTHOR'S NOTE:** This is for those who restore vehicles or who may work on restored vehicles. Some older VWs had a very simple self-adjuster. It basically was a wedge that was fitted narrow-end down through a slot on the brake strut. A small spring pulled downward on the wedge. As the lining wore off, the wedge was pulled down and the shoes adjusted outward, a simple system that worked fine if the technician knew that he had to push up on the wedge during installation of new shoes. Personal experience and great personal aggravation showed the impossibility of fitting the drum on with the wedge in the full down position. Also, do not forget to reconnect the small spring, as a friend who was a do-it-yourselfer found out. A loose wedge is strong enough to groove the drum and rip the lining in one or two stops.

Self-Adjuster Precautions. The self-adjuster mechanisms described in the preceding paragraphs are just a few common examples of such devices used on late-model drum brakes. When servicing these brakes, examine the self-adjuster parts closely and pay attention to how they are assembled. One previous paragraph mentioned that left- and right-hand parts exist for the left and right wheels of most systems. If parts are interchanged from side to side, the self-adjusters will not work.

Examine self-adjuster parts closely for wear and damage. Teeth wear off and adjusters freeze up. All threaded parts must move freely. Cable guides wear at the points where cables slide back and forth. Spring anchors can bend, stretch, or wear. Holes and slots wear where parts pivot or springs are anchored. Cables stretch with age and use. Many shop owners make it standard practice to replace self-adjuster cables whenever brakes are serviced and other parts when they show any sign of wear. Finally, if unfamiliar with a particular self-adjuster installation, service one wheel at a time and use the opposite brake assembly as a reference for parts installation.

Parking Brake Linkage

Almost all rear drum brake installations include mechanical parking brake linkage. The linkage basically consists of a cable, a lever, and a strut. The lever and strut spread the shoes against the drum when the cable pulls on the lever. The parking brake strut may contain part of the self-adjuster as explained in previous paragraphs. Chapter 9 covers parking brakes in detail.

Drum Brake Designs

Dozens of drum brake designs have been used over the decades of automobile manufacturing. Full-servo, partial-servo, nonservo, two-leading-shoe, two-trailing-shoe, and center-plane are just a few of the brake designs that are part of automobile history. Today, two designs account for almost all of the drum brakes installed on late-model vehicles:

1. **Leading-trailing brakes** (Figure 8-34), also called partial-servo or nonservo brakes
2. **Duo-servo brakes** (Figure 8-35), also called dual-servo or full-servo brakes

The following sections describe these brake designs and explain their operations.

Self-Energizing and Servo Actions

Two terms used to describe drum brake operation are "self-energizing" and "servo." They refer to the leverage developed on the brake shoe as it contacts the drum and the action of one shoe to help apply the other.

Brake shoes are described as leading or trailing and primary or secondary. Leading and trailing shoes are components of leading-trailing (nonservo) brakes. You might think that the **leading shoe** is the one that moves from the wheel cylinder's end that points in the direction of drum rotation. To identify the leading brake shoe, put your hand at the position of the wheel cylinder and then point in the direction of drum rotation: clockwise or counterclockwise. The first shoe that you point to is the leading shoe. Think about this for a moment and it is obvious that the leading shoe can be the front or the rear shoe, depending on whether the drum is rotating forward or in reverse and whether the wheel cylinder is at the top or the bottom of the backing plate.

Shop Manual
pages 376–382

In most cases of current drum leading-trailing brake systems, it is assumed that the leading shoe is always the shoe toward the front of the vehicle. This is assumed because the vehicle is normally braked while moving forward, and the wheel cylinder is at the top of the assembly.

Figure 8-34 A typical leading-trailing brake.

219

Figure 8-35 Typical duo-servo brake.

When the brakes operate, the wheel cylinder forces one end of the leading shoe outward against the drum. The other end of the shoe is forced back solidly against its anchor post or anchor block. The cylinder end of the leading shoe contacts the drum first and develops friction against the rotating drum. The drum friction actually pulls the shoe into tighter contact with the drum (Figure 8-36) and aids the hydraulic force of the cylinder to apply the brake shoe. As drum-to-lining contact increases, the rest of the lining is forced by the cylinder and pulled by friction against the drum to stop rotation. This is self-energizing action by the leading shoe.

Figure 8-36 Self-energizing operation of a brake shoe develops as friction against the rotating drum pulls the shoe into tighter contact with the drum.

Figure 8-37 The leading shoe is self-energizing, but the trailing shoe is non-self-energizing in a leading-trailing brake.

In a leading-trailing brake, the reaction of the **trailing shoe** to drum rotation is opposite to the reaction of the leading shoe. As the wheel cylinder tries to force the trailing shoe outward against the drum, rotation tries to force the shoe back against the cylinder. Eventually, hydraulic force overcomes drum rotation as the wheel slows, and the trailing shoe contributes to braking action. The trailing shoe, however, is said to be non-self-energizing (Figure 8-37).

In another brake design, the self-energizing action of one brake shoe can be used to help apply the other shoe. In a duo-servo brake, the **primary shoe** has the position of the leading shoe in a leading-trailing brake. The primary shoe is almost always the shoe toward the front of the vehicle and many times has less lining than the secondary shoe. The ends of both shoes opposite the wheel cylinder are not mounted on a rigid anchor attached to the backing plate as in a leading-trailing brake. The ends of the two shoes are linked to each other through the star wheel adjuster, and the ends of the shoes float in the drum.

As the primary shoe is applied by the wheel cylinder, it develops self-energizing action as in a leading-trailing brake. Instead of being forced against an anchor, however, the primary shoe is forced against the **secondary shoe** and applies leverage to force the secondary shoe against the drum. This is servo action (Figure 8-38), or the action of mechanically multiplying force. The secondary shoe is then forced against the drum by the wheel cylinder force at one end and the servo action of the primary shoe at the other.

Self-energizing action and servo operation are forces that naturally occur in internal-expanding drum brakes but do not exist in disc brakes. Engineers put these actions to work in different ways in different brake designs that are described and illustrated in the following sections.

Leading-Trailing Brakes Operation

Leading-trailing or nonservo brake design was common in four-wheel drum brake systems up to the late 1960s, but was steadily replaced by duo-servo four-wheel drum brakes as cars got heavier and

Figure 8-38 Self-energizing action for the primary shoe begins as the wheel cylinder forces it against the rotating drum (A). Reaction force of the primary shoe is transferred through the link to the secondary shoe (B). This transfer of force is servo action. The servo action of the primary shoe against the secondary shoe causes it also to become self-energizing (C). The top of the secondary shoe is forced against the anchor to complete the brake application.

faster. Duo-servo brakes are a more powerful drum brake design, and leading-trailing brakes almost vanished from the industry in the 1960s. They began to reemerge, however, as rear brakes on smaller and lighter cars with front disc brakes in the 1970s. Today, leading-trailing brakes are used as rear drum brakes on many FWD automobiles, as well as on some light trucks with front discs.

In a typical leading-trailing brake installation, the wheel cylinder is mounted at the top of the backing plate, and the cylinder pushrod or shoe links bear against the upper ends of the shoe webs. The lower end of each shoe bears against, or is held onto, an anchor block or anchor post toward the bottom of the backing plate. One or two strong return springs usually hold the lower ends of the shoes against the anchor, and another return spring usually is installed between the upper ends of the shoes to hold them together and hold them against the wheel cylinder pushrod. Various kinds of hallowedness, self-adjusters, and parking brake linkage as described earlier in this chapter also are installed.

The wheel cylinder of a leading-trailing brake acts equally on each brake shoe (Figure 8-39). The cylinder forces the top of each shoe outward toward the drum, and each shoe pivots on the anchor at the bottom. Each shoe thus operates separately and independently from the other.

Wheel cylinder force

Leading shoe

Trailing shoe

Anchor

◀ Front

Figure 8-39 Wheel cylinder force pushes both the leading and trailing shoes against the fixed anchor in a leading-trailing brake.

The leading shoe develops self-energizing action as described previously, however, and provides most of the braking force. The force of drum rotation works against the wheel cylinder force on the trailing shoe, so the trailing shoe is not self-energizing. The trailing shoe often is said to be nonenergized even though it is technically energized by the wheel cylinder.

The front shoe *is* the leading shoe for forward braking in a leading-trailing brake as described here. For reverse braking, however, the roles are reversed, and the rear shoe becomes the self-energized leading shoe.

When you compare a leading-trailing brake to a duo-servo brake described in the next section, it will be understood that a duo-servo brake provides greater braking force. One might then ask why engineers would install a "weaker" brake on an advanced, late-model automobile. Part of the reason is that FWD automobiles generate up to 80 percent of their braking force on the front wheels. Overly powerful duo-servo brakes at the rear could easily cause the rear wheels to lock as weight shifts forward and the rear end becomes very light during braking. Ironically, a more powerful drum brake could reduce overall braking performance.

Equally important, antilock brakes make overall brake balance at all four wheels a critical factor. To make brake lockup easier to control while maintaining the best total braking efficiency, engineers have actually had to "detune" the rear drum brakes on some FWD cars. For the sake of brake balance and overall efficiency, leading-trailing brakes will be used on the rear wheels of many vehicles for a long time to come.

Duo-Servo Brakes Operation

Just as do leading-trailing brakes, duo-servo brakes have a single, two-piston wheel cylinder mounted toward the top of the backing plate. A return spring holds each shoe against the cylinder pushrod. A duo-servo brake is distinguished from a leading-trailing brake by a single anchor post at the top of the backing plate. The top of each shoe web has a semicircular notch, and the return springs hold the notches tightly against the anchor post. The cylinder pushrod acts on the shoes a couple of inches below the anchor. The bottoms of the shoes are not anchored to the backing plate but are joined to each other by the adjuster link. Another spring holds the shoes tightly together and against the adjuster link. Think of the shoes of a duo-servo brake as hanging from the anchor post and floating within the drum at the bottom, and that is the key to their operation.

The forward shoe of a duo-servo brake is called the primary shoe, and its lining is shorter than the lining on the rear shoe, which is called the secondary shoe. As the wheel cylinder applies force against both brake shoes during forward braking, the cylinder force pushes the top of the primary shoe away from the anchor. The top edge or leading edge of the primary shoe contacts the rotating drum first and develops self-energizing action just as does the leading shoe of a leading-trailing brake. The self-energized primary shoe is both drawn and forced against the drum, but its lower end does not bear against a fixed anchor. Instead, the lower end of the primary shoe applies servo force to the bottom of the secondary shoe.

The secondary shoe then becomes self-energized, with the self-energizing action beginning at the bottom of the shoe. This combined servo and self-energized action actually works against the wheel cylinder, but servo force applied by the primary shoe to the secondary shoe is greater than the hydraulic force of the wheel cylinder alone. Servo force and the self-energizing action of the secondary shoe force its upper end against the anchor post and wrap the shoe tightly into the drum. The secondary shoe—with its longer and larger lining—thus applies most of the braking force for forward braking.

During reverse braking, the roles of primary and secondary shoe are reversed. The forward shoe with its smaller lining becomes the secondary shoe and receives servo action from the rear shoe. Even though the forward shoe with smaller lining must apply most of the braking force in reverse, it is not a problem because vehicle speed is usually much lower and the vehicle travels a shorter distance in reverse. Braking efficiency remains within safe limits.

Duo-servo brakes can apply a lot of braking power very efficiently, but excessive servo action can lead to brake grabbing and locking. Engineers, therefore, must balance several factors to take full advantage of servo power but maintain smooth braking. These factors include lining size on both shoes and placement of the lining on the primary shoe, which can be centered or mounted high or low on the shoe. Primary and secondary linings also can have different coefficients of friction to achieve smooth brake application.

Duo-servo drum brakes can develop a lot of braking force through combined hydraulic and mechanical action. From the 1950s through the mid-1960s duo-servo brakes with large diameter drums, wide linings, and cooling fins to help heat dissipation provided powerful braking for powerful cars. Disc brakes, however, provide even more powerful and efficient braking. By the mid-1960s disc brakes began to replace drum brakes on front wheels and on all four wheels of some high-performance cars. The 1976 revision to FMVSS 105 made disc brakes the most practical designs to achieve the brake system performance requirements. Duo-servo drum brakes are still used on the rear wheels of many vehicles, however, particularly RWD cars, trucks, and sport utility vehicles.

Summary

Terms to Know

Arcing

Backing plate

Bimetallic drum

Brake shoes

Cam-ground lining

Composite drum

Cup expander

❑ The basic parts of a drum brake are: a drum, a hub, brake shoes, a backing plate, a hydraulic wheel cylinder, shoe return springs, hold-down springs, an adjusting mechanism, and a parking brake.

❑ Front brake drums may be integrated with the hub, which contains the wheel bearings. Rear drums mount on a flange attached to the axle on RWD vehicles.

❑ The wheel cylinder has pressurized brake fluid applied to the pistons inside the cylinder. These pistons act on the brake shoes to move them into contact with the brake drum.

❑ The wheel cylinder and brake shoes are mounted on a solid backing plate, which is attached to the axle housing or steering knuckle.

- ❏ The brake shoes are a primary (or a leading) shoe and a secondary (or trailing) shoe.
- ❏ Brake linings are primarily made of semimetallic or synthetic materials that have replaced asbestos materials. Lining materials are attached to the shoes by either bonding or riveting.
- ❏ Brake shoes are held against their anchors and adjusters by springs.
- ❏ Servo brakes are systems in which one brake shoe increases the braking force of the other brake shoe.
- ❏ In a self-energizing brake, the rotation of the drum increases the braking force of the lining against the drum. The force of the brake drum is added to the force supplied by the wheel cylinder.
- ❏ A duo-servo brake is a servo brake in which servo action takes place whether the vehicle is moving forward or backward.
- ❏ A nonservo brake is a brake in which one shoe does not apply a force to the other shoe.
- ❏ In a leading-trailing brake, the leading shoe is self-energizing but the other is not self-energizing.
- ❏ Primary shoes and leading shoes face to the front of the vehicle.
- ❏ Secondary shoes and trailing shoes face to the rear of the vehicle.
- ❏ Self-adjusting brakes automatically adjust lining-to-drum clearances.

Terms to Know (continued)

Drum web
Duo-servo brake
Hold-down springs
Leading shoe
Leading-trailing brake
Overload spring
Pawl
Piston stop
Primary shoe
Return spring
Secondary shoe
Self-adjusters
Self-energizing operation
Shoe anchor
Star wheel
Table
Trailing shoe
Web

Review Questions

Short-Answer Essays

1. List the components of a typical drum brake assembly.
2. How is the drum brake protected from road splash?
3. Why do large vehicles usually use duo-servo brakes?
4. What happens to the drum during hard stops? Why?
5. Why are primary and secondary brake shoes not interchangeable?
6. What is a semimetallic lining made of? A synthetic lining?
7. What is the function of a self-adjuster? Why do we use them?
8. What is a self-energizing brake?
9. What is servo action?
10. How are the brake shoes returned to their released position when the brake pedal is released?

Fill in the Blanks

1. The component that the brake shoes react against to supply braking force is the _____.
2. Brake linings react against the brake _____ with friction that is dissipated as _____.
3. A _____ brake is a brake that increases the braking force of the other brake shoe.
4. The _____ at the end of the adjusting lever engages with the star wheel on the threaded adjuster.

5. The shoes in a duo-servo brake system are called the _____ shoe and the _____ shoe.

6. The hydraulic component that is mounted to the backing plate and moves the brake shoes is called the _____ _____.

7. When the car is moving in reverse, the _____ shoe provides most of the stopping force in a duo-servo system.

8. When the car is moving forward, the _____ shoe provides most of the stopping force in a leading-trailing-shoe system.

9. Most self-adjusters are connected to the _____ shoe.

10. The _____ _____ prevents road splash contamination from the steering knuckle side of the brake assembly.

Multiple Choice

1. *Technician A* says that for the same brake pedal force, duo-servo brakes apply a greater braking force than leading-trailing shoe brakes.
   Technician B says that the servo action of the duo-servo brakes takes place only when the car is moving forward.
   Who is correct?
   A. A only
   B. B only
   C. Both A and B
   D. Neither A nor B

2. *Technician A* says that the leading shoe lining receives the greatest wear on a leading-trailing shoe brake.
   Technician B says that leading-trailing brakes are nonservo brakes.
   Who is correct?
   A. A only
   B. B only
   C. Both A and B
   D. Neither A nor B

3. *Technician A* says that some leading-trailing shoe brakes with an adjustable parking brake self-adjust when the service brakes are applied.
   Technician B says that some brakes with an adjustable parking brake strut self-adjust when the parking brake is applied.
   Who is correct?
   A. A only
   B. B only
   C. Both A and B
   D. Neither A nor B

4. *Technician A* says that on a duo-servo brake, the force applied by the drum to the primary shoe aids the force applied by the wheel cylinder.
   Technician B says that on a leading-trailing brake, the force applied by the drum to the primary shoe opposes the force applied by the wheel cylinder.
   Who is correct?
   A. A only
   B. B only
   C. Both A and B
   D. Neither A nor B

5. Two technicians are discussing bimetallic aluminum brake drums:
   Technician A says that the bimetallic hub is lighter than a one-piece hub.
   Technician B says that the bimetallic brake hub contains an iron liner for the friction surface.
   Who is correct?
   A. A only
   B. B only
   C. Both A and B
   D. Neither A nor B

6. *Technician A* says that front brake drums are usually separate assemblies from the front hubs.
   Technician B says that special clips at the wheel studs hold the drum onto many front hubs.
   Who is correct?
   A. A only
   B. B only
   C. Both A and B
   D. Neither A nor B

7. *Technician A* says that the secondary shoe in duo-servo brakes usually has a thicker lining.
   Technician B says that the lining of a primary shoe may be made of a different material than the secondary shoe.
   Who is correct?
   A. A only
   B. B only
   C. Both A and B
   D. Neither A nor B

8. *Technician A* says that most self-adjusters are attached to the secondary shoe.
   Technician B says that some self-adjusters actuate only when the vehicle is moving backward.
   Who is correct?
   - **A.** A only
   - **B.** B only
   - **C.** Both A and B
   - **D.** Neither A nor B

9. *Technician A* says that the lining is generally thicker on the leading shoe on a leading-trailing brake.
   Technician B says that the trailing shoe is not self-energizing when braking in a forward direction.
   Who is correct?
   - **A.** A only
   - **B.** B only
   - **C.** Both A and B
   - **D.** Neither A nor B

10. *Technician A* says that self-adjusters perform lining-to-drum adjustments without driver intervention.
    Technician B says that all adjustments for self-adjusting brakes occur when the vehicle is moving in reverse and the brakes are applied.
    Who is correct?
    - **A.** A only
    - **B.** B only
    - **C.** Both A and B
    - **D.** Neither A nor B

Parking Brakes

Upon completion and review of this chapter, you should be able to:

❏ Explain the function of parking brakes.

❏ Identify the basic types of parking brake systems.

❏ Identify types of parking brake controls.

❏ Identify the types of cables used to operate the parking brakes.

❏ Identify and explain the operation of disc brake and drum brake parking brakes.

❏ Explain the operation of electric parking brakes.

Introduction

After the service brakes stop the moving car, the parking brakes hold it stationary. Parking brakes are often mistakenly called "emergency" brakes, but parking brakes are not intended to be used as an alternative to the service brakes to stop vehicles. The stopping power available from parking brakes is much less than from service brakes. Because the parking brakes work only on two wheels or on the driveline, much less friction surface is available for braking energy. In the rare case of total hydraulic failure, the parking brakes can be used to stop a moving vehicle, but their application requires careful attention and skill to keep the vehicle from skidding or spinning.

Parking Brake Operation

The parking brake system is generally not a part of the hydraulic braking system. It is either mechanically operated by cables and levers to apply the rear brakes, or it can be operated mechanically or by its own hydraulic system to activate a drum brake on the transmission or drive shaft.

Most parking brake systems use the service brake shoes or disc pads. Systems that use a separate set of shoes or pads, such as transmission or drive shaft parking brakes, are called independent parking brakes.

Parking brake actuators may be operated either by hand or by foot. Many small and medium-size vehicles use a hand-operated parking brake lever mounted in the console between the front seats (Figure 9-1). When the lever is pulled up, the parking brakes are applied. A ratchet-and-pawl mechanism acts to keep the brake lever applied. To release the lever and the brakes, a button on the lever is pressed and the lever is moved to unlock the ratchet. Some medium trucks and mobile construction equipment use the hydraulic service brakes as the parking brakes. With the vehicle/equipment stopped and the service brakes applied, an electric solenoid is activated. The solenoid closes the hydraulic lines between the wheels and master cylinders, effectively locking all wheels. The service brake pedal can be released until it is time to unlock the wheels.

Figure 9-2 shows a typical foot-operated pedal with a ratchet and pawl. Stepping on the pedal applies the brakes and engages the ratchet and pawl. A release handle and rod or cable is attached to the ratchet release lever. When the release handle is pulled, the pawl is lifted off the ratchet to release the brakes.

Many newer vehicles have a "press-to-release" feature on the parking brakes. To release the parking brakes, pressure (force) is applied to the parking brake pedal. This releases the locking ratchet and allows the pedal to return to the off or up position.

Figure 9-1 A typical lever-operated parking brake control.

Figure 9-2 A typical foot-operated parking brake with a mechanical release handle.

Some vehicles automatically disengage the parking brakes whenever the transmission is placed in drive or reverse; other vehicles release the brakes only when the transmission is placed in drive. The most common way to release the parking brakes automatically is with a vacuum motor (Figure 9-3). Vacuum is applied to the vacuum motor to move the release rod and release the brakes when the transmission is placed into gear. Figure 9-4 is a drawing of a simple vacuum circuit. The parking brake release lever can be operated manually if the automatic release mechanism fails.

This chapter explains the most common types of parking brake levers, handles, cables, and other linkage parts as well as warning lamps and switches. The final sections of this chapter describe typical drum, disc, and drive shaft parking brake assemblies and their operation.

Figure 9-3 A typical foot-operated parking brake with a vacuum release mechanism.

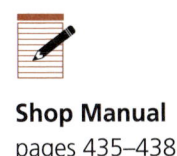

Figure 9-4 A simplified vacuum release circuit.

Parking Brake Controls—Levers and Pedals

The parking brakes on all late-model cars and light trucks are applied by a pedal or a lever, which is often called the **parking brake control.** Many older vehicles and a few current light- and medium-duty trucks have a handle under the instrument panel that is pulled to apply the parking brakes (Figure 9-5). Aside from the design and operation of the control handle, the linkage for this type of parking brake works the same as lever-operated or pedal-operated brakes. On some older vehicles, Chryslers in particular, the parking brakes should be applied before shifting the automatic transmission into park. Shifting into park without the parking brakes applied places the weight of the vehicle on the transmission parking gear making it very difficult to shift from park.

Shop Manual
pages 435–438

Figure 9-5 Older vehicles and some late-model trucks have a handle under the instrument panel to operate the parking brakes.

Most parking brakes use the service brake shoes or pads to lock the rear wheels after the vehicle is stationary. The parking brakes can be set most securely if the service brake pedal is pressed and held while the parking brake control lever or pedal is applied. The hydraulic system applies greater force to the shoes or pads than the parking brake mechanical linkage can apply. When the hydraulic system is used to set the brakes, the parking brake linkage simply takes up slack in the system and holds the shoes or pads tightly against the drums or rotors.

Levers

The control lever for lever-operated parking brakes usually is installed between the two front seats. As the lever is pulled upward, the ratchet mechanism engages to keep tension on the cables and hold the brakes applied. To release the brakes, the spring-loaded button in the end of the lever is pressed and held while the lever is lowered to the floor.

The lever-operated parking brakes on some Chevrolet Corvettes are examples of a design in which the lever drops back to the floor after the brakes are applied. The cables and linkage hold the brakes applied, but the lever returns to the released position. To actually release the parking brakes, you must pull up on the lever until you feel some resistance; then press and hold the button in the end of the lever while moving the lever back to the released position. The parking brake control lever on these Corvettes is located between the driver's seat and the door sill. If the control lever stayed in the upward position with the brakes applied, it would be difficult to climb in and out of the car.

Pedals

In a pedal-operated parking brake system, the pedal and its release mechanism are mounted on a bracket under the left end of the instrument panel. As the pedal is pushed downward by the driver's foot, the ratchet mechanism engages to keep tension on the cables and hold the brakes applied (Figure 9-6). A spring-loaded handle or lever is pulled or the pedal is pressed to release the brakes. A return spring moves the pedal to the released position. A rubber bumper is used to absorb the shock of the released parking brake pedal. If this bumper is missing, the pedal will break the warning light switch after a few operations.

FMVSS 105 requires that parking brakes must hold the vehicle stationary for 5 minutes on a 30 percent grade in both the forward and reverse directions (Figure 9-7). FMVSS 105 also specifies that the force needed to apply the parking brakes cannot exceed 125 pounds for foot-operated brakes or 90 pounds for hand-operated brakes. Some heavy full-size cars built in the late 1970s and early 1980s had trouble meeting the brake-holding requirements without exceeding the allowed maximum application force.

Manufacturers solved the problem with a pedal that had a very high leverage ratio but required two or three applications with the foot to set the brakes completely. The first pedal stroke partially applied the brakes, and the ratchet mechanism held the linkage in this position when the pedal was released. The second or third pedal stroke applied the brakes completely. A single pull on the release handle released the brakes.

Automatic Parking Brake Release

An automatic parking brake release mechanism is a convenience feature offered by some carmakers. It is used only with pedal-operated parking brakes on cars with automatic transmissions.

A vacuum motor is attached to the release mechanism. Vacuum is applied to the motor through a solenoid-operated valve that is actuated when the engine is running and the transmission is shifted into gear from park or neutral. The solenoid that controls the vacuum to the vacuum motor may be mechanically or electrically operated. On the manual system, a short link between the shift rod on top of the steering column is moved to open the vacuum port to the motor. On the electrical system,

Figure 9-6 The pedal ratchet is part of the parking brake pedal assembly.

30 percent grade

Figure 9-7 The parking brake must hold the vehicle on a 30 percent grade for 5 minutes in both the forward and reverse directions.

an electrical solenoid receives a positive or negative signal that causes the solenoid to open the vacuum port. As long as the transmission is in a drive gear, the parking brake cannot be locked in the applied mode. A rod connects the vacuum motor to the parking brake release lever. When vacuum is applied to the motor, it pulls the rod to release the brakes.

The automatic parking brake release is a supplement to the manual release handle, not a replacement for it. All parking brakes with this feature also have a manual release handle so that the brakes can be released in case the servo or its vacuum supply fails (Figure 9-8). The manual handle or lever is not easily visible from the driver's seat and is seldom used. Some drivers do not even know it exists. Although the automatic release mechanism is a convenience feature, it can help to prevent brake damage from driving with a partially engaged parking brake.

Figure 9-8 The small lever is to release the parking brake if the vacum release fails.

Warning Lamps

All brake systems on passenger cars and light trucks built since 1967 have a warning lamp on the instrument panel to indicate a failure in one-half of the hydraulic system. On many vehicles, the same warning lamp will light to indicate a low fluid level in the master cylinder. Most vehicles also use this lamp to indicate that the parking brake is applied. A normally open switch on the control linkage closes as the pedal is pressed or the lever is pulled. The lamp will not light, however, unless the ignition is on. Parking brake lamp switches are adjusted so that the lamp stays lit until the brake is released completely. In most vehicles, the red brake warning light will come on during key on, engine off, even if the fluid is correct and the parking brake is off. This is part of the "lamp check" or "prove out mode" to check the operation of the various warning light systems. The lamp should switch off when the engine is running if everything is correct.

Parking Brake Linkage

Shop Manual
pages 427–430

Parking brake linkage transmits force equally from the control pedal or lever to the shoes or pads at the rear wheels. There are as many different linkage designs as there are different vehicles, but all have the same job and work in basically the same way. The following paragraphs explain the cables, rods, levers, and equalizers or adjusters used in parking brake linkage.

Cables

Most parking brakes use cables to connect the control lever or pedal to the service brakes (Figure 9-9). Parking brake cables must transmit hundreds of pounds of force without jamming, breaking,

Figure 9-9 Typical parking brake cable installation showing an adjusting mechanism.

or stretching. Therefore, they are made of high-strength strands of steel wire that are tightly twisted together. The ends of the cables have different kinds of connectors that attach to other parts of the linkage. Some cables have threaded rods or clevises at their ends. Others have ball or thimble-shaped connectors that fit into holes and slots on other parts of the linkage.

The front cable connects the parking brake lever or pedal to the **equalizer,** which provides balanced braking force to each wheel. The equalizer is only a lever mounted on a pivot or a U-shaped grooved guide. Pulling the front cable moves the lever or guide. The lever or guide transmits the force equally to the two wheel cables (Figure 9-10).

Some vehicles have a three-part cable installation, which includes an intermediate cable that passes through the equalizer. Typically, one of the two rear cables is attached directly to this intermediate cable with a connector (Figure 9-11). The other rear cable is attached to the intermediate cable with some kind of cable adjuster, usually a turn buckle.

Some lever-operated parking brakes have a separate cable for each rear wheel attached to the control lever. Each cable is adjusted separately, and an equalizer is not necessary. The wheel cables run from each wheel all the way to the pedal or lever mechanism within the passenger compartment. Adjustment of the cables may be made at the pedal or lever.

Cable retainers and hooks maintain cable position on the rear axle, frame, and underbody of the vehicle (Figure 9-12). These retainers allow the cable to flex and move at their point of body attachment and help the equalizer to provide its equalizing action.

The equalizer may be referred to as the parking brake adjuster. In some cases, it is the point at which the parking brake is, in fact, adjusted.

Figure 9-10 This parking brake equalizer is under the center console.

Figure 9-11 Parking brake cable connector.

Figure 9-12 Cable retainers and clips hold the cables in position.

Most control cables and rear brake cables are partially covered with a flexible metal **conduit** or housing (Figure 9-13). The cable slides inside the conduit and is protected from chaffing or rubbing against the underside of the vehicle. One end of the cable conduit is fastened to a bracket on the underside of the vehicle with some type of retaining clip, and the other end is attached to the brake (Figure 9-14). Many cables are coated with nylon or plastic, which allows them to slide more easily through the conduit. The coatings help to reduce corrosion and contamination and make parking brake application easier.

AUTHOR'S NOTE: Conduits are common in other systems as well. Check the area around your house or shop and note the conduits that protect the electrical wiring. Parking brake conduits perform the same protective function for the cables.

Rods

The most common use of solid steel rods in parking brake linkage is in lever-operated systems to span a short distance in a straight line to an equalizer or intermediate lever. The linkage rod is usually attached to the control lever by a pin. The other end of the rod is often threaded to provide linkage adjustment.

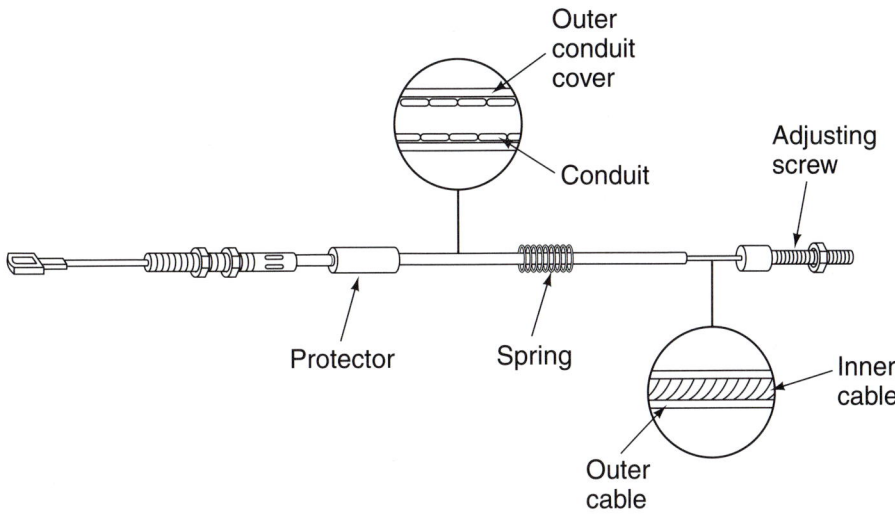

Figure 9-13 Typical cable and conduit.

Figure 9-14 A cable retainer secures the conduit to the backing plate.

Levers

Chapter 2 of this *Classroom Manual* explains how levers are used to multiply force in mechanical linkage. Mechanical leverage is necessary in parking brake linkage to make brake application easy for the driver. The parking brake pedals and control levers multiply the force applied by the driver. Many parking brake installations also have an **intermediate lever** under the vehicle body to increase the application force even more. The intermediate lever also is designed to work with the equalizer to ensure that force is applied equally to both rear wheels. Intermediate levers are most common on large cars and trucks that need greater force to apply the parking brakes.

Equalizers and Adjusters

Parking brakes that use the service brakes—either shoes or pads—to lock the wheels must apply equal force to each wheel. If application force is unequal, the parking brakes may not hold the vehicle safely. To meet this requirement, most parking brake linkage installations include an equalizer mechanism. The equalizer also usually contains the linkage adjustment point.

The simplest example of an equalizer is a U-shaped cable guide attached to a threaded rod. The rear cable (or an intermediate cable) slides back and forth on the guide to balance the force applied to each wheel. In some installations, the equalizer guide is attached to a lever to increase application force.

Another kind of equalizer is installed in a long cable that runs from the driver's position to one rear wheel. A shorter cable runs from the equalizer to the other wheel. When the parking brakes are applied, the long cable applies its brake directly and then continues to move forward after the shoes or pads lock the wheel. The continued forward motion pulls the equalizer and the shorter cable to lock the brake at the other wheel.

Rear Drum Parking Brakes

Rear drum parking brakes that use the rear service brakes to lock the wheels are the most common kind of parking brake system. Mechanical linkage that works with drum brake shoes is a relatively simple and economical design, and the self-energizing action of the brake shoes provides excellent holding power.

Figure 9-15 shows a typical parking brake installation with rear drum brakes. The brake cable runs through a conduit that goes through the backing plate. The cable end is attached to the lower end of the parking brake lever. The parking brake lever is hinged at the top of the web of the secondary or trailing shoe and connects to the primary or leading shoe through a strut. When activated, the lever and strut move the shoes away from both the anchor points and into contact

Front

Figure 9-15 A parking brake is applied by a lever working on the trailing or secondary shoe, depending on the system.

3. Lever works against link, and pivot forces secondary shoe against the drum.

2. Lever moves link against primary shoe and shoe against drum.

Spring

1. Cable pulls lever. Conduit

Figure 9-16 Parking brake lever and strut operation.

with the brake drum (Figure 9-16). When tension on the cable is released, the return springs move the shoes back to their unapplied positions.

The details of parking brake parts vary with different brake designs, but all work in the same way basically. Many parking brakes include various springs and clips to prevent rattles and to hold the parts in alignment.

Electrical Parking Brake Systems

Some may believe that electrically operated parking brakes were introduced only very recently. Electric parking brakes have been around since at least the late 1980s. They were first used on 1- to 2-ton trucks under the brand **Electro-Lok.**

The Electro-Lok was actually a simple mechanism that used one or two electrically activated solenoids mounted in the brake line(s) to the rear brakes and in some instances in the lines to the front brake also (Figure 9-17). When the driver stopped the vehicle and wanted to set the parking brakes, he/she applied the brake pedal firmly and with the pedal depressed, switched the

Toggle switch

Master
cylinder

Electro-Lok
solenoids

Hydraulic pressure held
by activated solenoids

Figure 9-17 On an Electro-Lok parking brake system, applied service brake fluid is prevented from returning to the master cylinder by one or more solenoids installed in the brake lines.

parking brake toggle (or push button) switch to "on." The solenoid(s) were then energized and blocked the brake fluid from returning from the wheel brake mechanisms. This provided a better parking brake because the high hydraulic pressure created by the brake pedal was trapped in the lines. To release the parking brakes, the driver applied force to the brake pedal and flipped the parking brake switch to "off," de-energizing the solenoid(s). It was not absolutely necessary to apply the brake pedal, but it was safer and reduced stress on the solenoids during plunger movement.

True electric parking brakes are being introduced today. The first is an extension of the ABS and active brake systems. When parking, the driver can switch on the parking brakes, and the hydraulic modulator applies fluid pressure to at least the rear wheels. The switch can also be activated by the movement of the transmission shift mechanism into park.

Some of the most recent electric parking brakes are available from Continental Teves North American, and Siemens. Both manufacturers offer electric parking brake systems that can be found on many top-line luxury and midsize models. Both systems dispense with lever/pedal controls and both will provide some form of emergency braking should the service brakes completely fail.

AUTHOR'S NOTE: Although a lot of us do not or cannot afford the top-line luxury models, electric parking brakes are on the way. It is only a matter of time before most manufacturers offer this system even down to the lower range models.

The Continental Teves system uses an electrical motor or solenoid that moves cables running to the rear wheels (Figure 9-18). It is designed for drum-in-hat or disc caliper parking brakes. The

Figure 9-18 This electric motor assembly pulls on the cables to apply the parking brakes.

system is an active parking brake that can be integrated into an electronic stability program (ESP) with appropriate interfaces. The system offers the advantages of automatically locking in park, releasing when engaging a drive gear, reducing or preventing rollback on slopes, increasing theft deterrent, and assisting in parking if integrated with a distance sensor system.

One version of the parking brake system has an electric mechanism at each rear wheel that applies the rear disc brake when activated (Figure 9-19). The two mechanisms are mounted just inside the rear calipers (Figure 9-20). The manual switch to activate and deactivate the system is mounted near the transmission's console shift lever or at a central point of the dash. An electrical signal can be mounted to sense transmission shift lever movement and engage/disengage the parking brake accordingly.

Figure 9-19 This parking brake is applied by an electric motor and cam operating directly on the caliper.

Figure 9-20 The electric motors are connected to the caliper and its adapter.

Rear Disc Parking Brakes

Shop Manual
pages 421–424

Two different types of parking brakes are used with rear disc brakes: auxiliary drum parking brakes and caliper-actuated parking brakes. Both are more complicated designs than parking brakes that are part of rear drum service brakes.

Auxiliary Drum Parking Brakes

Assuming that the driver does not drive with the parking brakes applied, the parking brake shoes within the drum will last the life of the car. However, they may get corroded over a period and should be checked when servicing the rear service brakes.

Fixed-caliper rear disc brakes, such as those used on early Corvettes, and some floating or sliding caliper rear disc brakes have a small drum cast into each rotor (Figure 9-21). A pair of small brake shoes is mounted on a backing plate that is bolted to the axle housing or the hub carrier. These parking brake shoes operate independently from the service brakes. They are applied by linkage and cables from the control pedal or lever. The cable at each wheel operates a lever and strut that apply the shoes in the same way that rear drum parking brakes work. These auxiliary drum parking brakes must be adjusted manually with star wheels that are accessible through the backing plate or through the outboard surface of the drum. They do not have self-adjusters.

Caliper-Actuated Parking Brakes

Eccentric means not round or concentric.

Most floating or sliding caliper rear disc brakes have components that mechanically apply the caliper piston to lock the pads against the rotors for parking. All caliper-actuated parking brakes have a lever that protrudes from the inboard side of the caliper. The caliper levers are operated by linkage and cables from the control pedal or lever. As with most brake assemblies and subassemblies, detail differences exist from one brake design to another. The two most common kinds of caliper-actuated parking brakes are the **screw-and-nut** type and the **ball-and-ramp** type. A few imported cars have a third kind that uses an **eccentric** shaft and a rod to apply the caliper piston. This type is not as common as the first two, however. An eccentric acts like a cam.

Figure 9-21 An auxiliary parking brake installation for rear disc brakes.

Figure 9-22 A GM screw-and-nut parking brake mechanism for rear disc brakes.

One portion of the shaft is oval shaped. As the shaft turns, the high part of the oval pushes the operating rod out of applying the brakes.

Screw-and-Nut Operation. General Motors' floating caliper rear disc brakes are the most common example of the screw-and-nut parking brake mechanism (Figure 9-22). The caliper lever is attached to an actuator screw inside the caliper that is threaded into a large nut. The nut, in turn, is splined to the inside of a large cone that fits inside the caliper piston. When the parking brake is applied, the caliper lever rotates the actuator screw. Because the nut is splined to the inside of the cone, it cannot rotate so it forces the cone outward against the inside of the piston. Movement of the nut and cone forces the piston outward. Similarly, the piston cannot rotate because it is keyed to the brake pad, which is fixed in the caliper. The piston then applies the inboard brake pad, and the caliper slides as it does for service brake operation and forces the outboard pad against the rotor.

An adjuster spring inside the nut and cone rotates the nut outward when the parking brakes are released to provide self-adjustment. Rotation of the nut takes up clearance as the brake pads wear.

Ball-and-Ramp Operation. Ford's floating caliper rear disc brakes are the most common example of the ball-and-ramp parking brake mechanism (Figure 9-23). The caliper lever is attached to a shaft inside the caliper that has a small plate on the other end. Another plate is attached to a thrust screw inside the caliper piston. The two plates face each other, and three steel balls separate them. When the parking brake is applied, the caliper lever rotates the shaft and plate. Ramps in the surface of the plate force the balls outward against similar ramps in the other plate. As the plates move farther apart, the thrust screw forces the piston outward. The thrust screw cannot rotate because it is keyed to the caliper. The piston then applies the inboard brake pad, and the caliper slides as it does for service brake operation and forces the outboard pad against the rotor.

Figure 9-23 A Ford ball-and-ramp parking brake mechanism for rear disc brakes.

When the caliper piston moves away from the thrust screw, an adjuster nut inside the piston rotates on the screw to take up clearance and provide self-adjustment. A drive ring on the nut keeps it from rotating backward.

Summary

Terms to Know

Ball and ramp

Conduit

Eccentric

Electro-Lok

Equalizer

Intermediate lever

Parking brake
control

Screw and nut

- ❏ The parking brake prevents the vehicle from moving when parked.
- ❏ The control device that applies the parking brakes may be operated by either hand lever or foot pedal. The release mechanism may be either a manual release or an automatic release using a vacuum servo controlled by the transmission gear selector.
- ❏ Equalizers are used to balance the forces applied to the parking brakes during application.
- ❏ Equalizer levers are used to multiply the effort of the driver applying the parking brakes.
- ❏ Rear drum brakes use a lever and strut to move the shoes into contact with the drum.
- ❏ Electro-Lok is a brand of electric parking brakes that uses solenoids to trap the applied hydraulic pressure in the lines.
- ❏ Electric parking brakes may operate conventional parking cables or directly apply the rear disc brakes.
- ❏ Integral disc parking brakes use the normal disc calipers as parking brakes to hold the vehicle while it is parked.
- ❏ Auxiliary drum parking brakes are contained inside the rotor of some rear disc brakes.

Review Questions

Short-Answer Essays

1. Explain the purpose of a parking brake.

2. Describe why a parking brake should not be used as an "emergency" brake.

3. How do the levers connected to equalizers on some brake systems multiply the driver's application effort?

4. What is the function of an equalizer in a parking brake system?

5. What is the construction of a parking brake cable?

6. Why are some parking brake cables plastic-coated?

7. How are integral disc parking brakes applied?

8. Describe how the rear drum brakes are applied to hold the vehicle when it is parked.

9. Describe how the internal-expanding shoe transmission-type brake works.

10. What is the difference between an integral parking brake and an independent parking brake?

Fill in the Blanks

1. Parking brakes may be operated by _____, _____, or _____.

2. A _____ _____ is used on some vehicles to automatically release the parking brake.

3. Parking brake cables are made up of a _____ of small wires to form the complete cable with the proper strength.

4. The _____ is a device that applies the same tension to each rear brake cable.

5. Another name for the cable housing of a parking brake cable is the cable _____.

6. The parking brake cable is attached to the _____ that is connected to the secondary shoe web on drum brakes.

7. The parking brake _____ connects the primary shoe to the secondary shoe.

8. A high-lead _____ is used to move the piston and apply the brakes.

9. A _____, lined with brake lining material, is wrapped around an external brake drum.

10. The Continental Teves electric parking brake system applies the parking brakes through movement of _____.

Multiple Choice

1. *Technician A* says that the parking brakes are provided with equal cable tension by the cable adjusters.
   Technician B says that the parking brakes are provided with equal cable tension by the equalizer.
   Who is correct?

   A. A only
   B. B only
   C. Both A and B
   D. Neither A nor B

2. *Technician A* says that the parking brakes are mechanically operated because mechanical brakes are much more effective than hydraulic brakes.
   Technician B says that parking brakes are mechanical because the parking brakes must operate separately from the service brakes.
   Who is correct?

   A. A only
   B. B only
   C. Both A and B
   D. Neither A nor B

3. *Technician A* says that the usual device that releases the parking brakes whenever the transmission gear selector is in drive or reverse is a vacuum motor.
   Technician B says that the device that releases the parking brakes is the driver.
   Who is correct?
   A. A only
   B. B only
   C. Both A and B
   D. Neither A nor B

4. *Technician A* says that the parking brakes are set by pulling the parking brake lever.
   Technician B says that parking brakes are set by depressing the parking brake foot pedal.
   Who is correct?
   A. A only
   B. B only
   C. Both A and B
   D. Neither A nor B

5. *Technician A* says that some electric parking brakes use solenoid(s) to trap pressurized fluid in the lines.
   Technician B says that some electric parking brakes may be applied by shifting the transmission into PARK.
   Who is correct?
   A. A only
   B. B only
   C. Both A and B
   D. Neither A nor B

6. *Technician A* says that a screw forces the disc pads against the rotor in a rear disc parking brake system.
   Technician B says that a ball-and-ramp assembly forces the disc pads against the rotor in a rear disc parking brake system.
   Who is correct?
   A. A only
   B. B only
   C. Both A and B
   D. Neither A nor B

7. Which of the following applies the rear caliper by cables for parking?
   A. Continental Teves
   B. Pedal control
   C. Lever control
   D. All of the above

8. *Technician A* says that another name for a rear disc-drum parking brake is auxiliary drum brake.
   Technician B says that another name for a rear disc-drum parking brake is integral brake.
   Who is correct?
   A. A only
   B. B only
   C. Both A and B
   D. Neither A nor B

9. *Technician A* says that it is important to remember that the parking brake is not part of the vehicle hydraulic brake system.
   Technician B says that it is important to remember that the parking brake is applied mechanically.
   Who is correct?
   A. A only
   B. B only
   C. Both A and B
   D. Neither A nor B

10. *Technician A* says that the effort of the driver when activating the foot pedal or hand lever is multiplied and is transferred on to the rear cables by the front cable.
    Technician B says that the effort of the driver when activating the foot pedal or hand lever is multiplied and is transferred on to the rear cables by the equalizer lever.
    Who is correct?
    A. A only
    B. B only
    C. Both A and B
    D. Neither A nor B

Electrical Braking Systems

Upon completion of this chapter, you should be able to:

❑ Define and understand the electronic terms commonly associated with electrical braking systems.

❑ Identify the components of a typical ABS.

❑ Explain the operation of a typical ABS.

❑ Describe the differences between integrated ABS and nonintegrated ABS.

❑ List the major components of two-wheel, four-wheel, and three- and four-channel ABS.

❑ List and explain the operation of the components for a TCS.

❑ Explain the integration of braking, steering, and suspension systems.

❑ Discuss the general operation of the Continental Teves Mark 20e system.

❑ Discuss the general operation of the Delphi DBC-7 system.

❑ Discuss the general operation of active brake systems.

❑ Discuss the general concept of electrical brake systems.

Introduction

Although 1987 was not the first year when antilock brake systems (ABSs) were used, it was the first year when they were installed on all light trucks and small vans. The systems were first used to reduce rear-wheel lockup during hard braking. The types of vehicles noted were the ones that were most likely to lock the rear wheels during braking. Since then rear-wheel antilock systems have evolved to today's initial offering of full electrical braking systems. Certain technical terms and components from ABSs have been incorporated into the newer systems and are discussed in the first section.

A BIT OF HISTORY

Ford Motor Company was the first U.S. carmaker to experiment with ABSs on a production car when it offered an antiskid option on the 1954 Lincoln Continental Mark II. It worked—sort of— but added too much weight to the car and cost too much, so it wound up in the technological trash can.

Common Components and Terms

Sensors

Shop Manual
pages 447–450

Sensors are electrical devices that measure some physical action and change that action to an electrical signal (Figure 10-1). They also are used to measure electrical voltage. Common actions measured by sensors are mechanical movement, temperature, pressure, and speed. Most sensors are simple to operate although the engineering design may be complex. The simplest are on–off switches, whereas others are actually small electrical generators.

Figure 10-1 The TPS is mounted to the side of the throttle body. As the throttle valve is moved by the accelerator cable, a wiper arm inside the TPS moves over a variable resistor wire. This changes the input 5-volt signal to a return signal between 0.5 volt and 4.5 volts.

Signals

Signals refer to the electrical signal sent from the sensor to the controller. They may be analog or digital (Figure 10-2). Digital signals are more easily "understood" by a computer so most analog signals are converted to digital by an ac/dc converter built into the computer. The ac voltage signal is infinitely variable; in other words, it flows smoothly as it changes voltage levels. The wheel speed signal is compared to the ones from different wheels, and the electronic brake control module (EBCM) determines which wheel is about to lock up or skid; the faster the wheel, the stronger the voltage signal produced.

The digital signal could be on or off, high or low, or yes or no. For simplicity in this discussion, the signal is either on or off. Digital signals can differ in the amount of on or off time and the rapidity of repeating signals (Figure 10-3). The time difference between on versus off is the duty cycle or frequency. This usually refers to the action of an actuator. Different speeds can be based on how often the on–off cycle occurs. A slow wheel may cause the sensor to be on or off each one-half millisecond, whereas a fast wheel may cause the signal to be on or off each one-tenth millisecond. In this case, the computer can compare wheel speeds and determine the proper command, if any, to be issued to the hydraulic modulator. Generally speaking, 12 volts represent an on signal, whereas zero volt is off. Digital wheel speed sensors (WSSs) are not used on many vehicles, but a WSS that produces a digital signal is available and is discussed later.

Actuators

Actuators are the workhorse of electrical/electronic systems. They perform some action based on commands from their controllers. Actuators are usually motors or solenoids. Some of the most common are cooling fans, fuel injectors, and transmission shift solenoids (Figure 10-4). In most cases, the controller can determine if the actuator is electrically and mechanically active by

AC VOLTAGE SIGNAL

DIGITAL SIGNAL

Figure 10-2 An analog signal switches from positive to negative to positive. A digital signal is usually positive or negative, but not both.

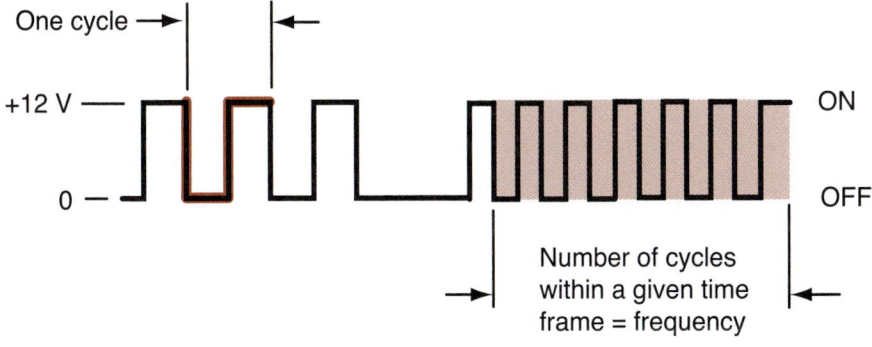

Figure 10-3 This signal indicates the amount of time on versus time off.

measuring the current flowing through that circuit. If the current is too high or too low, the controller may disable the actuator and/or illuminate a light in the dash.

Controller

A controller is a computer programmed to perform certain decisions based on sensor signals and to issue electrical commands to actuators (see Figure 10-4). Controllers are simple computers compared to the ones at work or school that can use multiple programs to perform different tasks. The automotive computer works within its own programmed parameters to control the engine, transmission, climate control, and other systems. Data are shared with other onboard computers to reduce the number of sensors and actuators needed to control multiple vehicle systems.

Figure 10-4 A fuel injector is a solenoid that will allow a specific amount of fuel to flow when commanded to open.

Multiplexing

Consider multiplexing to be a small Internet on the vehicle (Figure 10-5). Each computer has access to all data, but an individual computer will use only the data pertinent to its programming. The most common and most powerful controllers are the powertrain control module (PCM), the body control module (BCM), and the transmission control module (TCM). They control many smaller computer modules that have an even more specific function, such as adjusting the blower speed on an automatic climate control. There may be over fifty different controllers on a single vehicle, all networked.

CAN

The control area network (CAN) is the newest multiplex communication and it is being incorporated into all vehicles. Although this text does not discuss the CAN in detail, the reader needs to keep the CAN in mind as the various integrated systems are discussed.

Hydraulic Modulator

The hydraulic modulator is the electrical/hydraulic unit used on ABS, traction control system (TCS), and other braking systems to control hydraulic braking pressure to one or more wheels (Figure 10-6). The oldest could only release pressure, whereas the newest can release pressure or increase pressure to control wheel speed. Many times this pressure change is done without driver input.

Commands

Commands are nothing more than an electrical signal or output signal to an actuator. Usually a command is a negative polarity voltage that completes the actuator circuit. For example, a fuel injector always has 12 volts when the ignition switch is in the run or start position, but its circuit is not complete until the PCM grounds the circuit. When the circuit is complete the fuel injector can then function. Other actuators work in much the same manner.

Figure 10-5 Note the different modules connected in this wiring diagram. Also note the CAN+ and CAN– shown on most modules. This is a controlled area network (CAN).

Figure 10-6 A typical ABS hydraulic control modulator.

Antilock Brake System (ABS) and Vehicle Control

The ABS is not the end all to safe braking. The system was designed to provide some means of controlling wheel skid or **negative wheel slip.** If the driver overdrives the road condition or the capability of the vehicle, however, the ABS will not assist much in controlling the vehicle. Assuming the vehicle is operating within road and vehicle limits, the ABS is excellent in **directional stability** and **directional control.** Stability is achieved when the vehicle can be stopped in the shortest possible distance without wheel skid, whereas control refers to the fact that the driver can steer the vehicle during a panic or ABS stop. A locked wheel (sliding) has little, if any, traction and will not grip the road surface sufficiently for braking or steering. It should also be noted that the ABS will not necessarily stop the vehicle quicker than standard brake systems.

AUTHOR'S NOTE: One thing I have learned is that you cannot steer a vehicle if the front wheels are sliding. I missed a turn one day many years ago and slammed on the brakes of a 1969 VW Fastback. The front wheels locked and I steered to the right. The car continued to go straight ahead. I realized I was going to miss the turn anyway so I released the brakes. The wheels stopped sliding and the car immediately went into a hard right turn—I missed the turn and aged 10 years in less than 5 seconds. This experience has stayed with me for over 40 years: Do not try to steer the vehicle if the front wheels are locked and sliding. The ABS reduces this possible problem to almost zero.

ABS Types and General Operations

Shop Manual
pages 457–463

Basically when the ABS controller senses a wheel slowing faster than the other wheels, it opens a valve to bleed pressure from that wheel. It can command the valve to hold that new pressure, decrease the pressure further, or increase the pressure. The commands are based on the signals from the WSSs, and the commands are to the hydraulic modulator. Depending on the type of ABS, the system can control brakes to one axle, to one wheel, or to all four wheels.

Integrated and Nonintegrated ABS

The primary difference between these two systems is the placement of the hydraulic modulator. In the integrated system, the modulator is built with a master cylinder as one assembly (Figure 10-7). This led to some problems with the earlier versions. The earlier ABS modulators went through some growing pains that are typical of new systems, but these were particularly painful to some vehicle owners. Some modulator/master cylinder units cost as much as $3,000. This was understandable if the modulator failed because the owner could just ignore the ABS warning light, knowing that the standard brakes would work. The cost problem was not really with the modulator, however, but with a failed master cylinder. Because this is an integrated unit, a failed master cylinder that usually costs less than $75 now costs $3,000. For the most part, the weakest component was the master cylinder and the owner could not just ignore the problem. Generally, there were few problems with the system but the few caused some heated discussions between technicians and owners.

The nonintegrated system was a benefit to all concerned, from the manufacturer to the technician and especially the vehicle owner. The hydraulic modulator was separated from the master cylinder and usually was installed just below the master cylinder or somewhere on the left side of the engine compartment. This left the master cylinder as a separate unit that could be replaced alone. The cost of separating the two was nominal because the biggest cost was in engineering and a few extra feet of brake tubing and wires. The overall efficiency of the system was not affected.

Cap and connector

Fluid reservoir

Solenoid valve connector

Pressure warning switch

Solenoid valve block

Hydraulic pump

Figure 10-7 An integrated ABS combines the master cylinder, brake booster, and ABS components in a single unit.

Rear-Wheel ABS (RWAL or RABS)

This was the first true mass-production ABS installed on vehicles. The **RWAL/RABS** system was first installed on minivans and pickup trucks in late 1987. RWAL means rear-wheel antilock and is the most common term used by most manufacturers. RABS means rear antilock brake system and is the term commonly used by older Fords and some other manufacturers.

The system essentially controls wheel skid only on the rear wheels. Usually the speed sensor was located on the differential housing and measured the speed of the differential ring gear. A few manufacturers installed the sensor in the tail of the transmission and measured the drive shaft speed. In either case, only the speed of the two rear wheels was measured. The controller was a relatively simple device that processed the signal and commanded a hydraulic modulator commonly called an **isolation/dump valve** (Figure 10-8).

Figure 10-8 RWAL control valve.

On command the isolation/dump valve would block off or isolate the master cylinder from the rear brakes. As long as the brakes were applied and the vehicle was moving, the master cylinder remained isolated so additional fluid could not be applied to the rear brakes by the operator. At the same time the dump valve would open and allow a very small amount of fluid in the rear brake lines to enter an accumulator. This reduced the hydraulic pressure being delivered to the rear brakes and the brakes released enough to allow the wheels to turn. If the rear wheels sped up too much, the dump valve would essentially reverse and the gas-charged or spring-loaded accumulator would force a small amount of fluid back into the rear brakes. This constant dump/recharge is what causes the pulsation of the brake pedal during a panic or ABS stop. This could be disconcerting to a driver applying light foot force to the pedal on a gravel road where the rear wheels tend to slide easier.

The biggest drawback to the system was the placement of the isolation/dump valve under the vehicle body on the frame. Improvements in protection of the unit increased its durability, however. Another drawback resulted from a simple failure of the unit that could be confused as a master cylinder with an internal leak. This led to some unnecessary repairs. The *Shop Manual* explains this problem and the test procedure to isolate the hydraulic unit from the master cylinder.

Four-Wheel ABS

An ABS used for all four wheels is based on the same principle as the RWAL/RABS. It can control braking effect at each wheel through one of two systems: three-channel and four-channel. In the three-channel, the rear-wheel brakes are controlled just like the RWAL/RABS, but the two front wheels can be controlled independently of each other. To do this requires three WSSs and a larger, more sophisticated hydraulic modulator. Each front wheel has a speed sensor, whereas another is placed to measure the rear wheels. The hydraulic modulator basically has three "isolation/dump" valves in a single unit (see Figure 10-8). The basic principle of operation remained the same, but the different components improved drastically in design and serviceability. The controller's programming was increased to handle the additional workload and to effectively control the hydraulic module. The three-channel system works best with front disc and rear drum brake systems. This is mainly because the speed of release or application of drum brakes is a little slower than disc brakes.

The four-channel ABS works along the same lines as the three-channel, but all four wheels have a speed sensor with dedicated valves for each wheel. The hydraulic modulator is a full-fledged hydraulic unit housing four valves commanded by a stronger controller and usually with the controller integrated into the modulator. The time span for development between the first RWAL/RABS and the four-channel system allowed for better programming, better manufacturing, better materials, and a better fit into the service brake's hydraulic system. The four-channel works best with four-wheel disc but will work satisfactorily with a disc/drum combination. Its best feature is its adaptability to TCSs and active braking systems.

ABS Brands

ABS components are not manufactured by automotive manufacturers. They are designed and built for a specific vehicle line, but all function in much the same manner. Two of the different brands are discussed later. The primary manufacturers of ABSs are Bosch (and Delco-Bosch), Continental Teves, Kelsey-Hayes, Delphi Chassis, Bendix, Nippondenso, Nissan, and Sumitomo.

ABS Components

Shop Manual
pages 463–471

ABS components are additions to the standard braking hydraulic/mechanical components covered in earlier chapters. Common components are the controller, hydraulic modulator, WSSs, and brake switch.

Controllers

The antilock controller or microprocessor is the computer for the ABS. It is commonly referred to as the **electronic brake control module (EBCM)** although different manufacturers use different names. **Controller antilock brakes (CAB) module** is another common term for ABS controllers. If the controller is built into or attached to the hydraulic modulator, then the whole assembly is known as the integrated control unit (ICU). All controllers do basically the same thing: receive input signals, process the signals, store the signals for possible use later, and issue output commands. Most automotive computers, including the EBCM, are programmed with an adaptive memory capability or learning mode. This seems to mean that the computer can "learn," but that is not exactly true because computers do not have the ability to learn. Computers can process signals and determine exactly how this particular sensor or actuator, including the vehicle driver, responds to different conditions, however. This "learning" can be done only within the program parameters. For instance, if a new WSS is installed, the computer will determine the limits of the sensor. The new sensor's high/low signals may be a little (tenths of a volt) different from the old one. If the new signals are within set limits, the computer can accept them as the norm for that sensor. If not, then the computer activates the ABS warning light on the dash and disables the ABS entirely.

If the EBCM is incorporated with the computer on the hydraulic modulator, the unit is known as the **electrohydraulic unit.** This eliminates one controller and provides faster action with hydraulic pressure control even if the time difference is measured in microseconds.

As mentioned before, input signals may be analog or digital. Analog signals are produced by small ac electrical generators called **permanent magnet (PM) generators.** The most common ones on a vehicle are the WSS and vehicle speed sensor (VSS).

Wheel Sensors

The PM wheel sensor is composed of an ALNICO permanent magnet and a coil of copper wire. ALNICO is an abbreviation for the aluminum, nickel, and cobalt elements used in the metal alloy (Figure 10-9). The ALNICO magnet is very stable; that is, it loses only a very small percentage of its strength each year. This characteristic makes the ALNICO magnet ideally suited for its use in the wheel sensor because of its longevity.

A metal-toothed ring (tone ring or wheel) is placed in proximity to the permanent magnet and coil. One of the metal teeth on this ring attracts the magnetic lines of force of the magnet.

Figure 10-9 PM generator.

These magnetic lines of force are called **flux lines.** The flux lines must pass across the copper wires in the coil as they are attracted to the metal tooth of the **reluctor.** The principle of magnetic induction states that whenever a magnetic flux line crosses a conductive wire, it induces voltage in that wire. When that wire is in an electrical circuit, it generates electrical current.

The tooth rotates and aligns the valley between the teeth with the magnet. The distance between the metal of the reluctor and the permanent magnet is great enough to allow the flux lines to return to their original location. The flux lines reverse themselves as the magnetic pull weakens; as they travel in the reverse direction across the copper wire, they induce voltage once more. However, this time the voltage induced by the flux lines is the opposite of the initial induced voltage. Movement of the flux lines between the permanent magnet and the reluctor produces an ac voltage for use by the ABS (Figure 10-10).

The PM generator circuit was designed to self-test at POST. The technician needs to understand how this circuit works in case a wheel sensor code is stored in the diagnostic program of the microprocessor. Service technicians who do not understand the circuit are likely to start yanking expensive components, such as the microprocessor, the wiring, and the sensor, or, in a worst-case scenario, all three, and replacing these perfectly good components with identical parts. This is both expensive for the customer and labor intensive for the service technician.

The microprocessor has a regulated power supply that produces a stable direct current (dc) voltage. The voltage signal is usually 1.5 volts. That voltage begins its journey at the power supply inside the microprocessor and travels through a printed circuit until it arrives at an R_1 fixed-value resistor. The R_1 resistor value is usually 10,000 ohms. The current flow continues its forward movement from the microprocessor until it arrives at the PM generator. The resistance of the coil in the PM generator is usually 1,000 ohms. The current returns to the microprocessor power supply through the return wiring and printed circuit board. This completes the series circuit. A typical series circuit is illustrated in Figure 10-11.

Half wave   Half wave   = Full wave

Figure 10-10 Wheel sensor output.

Figure 10-11 PM generator circuit.

Anything that consumes electricity is called a load. Voltage drops occur across each resistance in a series circuit. The resistance may be a coil of wire or it may be a carbon fixed-value resistor. The total of the voltage drops in a circuit must equal the power supply. The **eyeball,** a voltage-monitoring detection circuit, sees the voltage drop. The eyeball expects to see a voltage of 136.3 millivolts. The technician can detect that voltage by test probing the body wiring with a voltmeter. If the eyeball sees no voltage, it assumes there is a problem. That problem could be with the power supply or with the printed circuit board at the resistor. The problem also could be with the body wiring to the PM generator. The generator coil could be shorted to ground.

Magnetoresistive Wheel Speed Sensors

The PM generator is a variable reluctance sensor in that the amount of voltage produced is directly related to the speed of the wheel. As the wheel slows, the strength of the voltage may not be detectable by the EBCM. So if the wheel begins to skid at a low speed, the ABS may not function correctly. Although this may be acceptable with standard ABSs, it is not acceptable with the active brake systems that are discussed later. Teves uses a magnetoresistive sensor as its WSS in the Teves Mark 20E ABS (Figure 10-12). The magnetoresistive sensor cannot produce a voltage, so it must be supplied with 12-volt power provided by the EBCM (Figure 10-13). This creates an electromagnet at the head of the sensor. The return voltage or signal is changed based on the resistance or disturbance of the magnetic field around the sensor and the relationship of a tone ring tooth to the sensor. This sensor has two integrated circuits (ICs) that increase or amplify the resistance change into a dc signal (digital) that is returned to the EBCM. The output signal of one IC is a constant 7 mA at 0.9 volt when no tooth on the tone ring is near the sensor. When a tooth aligns or enters the sensor's magnetic field, the second IC produces the same amount of current and voltage. As a result the return signal voltage is increased to 14 mA

Figure 10-12 This is a view of a magneto-resistive wheel speed sensor used with the Teves Mark 20E ABS.

Figure 10-13 The magneto-resistive sensor will return a signal of 14 mA, 1.65 volts with a tooth aligned with the sensor head. With no tooth aligned, the return signal is 7 mA, 0.9 volt.

Fast wheel ⎍⎍⎍⎍⎍⎍

Slow wheel ⎍ ⎍ ⎍ ⎍

Figure 10-14 A fast wheel will generate a rapid cycling or high frequency. A slow wheel will produce fewer cycles or a low frequency.

at 1.65 volts (Figure 10-14). Because the voltage output signal is not dependent on how fast a wheel turns, this sensor is more accurate and allows for the precise braking control demanded by active brake and active suspension systems. The speed of the wheel is calculated on how often or how fast the voltage changes from low voltage, 0.9 volt, to high voltage, 1.65 volts, or the cycle frequency. A fast wheel produces very quick repetitive cycles (high frequency), whereas slow cycles represent a slow wheel (low frequency).

At times, like when turning a corner, all four wheels may be traveling at different speeds. A four-wheel ABS has to take this speed difference into account on braking actions on curves and when cornering. Even simple RWAL/RABS have to allow for this condition. It is probable that most roads have more curves than straight sections. On the other hand, most panic stops are done when the vehicle is traveling straight ahead even if only for short distances.

CAUTION: Servicing a magnetoresistive sensor is different from servicing a PM generator sensor. Follow the manufacturer's instructions closely or damage to the sensor or electronic circuits could occur. Detail on testing and diagnosis is discussed in the *Shop Manual.*

CAUTION: Many magnetoresistive sensors have a mercury acceleration sensor internal to the sensor. Any sensor with mercury must be considered a hazardous material when stored or in use. It must be treated as hazardous waste when determined to be unserviceable.

Brake Switch. Most switch inputs to computer logic circuits are grounding devices. They complete the circuit to ground. The computer monitors a circuit with a voltage-dropping resistor. The ABS switch is different from these circuits because it sends a B+ voltage signal to the microprocessor when the brake pedal is depressed. This is a complex switch that contains two sets of contacts: one open circuit and one closed circuit (Figure 10-15).

Figure 10-15 Brake switch.

The normally open contacts in the brake switch close when the brake pedal is depressed. This sends B+ power to the stoplamp circuit, the center high-mounted stoplamp (CHMSL), and the ABS microprocessor. When the normally closed contacts open, they cancel the vehicle cruise control and torque converter lock-up features.

Motion Sensors

A final sensor that is coming of age on the vehicle is the motion sensor or **yaw** sensor. The yaw sensor is a lateral accelerometer that measures sideways motion and lean. Although this sensor is not found on earlier ABSs, it was first used on vehicles with automatic ride control systems. When first used with the ABS, the yaw sensor was used to determine forward, reverse, and sideways motion during acceleration and deceleration. It is now a major component in active suspension, active brakes, and active steering systems. Expect more and more use of the yaw sensor in future vehicles.

Earlier yaw sensors may contain mercury switches and must be treated under EPA guidelines. The newest yaw sensors are beam-type Hall-effect sensors that receive a 5-volt signal from the EBCM and return some portion of that voltage as a signal. Usually the portion returned is 2.5 volts, representing a zero gravity force in a lateral direction. In other words, the vehicle is completely level and moving straight ahead or sitting still. The variation from 2.5 volts is interpreted by the EBCM as the actual motion of the vehicle. The beam-type Hall-effect sensor produces an analog output signal and should not be tested when the vehicle is not still and level.

Pumps and Accumulators

The last major components of an ABS are internal hydraulic pumps and accumulators. Some ABSs have pumps to supply their own brake boost pressure instead of vacuum boosters.

High-Pressure Pumps and Accumulators. ABS boost pressure comes from a high-pressure electric pump. It charges a round ball, called the accumulator, with brake fluid from the reservoir. Anytime the accumulator pressure drops below a minimum point, a pressure switch grounds a power relay coil. This connects the high-pressure pump to the ignition B+ power. The ABS boost pressure operates the high-pressure pump when the key is turned on if the vehicle has not been started for a time and the pressure has bled below the minimum point.

Low-Pressure Return Pumps. An ABS without an accumulator or pressure switch but with an electrohydraulic unit that contains an electric motor is called a low-pressure system. The motor is connected to a low-pressure pump that quickly returns fluid to the master cylinder and the reservoir. On some systems, this pump is used to apply light pressure to an isolated hydraulic circuit of an individual wheel. This feature is used to correct traction problems during acceleration. If the vehicle is equipped with a lateral accelerometer, the ABS is programmed to eliminate fishtail during acceleration. It applies light brake pressure to the spinning drive wheel.

Dynamic Rear Proportioning (DRP) Value. The DRP replaces the standard hydraulic proportioning valve found on most of today's vehicles. It is an electronic operation within the hydraulic control unit and is used on many vehicles, particularly those with active brakes. The DRP uses active control and the existing ABS to regulate brake pressure to the base rear wheels. As with the hydraulic valve, this keeps most of the braking force on the front wheels during routine stops. The red brake warning light will illuminate if there is a hydraulic problem within the base brake system.

Lamps and Communications

The lamps are not essential to ABS operation; however, they are really the only contact between the vehicle's electronic package and the operator. They can also be used by the technician to retrieve DTCs if a scan tool is not available. Communications, on the other hand, is essential to the electronic package because without communications between the various components there would be no operation.

Warning Lamps. Most people understand that the red brake warning lamp is used by the foundation brake system. This lamp indicates problems to the driver. It comes on when brake fluid levels are low, when there is an unbalanced pressure in the hydraulic lines, and when the parking brakes are applied. This lamp also can be used as a brake pad wear indicator lamp. Some ABSs use the red warning lamp to indicate ABS problems when the amber ABS warning lamp or its circuit is nonoperational.

The amber warning lamp is used in several situations: to indicate that the microprocessor is testing the system; to indicate ABS operation; and to alert the driver about a malfunction in the ABS. In addition, the amber warning light may be used when the system is diagnosed for problems. It flashes a trouble code number. Some systems use flash diagnostics that are similar to those that were found on early engine controllers. A short pause separates the first and second digit of the code number. Each complete code is separated by a long pause. The RWAL system displays its failure code in a slightly different manner. The technician must total the number of flashes. This means that sixteen flashes represent failure code 16.

Some systems use a remote lamp driver to control the amber warning lamp. If the microprocessor is functional, it is able to command the remote lamp driver to turn the amber warning lamp off after the **POST.** If the microprocessor or its related remote lamp driver wiring is not functional, it is unable to turn off the amber warning light, so it remains on.

Communications. Some ABSs are strictly stand-alone units; that is, they do not interface with any other vehicle circuits. When these systems malfunction, they must be analyzed by interpreting the flash diagnostics. Or a special brake-out box and tester may be used to repair the system. Federal mandates led to the development of the OBD-II system. The OBD-II system was first used on some 1995 models. OBD-II made more data available to the service technician to aid in troubleshooting, diagnosis, and repair.

Some ABSs share data with other computers on the vehicle through a *local area network (LAN)*. This makes it possible for the technician to communicate with all the computers in the LAN through the use of a plug-in computer. This system allows the technician to use a hand-held scanner or shop computer not only to retrieve codes and records but also to command functions, read data lists, and take data snapshots. Some ABSs can hold as many as six failure codes for up to 100 ignition cycles.

Different carmakers use many variations in computer communication. The vehicle service manual is still the best source for specific vehicle repair information. It is the best source of information for code retrieval, data list interpretation, and testing.

Traction Control System (TCS)

Shop Manual
pages 476–487

The TCS is an outgrowth of four-wheel four-channel ABSs. Most of the same components are used in both systems with the major changes being in the controller and its programming. The TCS also interfaces with the engine (PCM) and transmission (TCM) control modules to request that specific actions be taken.

The TCS is designed to control wheel spin or positive wheel slip. If the vehicle is accelerated hard from a dead stop, the wheels tend to spin on the road surface assuming there is enough power available from the engine. This has led to some broken axles and other damage if the spinning wheel was on dirt and suddenly moved onto dry pavement. Although TCS can reduce wheel spin during hard acceleration, that is not its underlying purpose. It is designed to control wheel spin during snow, ice, wet pavement, and other road conditions.

Going back to the discussion on isolation/dump valves, it was noted that the accumulator in that valve could apply some pressure to the wheel brakes to reduce wheel speed. The TCS has taken that a step further by using a hydraulic modulator modified from the one used strictly for the ABS. If a wheel sensor indicates that a drive wheel's speed is greater than that of the other wheels, the TCS controller will command the hydraulic modulator to apply hydraulic fluid to that wheel. In effect, the brakes are applied. Because of the gear setup in the vehicle's differential, the power is transferred to the other drive wheel(s). When both or all four of the drive wheels approach the same speed, the TCS reduces or eliminates the pressure to the spinning wheel brakes. It can reapply the brakes if necessary or leave them off depending on the wheel speed.

AUTHOR'S NOTE: The TCS is an electronic version of the old, and still used, limited slip differential that was common on many production performance vehicles. This type of differential has some problems because of its mechanical nature, the locking clutches, and, in many cases, the type of lubricant. The TCS eliminates most of these mechanical problems and reacts much faster than its mechanical cousin.

If the wheel spin cannot be stopped by braking, or more than one drive wheel is spinning, the TCS controller will signal the PCM to reduce engine output by retarding spark or changing the air-fuel mixture. In the case of automatic transmissions, the TCS controller will signal the TCM to shift up to a higher gear, reducing the torque being supplied to the drive wheels. In either case, power is reduced to the drive wheels, and it is hoped the wheels can regain traction. It is possible that all three actions—engine reduction, higher gear ratio, and brake application—may be required to control positive wheel slip. This is especially helpful in snow and mud conditions.

The major differences between the ABS and the TCS are the newer, more variable hydraulic modulator and the controller. Some TCSs and ABSs use the same controller with more sophisticated programming (Figure 10-16) or there may be two interfaced controllers, one for the ABS and the other for the TCS.

System Integration

Since the early 2000s there has been a steady increase in the integration of steering, suspension, and braking. What started out as independent systems have evolved into one system that almost does it all regarding vehicle control. All of this is possible because of ongoing electronic research and application. Just like the computer at work and home, the vehicle is incorporating various electronic devices that can and do govern what the vehicle does in given conditions. Many times

Controller

ABS unit

Figure 10-16 The ABS controller on this unit is mated to the hydraulic modulator.

these actions are taken without driver input. Each of the systems mentioned has been installed independently on vehicles since the mid-1990s. The ABS, TCS, and **automatic ride control (ARC)** have been around in some form since the late 1980s, but they have been upgraded each year. Looking at today's electronic systems, these earlier versions were pretty basic.

ARC is an electronically controlled suspension system.

In 2000 models, integration of ABS/TCS and ARC began in earnest. They are used to control the vehicle during cornering and operating on certain road conditions. The ARC uses the yaw and ride height sensors to control the lean of the vehicle and the amount of absorption of road hazards. This is accomplished by magnetic shock absorbers and/or air springs. ABS/TCS on the other hand uses the same yaw sensor to measure body lean and cornering forces (lateral acceleration) to lightly apply the brakes on one or more wheels to slow the vehicle and slightly change the vehicle's body position, thus reducing the tendency of the vehicle to slide or flip over during hard cornering.

AUTHOR'S NOTE: One thing the reader should understand is that present technologies in the three systems under discussion do not truly replace the old mechanical systems. The main reason is lack of redundancy or backup. Remember that the power brake booster will provide boost for two or three stops if the booster fails. With absolute electrical systems, a loss of electrical power can seriously affect vehicle control in all three systems. As a result, the mechanical components remain in place and the electronics become more of a control or assist factor instead of a complete independent system. However, if a backup system can be developed and proved reliable it is only a matter of time before the mechanical components for brakes, steering, and suspension are greatly reduced.

Electronic or electric steering began in 1998 and 1999 with the most noticeable being on Ford F150 pickup trucks. Electrical steering at this point in time is more of an electronically controlled power steering assist. In the Ford truck mentioned, a sensor at the bottom of the steering column measures steering rotation in both degree of turn and speed of steering wheel rotation (Figure 10-17). Using these data and other information such as VSS, the electric steering computer regulates the flow of pressurized power fluid to the steering gear using a solenoid mounted on the power

Figure 10-17 A typical steering shaft (wheel) sensor that measures the amount of the turn and the speed of the turn.

steering pump (Figure 10-18). Other electrical steering assist can be done by attaching an electric motor to the steering shaft or gear that can apply force to the steering mechanism.

The ARC uses air springs that are just very tough air bags in place of mechanical springs (Figure 10-19). The bags are inflated or deflated to support the vehicle and to provide a smoother

Figure 10-18 This actuator is a solenoid used to control power steering fluid flow from the pump to the steering gear.

Figure 10-19 The air spring can be inflated or deflated to control body lean and ride height for better control and a better ride.

ride and better control of the vehicle. Shock absorbers using reversing polarity electronic magnets are being used in some vehicles to control vehicle bouncing instead of the standard hydraulic shocks. ARC systems can control body lean by adjusting the air pressure in individual air springs and raising or lowering the vehicle for better aerodynamics and easier entry and exit from the vehicle. Incorporating all systems, ARC, TCS, and ABS, into one interlaced system is a relatively simple procedure involving mostly electronic components and programming. As electric steering becomes more refined it will probably be merged with this system. The integration of the mentioned systems is quickly being accomplished as a vehicle stability system.

Shop Manual
pages 482–487

Continental Teves Mark 20 and Mark 20E

The Mark 20 is a nonintegral, four-wheel ABS in either three- or four-channel ABS with TCS capability. It was first offered in Europe in 1995 and is now available on many DaimlerChrysler, Ford, Honda, and BMW vehicles. The latest version is the Mark 20E.

Mark 20

The system combines the electric pump, solenoid, and controller into one assembly called an integrated control unit, or ICU (Figure 10-20). The ICU is an aluminum unit housing the valves, a pump, two low-pressure accumulators, an electric motor, and a control module. The valves' solenoids are part of the controller section of the ICU instead of the valve body. This section of the ICU is known as the CAB. The ICU may be located on or under the left front fender, under the master cylinder, on the right side of the crossmember, or behind the steering rack on DaimlerChyrsler vehicles. It is located in similar areas on other vehicles.

A three-channel, front/rear split uses six valves, an inlet and an outlet valve for each front wheel, and an inlet and an outlet valve for the rear wheels. The inlet valves are normally open and the outlet valves are normally closed. A four-channel system with a diagonally split hydraulic

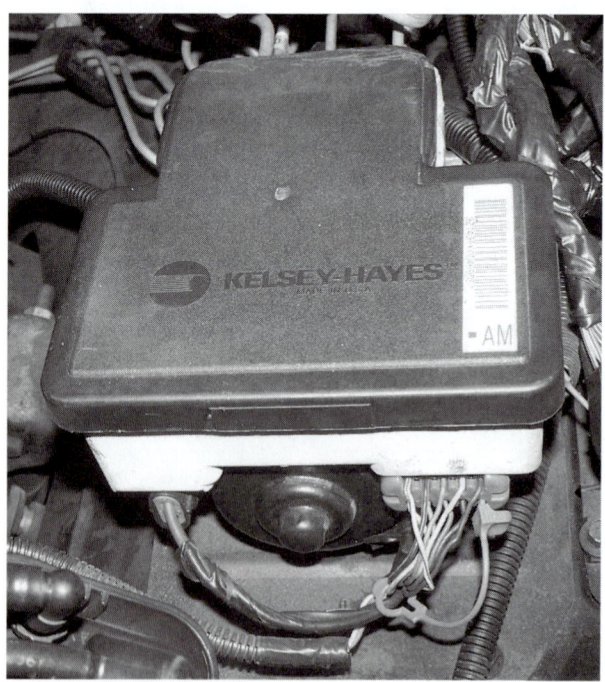

Figure 10-20 Shown is a combined ICU that includes the hydraulic modulator and the controller.

brake uses eight valves (four inlets and four outlets), one per hydraulic circuit (Figure 10-21). If the system is set up for TCS, two additional valves are added and used as isolation valves. The Mark 20 uses a brake pedal-mounted switch. The system will function without a signal from the switch, but the switch more quickly engages the ABS when the pedal is depressed. If equipped with a TCS, a switch on the dash allows the driver to switch off the TCS. Lights on the dash, TRAC ON and TRAC OFF, provide the driver with the status of the system (Figure 10-22). The ABS is not affected if the TCS is disengaged. If the TCS has been actively operating for a specific period, the controller will switch off the TCS to prevent brake heat buildup.

Figure 10-21 When the controller is separated from the modulator, the valve stems are visible. They are pointing up in this figure.

Figure 10-22 The style and location of indicator lamps vary for different vehicles, but the TCS always adds one or two more to the array.

The Mark 20 WSSs are PM sensors (Figure 10-23). The sensors measure wheel speed and send the appropriate signal to the CAB. The CAB compares the speeds of each monitored wheel and determines if one or more are about to lock up and lose braking traction. The Mark 20 has three phases of operation: isolation, decay, and build. If a wheel is slowing faster than the others, the CAB will command the inlet valve solenoid to close that valve but keeps the outlet valve closed (Figure 10-24). This isolation phase blocks additional pressure being applied to that wheel brake. If the wheel speed continues to decelerate faster than the other wheels, the system enters the decay phase. The inlet valve is kept closed while the outlet valve is open. This reduces the pressure by letting fluid in the brake line flow into the accumulator (Figure 10-25). The ABS pump is also switched on to transfer fluid from the accumulator to the braking or high-pressure side of the system. With the pressure reduced, the wheel should speed up, thereby reducing the tendency to lock up.

The build phase is the reapplication of fluid pressure to the wheel. Both valves are de-energized so the inlet valve returns to open position and the outlet to close position (Figure 10-26). This allows fluid that is being pumped from the accumulator to be directed into the brake line and reapplying the brake. The isolation, decay, and build phases are repeated as often as necessary

Figure 10-23 Wheel speed sensor.

Figure 10-24 The closed inlet valve blocks (isolates) the pressure from the master cylinder. The closed outlet valve prevents (holds) pressure from escaping from the brake line.

Inlet Outlet

From master cylinder → → To wheel cylinder

Accumulator and pump

Figure 10-25 In the decay phase, the inlet valve still blocks master cylinder pressure while the open outlet valve allows some brake line fluid to escape to the accumulator.

Inlet Outlet

From master cylinder → → To wheel cylinder

Accumulator and pump

Figure 10-26 With both valves in their normal state, pressure can flow from the master cylinder to the wheel.

to prevent wheel lockup. This very quick repetition of phases can cause a pulsation or severe vibration of the brake pedal or even the whole vehicle. During TCS operation the pump is used to apply hydraulic pressure to the wheel that is losing grip. After the vehicle is under control and the ABS returns to its normal mode, some residual pressure may remain in the affected brake line(s). This excess fluid is returned to the master cylinder in one of three ways. On a front/rear split the pump continues to operate for a few seconds, whereas a diagonally split system keeps the outlet valve open just long enough to drain the accumulator as the vehicle begins to move again. Systems programmed with TCS inlet (isolation) valves are cycled to deplete the excess pressure. Each of these actions will take only a second or two to drop the residual pressure to zero.

AUTHOR'S NOTE: Educating the owner of the vehicle often becomes an important aspect of doing business. Sometimes senior citizens with a new vehicle may complain of a serious braking noise and pedal movement during hard braking or when braking on a gravel road surface. They may not be aware of exactly how an ABS works. After questioning the owner on exactly what is happening and a complete inspection of the brake system reveals no faults, it may be best to explain to the customer exactly how an ABS works. It is also wise to inform them that if they are in a dangerous situation when panic braking is needed, they should not release the brake pedal until the danger is gone or the vehicle is stopped.

Mark 20E

The Mark 20E uses magnetoresistive WSSs, controls brake pressure proportioning on a front/rear split system, and is more efficient as ABS and TCS. The Mark 20E works almost the same as the Mark 20 with the exception of brake proportioning. The typical mechanical/hydraulic proportioning valve is replaced electronically with a process known as **electronically variable brake proportioning (EVBP).** The EVBP can control brake pressure to the rear wheels by cycling the rear outlet valves open and closed. The duration and frequency of the open/closed cycles regulate rear brake pressure. The Mark 20E also has ABS PLUS. This is a vehicle stability program and requires no extra hardware. When the stoplight switch is closed, the CAB monitors wheel speeds and can apply the brake to selective wheel(s) to help the driver control the vehicle. The CAB senses wheel speed, vehicle speed, brake application, and cornering forces to determine which brakes, if any, need to be applied.

Whereas PM WSSs generate an analog signal, the magnetoresistive sensor used with the Mark 20E generates a digital voltage signal that is more easily interpreted by the CAB (Figure 10-27). The CAB uses the speed of the generated signal changes to calculate wheel speed so that the EVBP functions properly and may also be used to detect an underinflated tire.

> **CAUTION:** Many magnetoresistive sensors have a mercury acceleration sensor internal to the sensor. Any sensor with mercury must be considered a hazardous material when stored or in use. It must be treated as hazardous waste when determined to be unserviceable.

Jeeps equipped with the Mark 20E system have an acceleration switch, or G-switch, used to change ABS operation according to traction conditions on the road surface. The acceleration switch has three internal mercury switches listed as G1, G2, and G3 (Figure 10-28). Under normal operations, all three switches are open. G1 and G2 sense forward motion, whereas G3 senses reverse motion. The combination of one or more switches being open or closed provides the CAB with the necessary information to determine what, if any, ABS or TCS action should occur.

Figure 10-27 Note the arrangement of this wheel sensor and the tone ring. The sensor can be replaced individually, but the tone ring is replaced by replacing the drive axle.

Arrow indicates front of
switch for proper mounting

Switch
part number

Mercury
switches

Figure 10-28 The accelaration sensor, or G-switch, used with the Teves Mark 20E ABS detects forward and rearward deceleration.

AUTHOR'S NOTE: The acceleration switch is marked as to installation. Positioning the switch incorrectly will cause some very uncomfortable conditions when driving the vehicle. See the *Shop Manual* for more on this.

Delphi DBC-7 ABS

Shop Manual
pages 476–482

The Delphi DBC-7 is used almost exclusively on General Motors vehicles. It is a nonintegral system that houses the valve body and controller into a single housing. The DBC-7 can be used on three- and four-channel ABS and can be adapted to TCS. Like the Teves 20E it is lighter and less costly than its predecessors. The DBC-7 is a major change from the Delco-VI it replaced.

The most obvious engineering change is the change from a motor pack hydraulic modulator back to the more standard modulator using solenoids (Figure 10-29). The DBC-7 is designed and works very similarly to the Teves Mark 20. The valves and pumps are in the valve block known as the **brake pressure modulator valve (BPMV).** Like the Mark 20, the valve solenoids and controller are mounted to the valve body and are called the EBCM. If the system is fitted with TCS, this control unit is listed as the **electronic brake traction control module (EBTCM).** The ABS relay, which is mounted separately in the Delco-VI, is now mounted within the EBCM or EBTCM. Each of these two units can be replaced as separate assemblies if necessary. In cavity 11 of the EBCM (EBTCM) wiring harness, a small vent tube has been installed to stop buildup of pressure or vacuum between the controlling unit and the valve housing.

In the BPMV are an inlet and an outlet valve for each brake hydraulic channel. Two accumulators are placed in the BPMV, one for each front/rear or diagonal hydraulic circuit. This is a total of eight valves and two accumulators for a four-channel system with six valves and two accumulators for a three-channel. Outlet valves are normally closed and inlets are normally open. The EBCM/EBTCM provides a ground to the solenoids to close the solenoid circuits.

Four-channel systems are used on General Motors passenger cars equipped with the DBC-7 ABS. Each BPMV is connected to a separate brake line and the lines are color coded: L/R is purple, R/R is yellow, L/F is red, and the R/F is green. The three-channel system like the one used on the GM Tracker has three outlets, one to each front wheel and the other to the rear. They are not color coded.

A brake pedal switch is used to signal that the brakes are applied. This signal is not needed for ABS functioning, but the signal will deactivate the TCS. Additional sensors are installed on the Tracker. One is an accelerometer sensor that measures the forward and reverse motions of the vehicle so the EBCM/EBTCM can adjust brake pressure according to how fast the vehicle is decelerating. The sensor is provided with a 5-volt reference signal and returns a portion of the 5 volts to the EBCM/EBTCM. If the voltage is high, then the pressure to the front brake can be reduced. This helps prevent nose diving at the front of the vehicle and provides better vehicle stability and control. The Tracker is also equipped with a switch so that the ABS operation can be changed when in four-wheel drive.

Figure 10-29 A motor pack hydraulic modulator used on the ABS-VI, the predecessor to the DBC-7.

Like all ABSs wheel speed is sensed by speed sensors. PM-type wheel sensors are used and may be integrated into the wheel bearing assembly (hub) or of the typical plug-in type. The integrated sensors provide a more concise signal that is read directly by the EBCM/EBTCM instead of being converted from ac to dc like the ABS VI systems. Integrated WSSs are used on higher-end GM FWD vehicles (Figure 10-30). Neither type of sensor is adjustable and all have a fixed air gap when installed.

The DBC-7 works like most ABSs in that it controls brake fluid pressure to a wheel(s) that is slowing faster than the other wheels. In the pressure hold phase, the inlet valve is closed to prevent further pressure buildup from the master cylinder and to hold the current brake line pressure. The pressure decrease phase opens the outlet valve to allow some fluid to flow from the affected brake line into the accumulator to relieve some of the pressure. This releases the brake and allows the wheel to turn faster. During pressure increase, the inlet valve is opened and the outlet valve closed, allowing master cylinder pressure to the affected brake line. The pump is switched on

Figure 10-30 Shown is a GM hub-and-bearing assembly with an integrated PM wheel speed sensor.

during the three phases to pump fluid from the accumulator back into the brake's hydraulic system. This cycle is repeated as long as the pedal is depressed and the ABS is required.

The DBC-7 uses **dynamic rear proportioning (DRP)** to control rear brake pressure during normal braking. This eliminates the mechanical/hydraulic proportioning valve and is much better at precise pressure control. The proportioning process is done by cycling the inlet and outlet valve(s) to the rear wheel(s) much the same as the ABS actions used to control wheel lockup.

The **tire inflation monitoring system (TIMS)** is accomplished by the EBCM/EBTCM comparing wheel speed during normal vehicle operation. An underinflated tire has a smaller rolling circumference, and if the inflation is 12 pounds or more below specification the speed difference between the wheels is sufficient to cause the EBCM/EBTCM to turn on the low tire pressure light in the dash. All 2007 and later vehicles sold in the United States must have some form of tire pressure monitoring system.

The DBC-7 has two possible traction control systems: **enhanced traction system (ETS)** and TCS. The TCS can control wheel slippage by signaling the PCM to reduce power and then applying brake pressure to the spinning wheel. This is a typical TCS operation using a program in the EBCM. The ETS can signal the PCM to reduce engine power output only by retarding spark timing, shutting down selected cylinders, leaning air-fuel mixture, or upshifting the transmission. One or any combination of these four actions may be commanded by the PCM. Of note is the fact that no braking action is taken in the ETS to control wheel spin.

If the EBCM senses wheel slip when the brakes are not applied, its first step is to request the PCM to reduce engine torque. The PCM can alter spark timing and open selected fuel injector circuits. The reduction in engine torque is signaled to the EBCM by the PCM. This signal is called the delivered torque signal.

If the wheel(s) continue to slip, the EBCM can apply the brakes to the affected wheel(s). This is done by closing the ABS inlet valve. The prime valve is opened and the pump is switched on to charge the accumulator. The EBCM then opens and closes the inlet and outlet valves to accomplish three phases: pressure hold, pressure increase, and pressure decrease. The different phases are cycled until the wheel slippage stops.

The TCS has two dash-mounted lights and an audible warning (Figure 10-31). One light is the TRAC OFF. When this lamp is lit, it means the TCS programming has been deactivated because of a sensed malfunction. At the same time the EBCM requests the BCM to sound an alarm by using the radio. The TRACTION ACTIVE light is lit when the EBCM module senses wheel slippage and begins to apply the brakes in an attempt to regain traction.

Figure 10-31 The traction control lights shown are typical of General Motors vehicles and are very similar in other vehicle brands.

Another major change between the DBC-7 and previous GM ABSs is the use of **Class 2 wiring.** Class 2 wiring is a multiplexing network. Each wired-in module can share information by different pulse signals used to identify each module on the net. Class 2 is faster than the older electronic communication and operates at a higher voltage. The communication speed is 10,400 bits per second (bps) versus the old 8,192 bps. The DBC-7 operates on 7 volts instead of 5 volts. Although these changes do not seem significant on the surface, when the speed of electrons is approximately 186,000 miles per second it is a huge change in the amount of information that can be transferred. The system does require a scan tool capable of reading Class 2 data.

Active Brake Systems

Active brake systems are a thing of the future as well as the present. The systems currently employed on vehicles are mostly first- and second-generation electronic systems. There are still a few bugs to correct, but not many. Most systems are known as vehicle stability control, or electronic stability program, or some version of those terms. The so-called active brake system was first installed on popular midrange passenger cars in 1999. Since then it has been installed on more models and refined to a good system with few failures. By 2008 all of the bugs should be found, corrected, and eliminated with more reliable components and programming. The system started out as a stand-alone system but in 2001 to 2006, it was merged or integrated with ARC. The basic concept behind active braking is to allow electronics to take over when the driver does something imprudent (driving too fast for a curve, for example) or the road conditions cause an unexpected loss of vehicle control. Active braking cannot replace the driver operational control, but it may help her/him out of some dangerous situations.

The basic active brake system has a pedal position sensor that is usually mounted to the vacuum brake booster. This switch measures the amount and speed of pedal travel. The signal is sent to a modified EBCM along with the yaw sensor signal from the ARC system. The two signals may indicate that the vehicle is leaning too much in a curve and the brakes to the inside wheels can be applied to slow the vehicle and act like a fulcrum to pull the vehicle around the arc. This is done without driver action. The more advanced active brake systems incorporate the signals from the WSS, VSS, yaw, steering, and other sensors that impact the directional control and stability of the vehicle.

The problem of control and stability may be the result of understeer or oversteer. Understeer is not enough steering motion compared to the arc of the curve. In other words, the vehicle heads for the ditch opposite to the turning motion. Oversteer is the opposite: too much steering for the arc. In understeer conditions the brakes are applied to the inside rear wheel. This creates a drag that pulls the vehicle toward the arc of the curve. In oversteer, the front outside brake is applied, pulling the vehicle out of its steering arc and reducing the chance of the rear of the vehicle fishtailing. In other cases indicating possible loss of vehicle control, either front or rear brakes may be applied to assist in vehicle control.

AUTHOR'S NOTE: If you want to experience an "active brake system" in action, take a disc/drum-equipped vehicle, adjust one of the rear-wheel brakes tightly with a lot of drag, find an empty smooth parking lot, and drive the vehicle at a slow speed. Remove your hands from the steering wheel and observe the actions of the vehicle. It will begin to drift or pull toward the side of the tightened brake. Apply the brake slightly and the drift/pull will increase noticeably. This is active braking, but in the electronic system the action is done more quickly with all brakes properly adjusted and functioning and no action from the driver. Many times the driver may never realize that the active brake system is at work. When integrated into the ARC, the ARC may inflate the air bag or change the magnetic shock absorber to raise the corner or side of the vehicle to shift its center of gravity toward the arc of the curve in conjunction with the operation of the braking system. Another action that may be signaled to the PCM is a reduction in engine power, slowing the vehicle.

AUTHOR'S NOTE: Some 2007 passenger vehicles offer an active brake system coupled with a front and rear radar-type system. If the front radar senses an object in the road ahead, it will illuminate a BRAKE light on the dash to alert the driver to apply the brakes. In some situations, the active brake will be applied electronically without driver brake input. This is extremely helpful in foggy or rainy conditions. The reverse radar can perform the same function, which is even more helpful when reversing when there are children near the vehicle.

It should be remembered, however, that no electronic system now available on vehicles can correct driver malfunction. Driving too fast for road conditions or exceeding the limits of the vehicle may still result in injury, damage, or death. As the old saying goes, if you want to play the game you have to pay the price. Sometimes the price is steep and electronics cannot stop it.

Electrical Braking Systems and Future Brake Technology

An electrical braking system is a system combining almost everything discussed in this text. The Mercedes brake-by-wire system and its problems were briefly mentioned earlier. A true electrical braking system will be the brake-by-wire, eliminating almost if not all of the mechanical and hydraulic systems. Braking will be done by electronically controlled hydraulic pressure, directly applying the wheels' mechanical braking components with motors or solenoids, or through the use of magnetic force. All braking will be based on signals indicating the driver's desire to slow or stop. Various sensors will provide vehicle and road conditions and various electronic commands will be issued to control hydraulic pressure through either a hydraulic modulator, motor/solenoids, or magnets within the wheel hubs. Magnetic brakes are not here yet for vehicles, but they are used on some amusement park rides and bullet trains. If the brake-by-wire system is finally realized, braking by magnetic force will be just around the corner. However, control of braking hydraulic pressure is now being done in all ABSs and active braking systems. If the redundancy problem can be resolved to the satisfaction of safety agencies, electrical braking or brake-by-wire will be viable and eventually will be installed on all vehicles. The author suspects, based on some research, that integrated braking, suspension, and steering will be common on most vehicles by about 2015.

There are two major problems involved with true electrical braking systems. One is redundancy or backup in the event of complete electric failure. That problem is solved or at least partially solved by the use of small master cylinders and standard hydraulic/mechanical brake components that will provide braking force even though the driver's efforts must be increased. This will be similar to the loss of a standard brake booster. In 2006 this seemed to be the way manufacturers were going, pending a better system. The other problem with electrical braking is present on today's vehicles equipped with many electronic systems: available electric power. There is a limit on the size of the ac generator and battery that is feasible to install on a vehicle. The cost and weight of large generators and battery packs make them unsuitable for a passenger car or light truck especially when the price of fuel, cost of repairs, and initial cost of the vehicle are considered. There are several options to help solve this power supply problem. The first and currently most available is the hybrid vehicle with its large drive battery and high-output ac generator. The vehicle initial cost is high but not unreasonable because this is a new technology; the vehicle price will come down as more vehicles are produced. The drive battery can deliver the power needed for all vehicle electronic systems now being offered and more. The energy needed to drive the ac generator can be reduced by using regenerative braking.

Regenerative braking is used when the vehicle has some means of a vehicle electric drive motor. During normal operation electric power is used to power the motor(s) that move the vehicle. When the brake switch is closed (brakes applied), the motor controller in effect reverses the magnetic field within the motor and it becomes a generator. This provides an electric current to help recharge the battery. All hybrid vehicles have an electric drive motor in some form so all can be designed to a low-output generator if certain conditions are met. Regenerative braking also has the effect of slowing the vehicle because the vehicle's moving energy is being used to drive the motor/generator. This prolongs brake life while reducing fuel consumption. Another advantage of regenerative braking is achieved when a vehicle moving down a hill has a higher speed than the drive motor is producing. The motor controller senses this and changes the motor to a generator, thereby slowing the vehicle speed and charging the battery. Consult Delmar Cengage Learning's publication *Hybrid, Electric, and Fuel-cell Vehicles* for complete details.

AUTHOR'S NOTE: In the late 1980s and early 1990s, some vehicle manufacturers tried regenerative braking on vehicles with the standard ac generator. When the brakes were applied, the ac generator was switched (regulated) to its full output capability. This slowed the engine, hence slowed the vehicle, and charged the 12-volt battery. However, this was an expensive generator with an expensive program and was soon dropped. To the best of my knowledge, this system had no more failures than the standard charging systems. At least there were no failures reported that could be directly related to regenerative braking.

Summary

❑ A sensor may measure electrical energy directly or convert a mechanical action into electrical energy.

❑ Signals may be analog or digital; digitals are the most easily understood by controllers.

❑ Frequency is the number of times a cycle occurs within a given time frame.

❑ Duty cycle indicates the amount of time on versus time off.

❑ Actuators are normally used to change electrical energy into mechanical action.

❑ A controller is the computer module for a particular device, system, or group of systems.

❑ Multiplexing is the term applied to the communications on the vehicle computer system.

❑ Control area network (CAN) is the newest form of multiplexing for vehicles.

❑ The hydraulic modulator is the actuator that directly controls brake fluid pressure during electronically controlled braking.

❑ Commands are electrical signals generated by the controller and sent to an actuator to perform an action.

❑ An ABS offers drivers a braking system that can help control and stabilize their vehicles.

❑ Integrated and nonintegrated ABSs usually indicate the placement of the hydraulic modulator in relation to the master cylinder and power booster.

❑ RWAL/RABS provides a means to control wheel lockup on the rear wheels only.

❑ Four-wheel ABSs come in two configurations: three-channel and four-channel.

❑ Controllers may be known as electronic brake control module (EBCM), controller antilock brake (CAB), or similar designations desired by the ABS manufacturers.

- ❏ An electrohydraulic unit is the combined assembly of a controller and hydraulic modulator.
- ❏ Permanent magnet (PM) sensors generate an ac voltage based on the speed of the wheel and are the most common wheel speed sensors (WSS).
- ❏ Magnetoresistive sensors have a low and a high voltage and current signal regardless of wheel speed.
- ❏ Wheel speed is calculated based on the frequency of change between high- and low-voltage signals generated by the magnetoresistive sensor.
- ❏ A yaw sensor is a motion sensor that can sense and signal a change in the movement of the vehicle.
- ❏ A beam-type Hall-effect yaw sensor has a return signal of 2.5 volts, which represents a level vehicle.
- ❏ ABSs may be equipped with high-pressure pumps and accumulators.
- ❏ The mechanical proportioning valve has been replaced on many vehicles with electrically controlled proportioning programming known as dynamic rear proportioning (DRP).
- ❏ Common TCS/ABS components are WSS, ICU, and the mechanical/hydraulic service brake assemblies.
- ❏ ARC, TCS, and ABS have been integrated into many vehicles as a stability system.
- ❏ The Teves Mark 20 and 20E are three-channel and four-channel ABSs with the adaptability to add TCS.
- ❏ The Mark 20/20E ABSs have two valves per hydraulic circuit plus two additional valves if TCS equipped.
- ❏ The Mark 20 ABS uses PM wheel speed sensors, whereas the 20E uses magnetoresistive sensors.
- ❏ The phases of operation with the Mark 20/20E are isolate, decay, and build.
- ❏ The Delphi DBC-7 ABS works similarly to the Teves Mark 20E.
- ❏ The two main components of the DBC-7 are the brake pressure modulator valve (BPMV) and the EBCM.
- ❏ The DBC-7 with TCS has two additional valves and an electronic brake traction control module (EBTCM).
- ❏ The DBC-7 uses integrated PM sensors for wheel speed.
- ❏ The DBC-7 ABS provides programming for dynamic rear proportioning, tire inflation monitoring system, and two TCS programs named enhanced traction system (ETS) and traction control system.
- ❏ Class 2 wiring is used within the DBC-7 system for shared communications between networked controllers.
- ❏ Active braking systems may apply the brakes to one or more wheels for vehicle directional control without driver input.
- ❏ Electrical braking systems will use computer commands to either apply and/or control brake fluid pressures or will directly apply the wheels' mechanical brake components.

Terms to Know (continued)

Electronically variable brake proportioning (EVBP)

Enhanced traction system (ETS)

Eyeball

Flux lines

Isolation/dump valve

Negative wheel slip

Permanent magnet (PM) generator

Regenerative braking

Reluctor

RWAL/RABS

Tire inflation monitoring system (TIMS)

Yaw

Review Questions

Short-Answer Essays

1. Describe the general operation of a PM wheel speed sensor.

2. Describe the general operation of a magnetoresistive wheel speed sensor.

3. Explain how a RWAL/RABS generally controls wheel lockup.

4. Describe the construction differences between the integrated ABS and the nonintegrated ABS.

5. Describe the brake valves in an ICU without TCS.

6. Explain the major difference between the Delphi DBS-7 and its predecessor, the ABS-VI.

7. Explain the difference between positive wheel slip and negative wheel slip.

8. Define directional control and directional stability.

9. Define multiplexing.

10. Discuss generally active braking systems.

Fill in the Blanks

1. The basic active brake system has a _____ sensor that is usually mounted to the vacuum brake booster.

2. The DBC-7 operates on _____ volts instead of _____ volts.

3. The DBC-7 has two possible traction control systems: _____ and traction control system (TCS).

4. An underinflated tire has a _____ rolling circumference than one that is properly inflated.

5. _____ is used to control rear brake pressure during normal braking.

6. The Mark 20E system has an _____ or _____ used to change ABS operation according to traction conditions on the road surface.

7. On a Teves Mark 20 ABS the _____ is the reapplication of fluid pressure to the wheel.

8. On a Mark 20 ABS the valves' _____ are part of the controller section of the ICU instead of the valve body.

9. With a Ford electric steering system a sensor at the bottom of the steering column measures _____ in both degree of turn and the _____ of steering wheel rotation.

10. The traction control system is designed to control _____ or _____.

Multiple Choice

1. *Technician A* says that an ac volt signal continually goes from low to high to low values.
   Technician B says that a dc volt signal may be yes/on, high/low, or on/off.
   Who is correct?
   - **A.** A only
   - **B.** B only
   - **C.** Both A and B
   - **D.** Neither A nor B

2. Voltage signals are being discussed:
   Technician A says that digital cycles can be used to determine the duty cycle of a device.
   Technician B says that an analog signal is more easily interpreted by a controller.
   Who is correct?
   - **A.** A only
   - **B.** B only
   - **C.** Both A and B
   - **D.** Neither A nor B

3. *Technician A* says that an actuator converts mechanical action to electric energy.
   Technician B says that a solenoid uses electric current to perform a mechanical action.
   Who is correct?
   - **A.** A only
   - **B.** B only
   - **C.** Both A and B
   - **D.** Neither A nor B

4. Actuators and sensors are being discussed:
   Technician A says that a too low current may mean that an actuator or sensor is not functioning properly.
   Technician B says that the controller can monitor actuator and sensor operation electronically.
   Who is correct?
   - **A.** A only
   - **B.** B only
   - **C.** Both A and B
   - **D.** Neither A nor B

5. Command signals are being discussed:
   Technician A says that all commands are performed by closing the circuit on the positive side.
   Technician B says that most of the command signals consist of ac current applied at a specific rate of repetition.
   Who is correct?
   - **A.** A only
   - **B.** B only
   - **C.** Both A and B
   - **D.** Neither A nor B

6. RWAL is a brake term generally used by
   - **A.** most vehicle manufacturers.
   - **B.** Ford Motor Company.
   - **C.** Asian imports.
   - **D.** European imports.

7. RWAL is being discussed:
   Technician A says that some RWD vehicles have the WSS mounted to the transmission and that it measures output shaft speed.
   Technician B says that this system can control braking on individual rear wheels.
   Who is correct?
   - **A.** A only
   - **B.** B only
   - **C.** Both A and B
   - **D.** Neither A nor B

8. RABS is being discussed:
   Technician A says that this system uses an isolation/dump valve assembly that works completely different from the one in all RWAL.
   Technician B says that the isolation/dump valve operation is dependent on signals generated by the VSS mounted on the RWD transmission.
   Who is correct?
   - **A.** A only
   - **B.** B only
   - **C.** Both A and B
   - **D.** Neither A nor B

9. Active brake systems are being discussed:
   Technician A says that this system is used for directional control.
   Technician B says that the system is used for vehicle stability.
   Who is correct?
   - **A.** A only
   - **B.** B only
   - **C.** Both A and B
   - **D.** Neither A nor B

10. *Technician A* says that the speed of a wheel is calculated by the on/off signal generated by a PM sensor.
    Technician B says that at times all wheels may be rotating at different speeds.
    Who is correct?
    - **A.** A only
    - **B.** B only
    - **C.** Both A and B
    - **D.** Neither A nor B

Glossary
Glosario

Note: Terms are highlighted in color, followed by Spanish translation in bold.

ABS event   A rapid reduction in speed where one or more wheels begin to lock up.

evento ABS   Reducción rápida de velocidad en la que una o más ruedas comienzan a bloquearse.

acceleration sensor   A sensor that provides information about the rate of forward or reverse acceleration or deceleration; also known as a G-switch.

sensor de aceleración   Sensor que proporciona información sobre el nivel de aceleración o desaceleración hacia delante o hacia atrás; también denominado conmutador G.

acceleration slip regulation (ASR)   The name of the traction control system manufactured by Bosch.

regulación de deslizamiento en aceleración (ASR)   Nombre del sistema de control de tracción fabricado por Bosch.

accumulator   A container that stores hydraulic fluid under pressure. It can be used as a fluid shock absorber or as an alternate pressure source. A spring or compressed gas behind a sealed diaphragm provides the accumulator pressure. In an ABS, the accumulator is a gas-filled chamber that acts as both a storage container for system fluid and a reserve pressure chamber to provide smooth antilock operation and dampen pressure pulses from the system pump.

acumulador   Recipiente que contiene líquido hidráulico bajo presión. Se puede utilizar como líquido amortiguador o como fuente de presión alternativa. Un resorte o gas comprimido detrás de un diafragma sellado proporciona presión al acumulador. En un sistema ABS, el acumulador es una cámara llena de gas que actúa a la vez como recipiente de almacenamiento de líquido para el sistema y como cámara de reserva de presión para proporcionar una operación suave de antibloqueo y amortiguar los pulsos de presión que provienen de la bomba del sistema.

actuator   Any device that receives an output signal or command from a computer and does something in response to the signal.

impulsor   Cualquier dispositivo que reciba un comando o una señal de salida de un computador y haga algo como respuesta a dicha señal.

adjustable pedal system (APS)   Mechanical devices capable of moving the brake, accelerator, and clutch pedal forward (up) and backward (down) to accommodate different drivers. Usually computer controlled based on the driver's input.

sistema de pedal adaptable (APS)   Dispositivo mecánico capaz de mover los pedales del freno, el acelerador y el clutch o embrague hacia adelante (arriba) y hacia atrás (abajo) para acomodar a los diferentes conductores. Usualmente está controlado por una computadora en el mando de entrada del conductor.

ampere (A)   The unit for measuring electric current. One ampere equals a current flow of 6.28×1018 electrons per second.

amperio (A)   Unidad de medida de la corriente eléctrica. Un amperio equivale a un flujo de corriente de $6,28 \times 1018$ electrones por segundo.

analog   A signal that varies proportionally with the information that it measures. In a computer, an analog signal is voltage that fluctuates over a range from high to low.

analógica   Señal que varía proporcionalmente con la información que mide. En un computador, una señal analógica es la tensión que fluctúa en un margen de alto a bajo.

antilock brake control module (ABCM)   The computer that controls the ABS operation. May be used on some systems for traction control.

módulo de control antibloqueo de frenos (AC)   Computador que controla el funcionamiento de los frenos ABS. En algunos sistemas se puede utilizar para controlar la tracción.

antilock brake system (ABS)   A service brake system that modulates hydraulic pressure to one or more wheels as needed to keep those wheels from locking during braking. An antilock brake system can improve vehicle control during hard braking and eliminate or reduce the tendency for the vehicle to skid.

sistema antibloqueo de frenos (ABS)   Sistema de frenos de servicio que modula la presión hidráulica a una o más ruedas cuando es necesario para evitar que éstas se bloqueen al frenar. Un sistema antibloqueo de frenos puede mejorar el control del vehículo durante un frenado brusco y eliminar o reducir la tendencia del vehículo a patinar.

aramid fibers   A family of synthetic materials that are stronger than steel but weigh little more than half what an equal volume of fiberglass weighs.

fibras de aramida   Familia de materiales sintéticos más fuertes que el acero pero que pesan poco más de la mitad que un mismo volumen de fibra de vidrio.

arcing   The process of grinding or forming drum brake linings to conform to the drum diameter and provide clearance where needed.

formación de arco   Proceso de limar o formar los forros de frenos de tambor para que se adapten al diámetro del tambor y den tolerancia donde sea necesario.

asbestos   The generic name for a silicate compound that is very resistant to heat and corrosion. Its excellent heat dissipation abilities and coefficient of friction make it ideal for automotive friction materials such as clutch and brake linings. Asbestos fibers are a serious health hazard if inhaled.

amianto   Nombre genérico de un compuesto de silicatos muy resistente al calor y a la corrosión. Su excelente capacidad de disipación de calor y su coeficiente de fricción lo hacen ideal para materiales de fricción para automoción, como el embrague y los forros de frenos. Las fibras de amianto constituyen un serio peligro para la salud si se inhalan.

asbestosis   A progressive and disabling lung disease caused by inhaling asbestos fibers over a long period of time

asbestosis   Enfermedad pulmonar progresiva que provoca la incapacidad de la persona a causa de la inhalación de fibras de amianto.

aspect ratio   The ratio of the cross-sectional height to the cross-sectional width of a tire expressed as a percentage.

relación entre dimensiones   Relación entre la altura y la anchura de la sección transversal de un neumático, expresada en un porcentaje.

atmospheric pressure   The weight of the air that makes up the Earth's atmosphere.

presión atmosférica   Peso del aire que constituye la atmósfera de la Tierra.

atmospheric suspended A term that describes a power brake vacuum booster in which atmospheric pressure is present on both sides of the diaphragm when the brakes are released. An obsolete type of vacuum booster.

de suspensión atmosférica Término que describe un reforzador de vacío para frenos de potencia en el cual hay presión atmosférica a ambos lados del diafragma cuando se sueltan los frenos. Tipo obsoleto de reforzador de vacío.

automatic ride control (ARC) An electronically controlled suspension system designed to improve vehicle ride and increase vehicle stability.

control automático del movimiento (ARC) Sistema de suspensión controlado electrónicamente diseñado para mejorar el manejo y la estabilidad del vehículo.

Automotive Friction Material Edge Code A series of codes on the side of a brake lining (disc or drum) that identifies the manufacturer, the lining material, and coefficient of friction. These codes are for lining identification and comparison; they do not indicate quality.

Código de Borde para Material de Fricción en Automóviles Serie de códigos situados en el lado de un forro de frenos (de disco o de tambor) que identifica el fabricante, el material del forro y el coeficiente de fricción. Estos códigos identifican y comparan los forros; no indican calidad.

backing plate The mounting surface for all other parts of a drum brake assembly except the drum.

placa de refuerzo Superficie en que se montan todas las partes de un freno de tambor, excepto el tambor.

ball and ramp A common kind of caliper-actuated parking brake.

bola y rampa Tipo común de freno de estacionamiento accionado por calibre.

banjo fitting A round, banjo-shaped tubing connector with a hollow bolt through its center that enables a brake line to be connected to a hydraulic component at a right angle.

ajuste de banjo Conector redondo, en forma de banjo, atravesado en el centro por un perno hueco, que permite conectar en ángulo recto una línea de frenos a un componente hidráulico.

bearing cage The steel component that holds the rollers together in a tapered roller bearing.

jaula de cojinetes Componente de acero que mantiene juntas las bolillas en un cojinete de bolillas cónicas.

bearing cone The inner race of a tapered roller bearing; usually an integral assembly with the rollers and the cage.

cono del cojinete Canaleta interior de un cojinete de bolillas cónicas; por lo general, conjunto integrado de bolillas y jaula.

bearing cup The outer race of a tapered roller bearing; usually pressed into the wheel hub.

taza del cojinete Canaleta exterior de un cojinete de bolillas cónicas; por lo general está comprimida en el buje de la rueda.

belted bias ply tire Tire construction that incorporates the belts used in radial ply tires with the older bias ply construction.

neumático encintado de capas contrapuestas Construcción de neumáticos que incorpora las cintas usadas en los neumáticos de capas radiales a la construcción más antigua de capas contrapuestas.

bias ply tire Tire construction in which the cords in the body plies of the carcass run from bead to bead at an angle from 26 to 38 degrees instead of 90 degrees as in a radial ply tire.

neumático de capas contrapuestas Construcción de neumáticos en la cual los cables de las capas del cuerpo de la carcasa van de una a otra nervadura en un ángulo de 26 a 38 grados, en vez de uno de 90 grados como en los neumáticos radiales.

bimetallic drum A composite brake drum made of cast iron and aluminum.

tambor bimetálico Tambor de frenos hecho de hierro fundido y aluminio.

binary system The mathematical system that uses only the digits 0 and 1 to present information

sistema binario Sistema matemático que utiliza sólo los dígitos 0 y 1 para presentar información.

binder Adhesive or glue used in brake linings to bond all the other materials together.

aglomerante Adhesivo o pegamento que se utiliza en los forros de frenos para unir todos los demás materiales.

bit A binary digit (0 or 1). Bit combinations are used to represent letters and numbers in digital computers. Eight bits equal one byte.

bit Dígito binario (0 o 1). Las combinaciones de bits se emplean para representar letras y números en los computadores digitales. Ocho bits equivalen a un byte.

bleeding A service procedure that removes air from the hydraulic system.

extracción de vapor Procedimiento de mantenimiento que extrae aire del sistema hidráulico.

bonded lining Brake lining attached to the pad or shoe by high-strength, high-temperature adhesive.

forro adherido Forro de freno unido a la pastilla o zapata por un adhesivo muy fuerte, de alta temperatura.

brake assisted (BA) A sensor mounted on the vacuum brake booster to detect brake pedal motion and speed of motion for faster brake application.

freno, asistido (BA) Sensor instalado en el sistema booster del freno para detectar el movimiento del pedal y la velocidad del movimiento para una más rápida aplicación del freno.

brake caliper The part of a disc brake system that converts hydraulic pressure back to mechanical force that applies the pads to the rotor. The caliper is mounted on the suspension or axle housing and contains a hydraulic piston and the brake pads.

calibre del freno Parte de un sistema de frenos de disco que vuelve a convertir la presión hidráulica en fuerza mecánica que aplica las pastillas al rotor. El calibre va montado en el alojamiento del eje o la suspensión, y contiene un pistón hidráulico y las pastillas de freno.

brake fade The partial or total loss of braking power caused by excessive heat, which reduces friction between the brake linings and the rotors or drums.

pérdida de freno Pérdida parcial o total de la potencia de freno debido a un calor excesivo, el cual reduce la fricción entre los forros de los frenos y los rotores o tambores.

brake pad The part of a disc brake assembly that holds the lining friction material that is forced against the rotor to create friction to stop the vehicle.

pastilla de freno Parte de un conjunto de frenos de disco que aloja el material de fricción del forro que se fuerza contra el rotor para crear la fricción que detendrá el vehículo.

brake pressure modulator valve (BPMV)   A valve assembly used in the Teves Mark 20 ABS to control braking of the wheels. Commonly referred to as the hydraulic modulator.

válvula moduladora de la presión de los frenos (BPMV)   Válvula que se usa en el sistema de ABS del Teves Mark 20 para controlar el frenado de las ruedas. Comúnmente se le llama modulador hidráulico.

brake shoes   The curved metal parts of a drum brake assembly that carry the friction material lining.

zapatas de freno   Partes metálicas curvas de un conjunto de frenos de tambor que llevan el forro de material de fricción.

buffer   An isolating circuit used to avoid interference between a driven circuit and its driver circuit. Also a storage device, or circuit, that compensates for a difference in the rate of data transmission. A buffer can absorb data transmitted faster than a receiving circuit or device can respond.

separador de interferencias   Circuito aislante que se emplea para evitar posibles interferencias entre un circuito controlado y su circuito controlador. También un dispositivo, o circuito, de almacenamiento que compensa una diferencia en la velocidad de transmisión de datos. Un separador de interferencias puede absorber datos transmitidos más rápidamente de lo que puede responder un circuito o dispositivo receptor.

caliper   The major component of a disc brake system. It houses the piston(s) and supports the brake pads.

calibre   Componente principal de un sistema de frenos de disco. Contiene el pistón (o pistones) y soporta las pastillas de freno.

caliper support   The bracket or anchor that holds the brake caliper.

soporte del calibre   Mordaza o ancla que aloja el calibre de freno.

camber   The inward or outward tilt of the wheel measured from top to bottom and viewed from the front of the car.

inclinación   Inclinación de la rueda hacia dentro o hacia fuera, medida de arriba abajo y vista desde la parte frontal del vehículo.

cam-ground lining   A brake shoe lining that has been arced or formed so that it is thinner at the ends than at the center, and the lining surface is not a portion of a circle with a constant radius.

forro elíptico   Forro de zapata de freno que se ha arqueado o formado de tal manera que es más delgado en los extremos que en el centro, y cuya superficie de recubrimiento no forma parte de un círculo de radio constante.

carcass   The steel beads around the rim and layers of cords or plies that are bonded together to give a tire its shape and strength.

carcasa   Nervaduras de acero alrededor de la llanta y las capas de cordones, o capas que están unidas para dar forma y resistencia a un neumático.

casing   Layers of sidewall and undertread rubber added to a tire carcass.

cubierta   Capas de caucho añadidas a la carcasa del neumático.

caster   The backward or forward angle of the steering axis viewed from the side of the car.

inclinación del eje   Inclinación hacia delante o hacia atrás del eje de dirección visto desde el costado del auto.

central processing unit (CPU)   The calculating part of any computer that makes logical decisions by comparing conditioned input with data in memory.

unidad central de proceso (UCP)   Parte de cálculo de un computador que llega a decisiones lógicas comparando los datos de entrada condicionados con los datos que tiene en la memoria.

ceramic pads   Pads consisting of a combination of ceramic material and copper or some other metal fibers.

cojines de freno de cerámica   Cojines que consisten en una combinación de materiales y del cobre de cerámica, o algunas otras fibras del metal.

channel   Individual legs of the hydraulic system that relay pressure from the master cylinder to the wheel cylinder.

canal   Terminales del sistema hidráulico que transmiten presión desde el cilindro maestro al cilindro de la rueda.

check valve   Valve that allows fluid or air to flow in one direction but not in the opposite direction.

válvula de control   Válvula que permite el paso de líquidos o aire en un sólo sentido contrario.

chlorinated hydrocarbon solvents   A class of chemical compounds that contain various combinations of hydrogen, carbon, and chlorine atoms; best known as a class of cleaning solvents.

solventes clorohidrocarbonados   Clase de compuestos químicos que contienen diferentes combinaciones de átomos de hidrógeno, carbono y cloro; más conocidos como un tipo de solventes limpiadores.

Class 2 wiring   A multiplexing network in which each wired-in module can share information through different pulse signals that are used to identify each module on the net.

cableado Clase 2   Red múltiple en la cual cada módulo conectado pueden compartir información mediante diferentes señales de pulsación que se usan para identificar cada módulo en la red.

coefficient of friction   A numerical value that expresses the amount of friction between two objects, obtained by dividing tensile force (motion) by weight force. A coefficient of friction can be either static or kinetic.

coeficiente de fricción   Valor numérico que expresa la cantidad de fricción que hay entre dos objetos, obtenido mediante la división de la fuerza de tracción (movimiento) por la del peso. Los coeficientes de fricción pueden ser estáticos o dinámicos.

cold inflation pressure   The tire inflation pressure after a tire has been standing for 3 hours or driven less than 1 mile after standing for 3 hours.

presión de inflado en frío   Presión de inflado del neumático después de haber estado en reposo durante tres horas o haber recorrido menos de una milla después del mismo tiempo de reposo.

combination valve   A hydraulic control valve with two or three valve functions in one valve body.

válvula de combinación   Válvula de control hidráulico con dos o tres funciones de paso en el mismo cuerpo de válvula.

command   An electrical signal or output signal from a computer (controller) to an actuator.

comando   Señal eléctrica o señal de salida de una computadora (servo-regulador) a un servo-motor.

composite drum   A drum made of different materials, usually cast iron and steel or aluminum, to reduce weight. The friction surfaces and the hubs are cast iron, but supporting parts are made of lighter metal.

tambor compuesto   Tambor hecho de diferentes materiales, normalmente de hierro fundido y acero o aluminio, para reducir su peso. Las superficies de fricción y los bujes son de hierro fundido, pero las partes de soporte se hacen de metal más ligero.

composite rotor   A rotor made of different materials, usually cast iron and steel, to reduce weight. The friction surfaces and the hubs are cast iron, but supporting parts are made of lighter steel stampings.

rotor compuesto   Rotor construido con diferentes materiales, normalmente hierro fundido y acero para reducir su peso. Las superficies de fricción y los bujes son de hierro fundido, pero las partes de soporte son estampaciones de acero más ligero.

conduit A flexible metal housing or jacket that houses the parking brake cables to protect them from dirt, rust, abrasion, and other damage.

conducto Alojamiento de metal flexible o forro que recubre los cables del freno de estacionamiento para protegerlos de la suciedad, el polvo, la abrasión y otros daños.

control valve assembly The block of metal that contains the hydraulic passages and electric solenoids used to direct brake fluid during an ABS event.

conjunto de válvulas de control Bloque metálico que contiene los pasos hidráulicos y los solenoides eléctricos que se usan para dirigir el líquido de frenos en un evento ABS.

controller A computer programmed to perform certain decisions based on sensor signals and issue electrical commands to actuators.

servo-regulador Computadora programada para tomar ciertas decisiones de acuerdo a señales de un sensor y envíar ordenes eléctricas a los servo-motores.

controller, antilock brake (CAB) The computer that controls the ABS operation.

controlador, antibloqueo de freno (CAB) Computador que controla el funcionamiento de los frenos ABS.

cup expander A metal disc that bears against the inner sides of wheel cylinder seals to hold the seal lips against the cylinder bore when the brakes are released. This keeps air from entering the cylinder past the retracting pistons and seals.

cubeta de expansión Disco metálico que se ajusta a los costados internos de las juntas del cilindro de las ruedas para sujetar los bordes de la junta contra el hueco del cilindro al aplicar los frenos. Así se impide que el aire que entra en el cilindro pase más allá de los pistones retráctiles y las juntas.

cup seal A circular rubber seal with a depressed center section surrounded by a raised sealing lip to form a cup. Cup seals often are used on the front ends of hydraulic cylinder pistons because they seal high pressure in the forward direction of travel but not in the reverse.

cubeta de obturación Junta de goma circular con una sección central hundida rodeada por un borde de junta saliente que forma una copa. Las cubetas de obturación se suelen usar en los extremos frontales de los pistones de los cilindros hidráulicos porque impiden el paso de alta presión hacia delante pero no en el sentido contrario.

curing agent A class of materials used in brake linings to accelerate the chemical reaction of the binders and other materials.

agente endurecedor Una clase de material que se utiliza en los forros de frenos para acelerar la reacción química de los aglomerantes y otros materiales.

cycle The microprocessor action of turning solenoids on or off.

ciclo Acción del microprocesador al activar o desactivar los solenoides.

Department of Transportation (DOT) The U.S. government executive department that establishes and enforces safety regulations for motor vehicles and for federal highway safety.

Departamento de Transportes (DOT) Departamento ejecutivo del Gobierno de los EE.UU. que establece y hace cumplir las normas de seguridad para los vehículos a motor y para la seguridad vial federal.

diaphragm A flexible membrane, usually made of rubber, that isolates two substances or areas from each other. A rubber diaphragm isolates brake fluid in the master cylinder reservoir from the air. A diaphragm separates the two chambers of a power brake vacuum booster.

diafragma Membrana flexible, por lo común de goma, que aísla una sustancia o una zona de otra. Un diafragma de goma aisla del aire el líquido de frenos en el depósito del cilindro maestro. Un diafragma separa las dos cámaras de un reforzador de vacío en frenos de potencia.

digital A signal that is either on or off and that is translated into the binary digits zero and one. In a computer, a digital signal is voltage that is either low or high or current flow that is on or off.

digital Señal que está activada o desactivada y que se traduce por los dígitos binarios cero y uno. En un computador, una señal digital es una tensión alta o baja, o un flujo de corriente que está abierto o cerrado.

digital ratio adapter controller (DRAC) Used by the microprocessor to change analog signals to digital signals.

controlador adaptador de relación digital (DRAC) Usado por el microprocesador para transformar las señales analógicas en digitales.

digitized The process of converting an analog voltage signal to a digital equivalent that the computer can understand.

digitalizado Proceso de conversión de una señal de tensión analógica a una señal equivalente que el computador pueda entender.

directional control The ability to steer the automobile while stopping.

control direccional Capacidad de gobernar el automóvil al detenerlo.

directional stability The ability to maintain a straight line stopping action.

estabilidad de dirección Capacidad de conservar la línea recta al detener el vehículo.

disc brake A brake in which friction is generated by brake pads rubbing against the friction surfaces on both sides of a brake disc or rotor attached to the wheel.

freno de disco Freno en el que la fricción se genera al rozar las pastillas de freno contra las superficies de fricción a ambos lados de un disco o rotor de freno que está unido a la rueda.

DOT 3, DOT 4, and DOT 5 U.S. Department of Transportation specification numbers for hydraulic brake fluids.

DOT 3, DOT 4 y DOT 5 Números de especificación del Departamento de Transportes de los EE.UU. para los líquidos de frenos hidráulicos.

double flare A type of tubing flare connection in which the end of the tubing is flared out, then is formed back on to itself.

doble ensanche Tipo de conexión de tubos en la que el extremo del tubo se acampana y luego se vuelve a doblar.

drum brake A brake in which friction is generated by brake shoes rubbing against the inside surface of a brake drum attached to the wheel.

freno de tambor Freno en el que la fricción la generan zapatas que rozan contra la superficie interior de un tambor de freno unido a la rueda.

drum web The closed side of a brake drum.

membrana del tambor Lado cerrado de un tambor de freno.

duo-servo brake A drum brake that develops self-energizing action on the primary shoe, which in turn applies servo action to the secondary shoe to increase its application force. Brake application force is interrelated for the primary and the secondary shoes. Also called a dual-servo or a full-servo brake.

freno servoduo Freno de tambor que desarrolla acción autónoma sobre la zapata primaria, que a su vez aplica servoacción a la zapata secundaria para aumentar su fuerza de aplicación. La fuerza de aplicación de freno es interrelacionada para las zapatas primarias y las secundarias. También se conoce como freno servo dual o totalmente asistido.

duty cycle The percentage of time that a solenoid is energized during one complete on–off cycle during pulse-width modulation.

ciclo de trabajo Porcentaje de tiempo que recibe energía un solenoide en lo que dura un ciclo completo de activación–desactivación durante la modulación de amplitud de pulso.

dynamic range The operating range of a sensor.

alcance dinámico Alcance de operación de un sensor.

dynamic rear proportioning An electric valve within an ABS used to control fluid to the rear brake in an action similar to the mechanical/hydraulic proportioning valve.

el proporciónar dinámico de la parte posterior Una válvula eléctrica dentro de un ABS controlaba el líquido al freno posterior en la acción similar a la válvula mecánica/hidráulica que proporcionaba.

eccentric Not round, or concentric.

excéntrico Que no es redondo ni concéntrico.

electronic brake control module (EBCM) A term used to designate an ABS control module or the ABS computer.

módulo de control electrónico de freno (EBCM) Término que se usa para designar un módulo de control ABS o una computadora ABS.

electronic brake system (EBS) A hydraulic modulator used in conjunction with the brake assist sensor.

sistema de frenado electrónico (EBS) Modulador hidráulico que se usa en conjunto con el sensor de asistencia de frenado.

electrohydraulic brake An electrically controlled hydraulic braking system that is being used to combine ABS and TCS into one system.

freno electrohidráulico Sistema de control hidráulico controlado eléctricamente que se usa para combinar el ABS y el TCS en un sistema.

electrohydraulic unit The microprocessor and hydraulic unit are combined in one unit.

unidad electrohidráulica Microprocesador y unidad hidráulica combinados en un solo elemento.

Electro-Lok A brand of electric parking brakes. In this type of brakes, the applied hydraulic pressure is trapped in the lines by solenoids controlled by the vehicle operator.

Freno eléctrico Marca de frenos eléctricos para estacionarse. En este tipo de frenos, la presión hidráulica aplicada queda regulada en las líneas por los solenoides que controla el conductor del vehículo.

electromagnetic induction The generation of voltage in a conductor by relative motion between the conductor and a magnetic field.

inducción electromagnética Generación de tensión en un conductor mediante un movimiento relativo entre el conductor y un campo magnético.

electromagnetic interference (EMI) A magnetic force field that influences a signal being sent to the microprocessor.

interferencia electromagnética (EMI) Campo de fuerza magnética que influye en una señal enviada al microprocesador.

electronically variable brake proportioning (EVBP) An ABS software program that proportions brake pressure to the rear wheels during braking.

distribuidor electrónico variable de la frenada (EVBP) Programa de software para el sistema de ABS que distribuye la presión de frenado a las ruedas traseras durante el frenado.

enhanced traction system (ETS) A traction control system used with the DBC-7 ABS.

sistema de tracción mejorada (ETS) Sistema de control de tracción que se usa con el ABS del DBC-7.

Environmental Protection Agency (EPA) The U.S. government executive department that establishes and enforces regulations to protect and preserve the physical environment, best known for regulations relating to air quality.

Agencia Protectora del Medio Ambiente (EPA) Departamento gubernamental de EE.UU. que establece y hace cumplir las normas que protegen y preservan el entorno físico; más conocido por las normas que regulan la calidad del aire.

equalizer Part of the parking brake linkage that balances application force and applies it equally to each wheel. The equalizer often contains the linkage adjustment point.

compensador Parte del acoplamiento del freno de estacionamiento que equilibra la fuerza ejercida y la aplica por igual en cada rueda. Con frecuencia, el compensador incluye el punto de ajuste del acoplamiento.

erasable programmable read-only memory (EPROM) Computer memory program circuits that can be erased and reprogrammed. Erasure is done by exposing the integrated circuit chip to ultraviolet light.

memoria de sólo lectura borrable y programable (EPROM) Circuitos de programa de memoria del computador que se pueden eliminar y reprogramar. La eliminación es posible exponiendo a la luz ultravioleta el chip de circuito integrado.

eyeball A voltage monitoring circuit.

ojo Circuito de control de tensión.

Federal Motor Vehicle Safety Standards (FMVSS) U.S. government regulations that prescribe safety requirements for various vehicles, including passenger cars and light trucks. The FMVSS regulations are administered by the U.S. Department of Transportation (DOT).

Normas Federales de Seguridad de Vehículos a Motor (FMVSS) Normas gubernamentales de EE.UU. que dictan los requisitos de seguridad para diversos vehículos, incluyendo los automóviles de pasajeros y los camiones ligeros. El Departamento de Transportes de los EE.UU. (DOT) es el que administra las normas FMVSS.

filler A class of materials used in brake linings to reduce noise and improve heat transfer

relleno Tipo de material usado en los forros de frenos para reducir el ruido y mejorar la transferencia de calor.

fittings The term applied to all plumbing connections on the car or in the house.

ajustes Término aplicado a las conexiones de tuberías del auto o de la casa.

fixed caliper brake A brake caliper that is bolted to its support and does not move when the brakes are applied. A fixed caliper must have pistons on both the inboard and the outboard sides.

freno de calibre fijo Calibre de freno que se fija al soporte con un perno y no se mueve al aplicar los frenos. Un calibre fijo debe tener pistones tanto en el lado exterior como en el interior.

fixed rotor A rotor that has the hub and the rotor cast as a single part.

rotor fijo Rotor en el que el buje y el rotor forman una única pieza.

fixed seal A seal for a caliper piston that is installed in a groove in the caliper bore and that does not move with the piston.

junta fija Junta para un pistón de calibre que se instala en una hendidura del hueco del calibre y que no se mueve con el pistón.

floating caliper   A caliper that is mounted to its support on two locating pins or guide pins. The caliper slides on the pin in a sleeve or bushing. Because of its flexibility, this kind of caliper is said to float on its guide pins.

calibre flotante   Calibre que se monta al soporte mediante dos pasadores de posición o pasadores guía. El calibre se desliza dentro del pasador a través de un manguito o casquillo. Debido a su flexibilidad, se dice que este tipo de calibre flota en los pasadores guía.

floating rotor   A rotor and hub assembly made of two separate parts.

rotor flotante   Conjunto de rotor y buje en dos piezas distintas.

flux lines   Lines of magnetism.

líneas de flujo   Líneas de magnetismo.

force   Power working against resistance to cause motion.

fuerza   Energía que trabaja contra la resistencia para producir movimiento.

free play   The distance the brake pedal moves before the master cylinder primary piston moves.

holgura   Distancia recorre el pedal del freno antes de que se mueva el pistón primario del cilindro maestro.

frequency   The number of times, or speed, at which an action occurs within a specified time interval. In electronics, frequency indicates the number of times that a signal occurs or repeats in cycles per second. Cycles per second are indicated by the symbol "hertz" (Hz).

frecuencia   Número de veces, o velocidad, a la que ocurre una acción dentro de un intervalo de tiempo especificado. En electrónica, la frecuencia indica el número de veces que una señal se da o se repite en ciclos por segundo. Los ciclos por segundo se indican con el símbolo "herzios" (Hz).

friction modifier   A class of materials used in brake linings to modify the final coefficient of friction of the linings.

modificador de rozamiento   Tipo de material usado en los forros de frenos para modificar el coeficiente final de fricción de las envolturas.

friction   The force that resists motion between the surfaces of two objects or forms of matter.

fricción   Fuerza que se opone al movimiento entre las superficies de dos objetos o formas de materia.

fulcrum   The pivot point of a lever.

punto de apoyo   Punto de apoyo de una palanca.

gas fade   Brake fade due to hot gas and dust, which reduce friction between the drum or rotor during prolonged hard braking.

pérdida de frenado por gas   Disminución progresiva de la acción de frenado debido al gas caliente y al polvo que reducen el rozamiento entre el tambor o el rotor durante un frenado fuerte prolongado.

geometric centerline   A static dimension represented by a line through the center of the vehicle from front to rear.

línea media geométrica   Dimensión estática representada por una línea en el centro del vehículo desde el frente hasta la parte de atrás.

gross vehicle weight rating (GVWR)   Total weight of a vehicle plus its maximum rated payload.

peso bruto del vehículo (GVWR)   Peso total de un vehículo más la carga de régimen máxima.

Hall-effect switch   A device that produces a voltage pulse dependent on the presence of a magnetic field. Hall-effect voltage varies as magnetic reluctance varies around a current-carrying semiconductor.

conmutador de efecto Hall   Aparato que produce una variación rápida de tensión dependiente de la presencia de un campo magnético.

La tensión de efecto Hall varía al cambiar la resistencia magnética de alrededor de un semiconductor que lleva corriente.

hardware   The electrical and mechanical parts of a computer system. Hardware includes resistors, diodes, capacitors, transistors, and other electronic parts mounted on a circuit board. Hardware also can include sensors and actuators in a system.

hardware   Partes eléctricas y mecánicas de un computador. El hardware incluye resistores, diodos, capacitores, transistores y otras partes electrónicas montadas en un tablero de circuitos. El hardware incluye también los sensores e impulsores del sistema.

height-sensing proportioning valve   A proportioning valve in which hydraulic pressure is adjusted automatically according to the vertical movement of the chassis in relation to the rear axle during braking; sometimes also called a weight-sensing proportioning valve.

válvula dosificadora de detección de altura   Válvula dosificadora en la que la presión hidráulica se ajusta automáticamente según el movimiento vertical del chasis en relación con el eje trasero durante el frenado; a veces también se la denomina válvula dosificadora de detección de peso.

hertz   A unit of frequency equal to one cycle per second.

hertzio   Unidad de frecuencia que equivale a un ciclo por segundo.

hold-down springs   Small springs that hold drum brake shoes in position against the backing plate while providing flexibility for shoe application and release.

resortes de sujeción   Pequeños resortes que mantienen las zapatas de frenos de tambor en posición contra la placa de refuerzo, a la vez que dan flexibilidad para aplicar y soltar la zapata.

hybrid-electric vehicle (HEV)   A vehicle that uses an internal combustion engine to charge a battery, which supplies electrical power to one or more direct current motors.

los vehículos híbridos (HEV)   Vehículo que utilizan un motor de combustión interna para cargar una batería, que provee corriente eléctrica a unos o más motores de la corriente directa.

hydraulic modulator   The common term for an ABS computer-controlled set of valves used to control brake pressure to various wheels.

modulador hidráulico   Término común para el conjunto de válvulas controladas por computadora del ABS que se usan para controlar la presión del frenado a las distintas ruedas.

hydraulic system mineral oil (HSMO)   A brake fluid made from a mineral oil base, used by a few European carmakers. DOT specifications do not apply to HSMO fluids, and HSMO fluids cannot be mixed with DOT fluids.

aceite mineral del sistema hidráulico (HSMO)   Líquido de frenos a base de aceite mineral que utilizan algunos fabricantes de automóviles europeos. Las especificaciones de los líquidos DOT no se aplican a los líquidos HSMO, y éstos últimos no se pueden mezclar con los primeros.

hydro-boost   A hydraulic power brake system that uses the power steering hydraulic system to provide boost for the brake system.

reforzador hidráulico   Un sistema hidráulico de frenos de potencia que utiliza el sistema hidráulico de dirección de potencia para alimentar el sistema de frenado.

hydroplane   The action of a tire rolling on a layer of water on the road surface instead of staying in contact with the pavement. Hydroplaning occurs when water cannot be displaced from between the tread and the road.

aquaplaning   Acción de un neumático que rueda sobre una capa de agua sobre la superficie vial en lugar de mantenerse en contacto con el pavimento. El "aquaplaning" se produce cuando no se puede desplazar el agua entre los dibujos del neumático y la calle.

hygroscopic   The chemical property or characteristic of attracting and absorbing water, particularly out of the air. Polyglycol brake fluids are hygroscopic.

higroscópico   Que posee la propiedad química o característica de atraer y absorber agua, en especial del aire. Los líquidos de frenos con poliglicol son higroscópicos.

inertia   The tendency of an object in motion to keep moving and the tendency of an object at rest to remain at rest.

inercia   Tendencia de un objeto en movimiento a seguir moviéndose y la de un objeto en reposo a permanecer en reposo.

integrated ABS   An antilock brake system in which the ABS hydraulic components, the standard brake hydraulic components, and a hydraulic power booster are joined in a single, integrated hydraulic system.

ABS integrados   Sistema antibloqueo de frenos en el que los componentes hidráulicos de los frenos ABS, los componentes hidráulicos de los frenos estándar y un reforzador hidráulico de potencia se unen en un único sistema hidráulico integrado.

integrated circuit (IC)   A complete electronic circuit of many transistors and other devices, all formed on a single silicon semiconductor chip.

circuito integrado (IC)   Completo circuito electrónico de muchos transistores y otros dispositivos, todos formados sobre un único chip semiconductor de siliconas.

intermediate lever   Part of the parking brake linkage under the vehicle that increases application force and works with the equalizer to apply it equally to each wheel.

palanca intermedia   Parte del acoplamiento del freno de estacionamiento, situada debajo del vehículo, que aumenta la fuerza de aplicación y que trabaja con el compensador para aplicarla por igual a cada rueda.

ISO flare   A type of tubing flare connection in which a bubble-shaped end is formed on the tubing; also called a bubble flare.

ensanche ISO   Tipo de conexión con ensanche del tubo en el cual un extremo toma forma de burbuja; también se le llama ensanche de burbuja.

isolate, hold, and dump   Actions commanded by the microprocessor to regulate hydraulic pressure to an individual wheel brake.

aislar, mantener y descargar   Acciones solicitadas por el microprocesador para regular la presión hidráulica que va al freno de una rueda individual.

isolation/dump valve   A term given to a brake pressure control valve in an RWAL or RABS.

válvula de vertedero/aislamiento   Término dado a la válvula que controla la presión del frenado en RWAL o RABS.

keep-alive memory (KAM)   Random-access memory that is retained by keeping a voltage applied to the circuits when the engine is off.

memoria estable (KAM)   Memoria de acceso aleatorio que se conserva manteniendo la aplicación de tensión a los circuitos cuando el motor está parado.

kinetic energy   The energy of mechanical work or motion.

energía cinética   Energía del trabajo mecánico o movimiento.

kinetic friction   Friction between two moving objects or between one moving object and a stationary surface.

fricción cinética   Fricción entre dos objetos en movimiento, o entre uno en movimiento y una superficie estacionaria.

lands   The raised surfaces on a valve spool.

partes planas   Superficies elevadas en un carrete de válvula.

lateral runout   A side-to-side variation or wobble as the tire and wheel are rotated.

desviación lateral   Variación u oscilación de un lado a otro cuando se hacen girar el neumático y la rueda.

lathe-cut seal   A fixed seal for a caliper piston that has a square or irregular cross section; not round like an O-ring.

junta torneada   Junta fija para un pistón de calibre que tiene una sección transversal cuadrada o irregular, no redonda como en un toroide.

lateral accelerometer   Vehicle sensor used to measure the speed of the vehicle's lateral movement during operation.

acelerómetro lateral   Un acelerómetro lateral es un sensor del vehículo usado para medir la velocidad del movimiento lateral del vehículo durante la operación.

leading shoe   The first shoe in the direction of drum rotation in a leading-trailing brake. When the vehicle is going forward, the forward shoe is the leading shoe, but the leading shoe can be the front or the rear shoe depending on whether the drum is rotating forward or in reverse and whether the wheel cylinder is at the top or the bottom of the backing plate. The leading shoe is self-energizing.

zapata tractora   Primera zapata en la dirección de giro del tambor en un freno de tracción-remolque. Cuando el vehículo avanza, la zapata delantera es la tractora, pero la zapata tractora puede ser la delantera o la trasera dependiendo de si el tambor está girando hacia delante o hacia atrás y de si el cilindro de la rueda está en la parte superior de la placa de refuerzo o en la inferior. La zapata tractora es autónoma en cuanto a energía.

leading-trailing brake   A drum brake that develops self-energizing action only on the leading shoe. Brake application force is separate for the leading and the trailing shoes. Also called a partial-servo or a nonservo brake.

freno tracción-remolque   Freno de tambor que desarrolla una acción autónoma sólo sobre la zapata tractora. La fuerza de aplicación de freno es independiente para las zapatas tractoras y para las de remolque. También se conoce como freno parcialmente asistido o freno no asistido.

leverage   The use of a lever and fulcrum to create a mechanical advantage, usually to increase force applied to an object. The brake pedal is the first point of leverage in a vehicle brake system.

transmisión por palancas   Utilización de una palanca y un punto de apoyo para crear una ventaja mecánica, generalmente aumentar la fuerza que se aplica a un objeto. El pedal del freno es el primer punto de transmisión por palancas en un sistema de frenos de vehículos.

linearity   The expression of sensor accuracy throughout its dynamic range.

linealidad   Expresión de la exactitud de un sensor en todo su alcance dinámico.

lining fade   Brake fade due to a loss of brake lining coefficient of friction caused by excessive heat.

pérdida por envolturas   Pérdida de frenado debido a una disminución del coeficiente de fricción de los forros de freno a causa de un calor excesivo.

lockup   A condition in which a wheel stops rotating and skids on the road surface.

bloqueo   Condición en la cual una rueda deja de girar y patina sobre la superficie vial.

magnetoresistive sensor   A sensor that measures resistance in a magnetic field; commonly used as wheel speed sensor in ABS.

sensor magnético-resistente   Sensor que mide la resistencia en un campo magnético; se usa comúnmente como sensor de velocidad de la rueda en el ABS.

mass   The measure of the inertia of an object or form of matter or its resistance to acceleration. Also the molecular density of an object.

masa   Medida de la inercia de un objeto o forma de materia, o su resistencia a la aceleración. También la densidad molecular de un objeto.

master cylinder   The liquid-filled reservoir in the hydraulic brake system or clutch where hydraulic pressure is developed when the driver depresses a foot pedal.

cilindro principal   El cilindro principal es el depósito llenado del líquido en el sistema de frenos hidráulico donde se desarrolla la presión hydráulica cuando el conductor presiona un pedal con su pie.

material safety data sheets (MSDS)   Information sheets issued by the manufacturers of hazardous materials. An MSDS provides detailed information on dangerous ingredients, corrosiveness, reactivity, toxicity, fire and explosion data, health hazards, spill and leak procedures, and special precautions. Federal law requires that an MSDS be available for each hazardous material in the workplace.

hojas de datos de seguridad de materiales (MSDS)   Hojas de información proporcionadas por los fabricantes de materiales peligrosos. Una MSDS ofrece información detallada sobre ingredientes peligrosos, corrosión, capacidad de reacción, toxicidad, datos sobre incendios y explosiones, peligros para la salud, procedimientos en caso de derrame o goteo de estos materiales, y precauciones especiales a tomar. La ley federal exige que haya una MSDS para cada material peligroso en el lugar de trabajo.

mechanical fade   Brake fade due to heat expansion of a brake drum away from the shoes and linings. Mechanical fade does not occur with disc brakes.

pérdida mecánica de frenado   Pérdida de frenado debido a la dilatación térmica de un tambor de freno, que lo separa de las zapatas y los forros. La pérdida mecánica no se produce con los frenos de disco.

metallic lining   Brake friction material made from powdered metal that is formed into blocks by heat and pressure.

forro metálico   Material de fricción para frenos hecho de metal en polvo que se transforma en bloques por medio de calor y presión.

metering valve   A hydraulic control valve used primarily with front disc brakes on RWD vehicles. The metering valve delays pressure application to the front brakes until the rear drum brakes have started to operate.

válvula de dosificación   Válvula de control hidráulica que se usa principalmente con frenos delanteros de disco en vehículos RWD. La válvula de dosificación demora la aplicación de presión a los frenos delanteros hasta que hayan comenzado a funcionar los frenos traseros de tambor.

microprocessor   A digital computer or processor built on a single integrated circuit chip. A microprocessor can perform functions of arithmetic logic and control logic. It is the basic building block of a microcomputer system.

microprocesador   Computador digital o procesador construido sobre un único chip de circuitos integrados. Un microprocesador puede realizar funciones de lógica aritmética y lógica de control. Es el elemento básico de construcción de un sistema de microprocesador.

mold-bonded lining   A pad assembly made by applying adhesive to the pad and then pouring the uncured lining material onto the pad in a mold. The assembly is cured at high temperature to fuse the lining and adhesive to the pad.

forro adherido al molde   Conjunto de pastillas formado por la aplicación de adhesivo a las pastillas y vertiendo luego el material de envoltura no vulcanizado sobre éstas en un molde. El conjunto se vulcaniza a una temperatura elevada para fundir el forro y el adhesivo a la pastilla.

momentum   The force of continuing motion. The momentum of a moving object equals its mass times its speed.

impulsor   Fuerza de un movimiento continuo. El impulso de un objeto en movimiento es igual a la masa por la velocidad.

multiplexing   The network used by multiple computers so that only one component is transmitting at a time.

multiplexación   La red usada por las computadoras múltiples de modo que solamente una componente esté transmitiendo a la vez.

National Highway Transportation and Safety Agency (NHTSA)   A federal agency assigned to develop regulations for highway safety including vehicle safety features.

Agencia Nacional de Transporte y Seguridad en Carreteras (NHTSA)   Agencia federal asignada a desarrollar regulaciones para la seguridad en las carreteras incluyendo detalles de seguridad vehicular.

negative wheel slip   Wheel lockup that occurs when too much braking force is applied to a wheel and the tire skids on the pavement. ABS controls negative wheel slip by modulating (decreasing and increasing) the hydraulic pressure to the wheel or wheels that is/are skidding.

deslizamiento negativo de la rueda   Bloqueo de la rueda que tiene lugar cuando se aplica un frenado demasiado fuerte a la rueda y la llanta patina sobre el pavimento. Los frenos ABS controlan el deslizamiento negativo de la rueda regulando (aumentando y disminuyendo) la presión hidráulica de la rueda o las ruedas que patina/n.

network   The channel through which several computers share information.

red   Canal a través del cual diversos computadores comparten la información.

nonintegrated ABS   An antilock brake system in which the ABS hydraulic components are attached to, but separate from, the normal brake hydraulic system and power booster.

ABS no integrado   Sistema antibloqueo de frenos en el que los componentes hidráulicos del ABS se conectan, pero estando separados, al sistema hidráulico normal de frenos y al reforzador de potencia.

O-ring   A circular rubber seal shaped like the letter "O."

toroide   Junta de goma circular con la forma de la letra "O".

Occupational Safety and Health Administration (OSHA)   A division of the U.S. Department of Labor that establishes and enforces workplace safety regulations.

Administración de Seguridad y Salud en el Trabajo (OSHA)   División del Departamento de trabajo de los EE.UU. que establece y hace cumplir las normas de seguridad en el lugar de trabajo.

ohm   The unit used to measure the amount of electrical resistance in a circuit or an electrical device. One ohm is the amount of resistance present when one volt forces one ampere of current through a circuit or a device. Ohm is abbreviated with the Greek letter omega (Ω).

ohmio Unidad usada para medir la cantidad de resistencia eléctrica de un circuito o de un aparato eléctrico. Un ohmio es la cantidad de resistencia presente cuando un voltio hace pasar un amperio de corriente a través de un circuito o de un aparato. El ohmio se abrevia con la letra griega omega (Ω).

Ohm's law The mathematical formula for the relationships between voltage, current, and resistance; often stated simply as E (voltage) = I (current) × R (resistance).

ley de Ohm Fórmula matemática que expresa las relaciones entre tensión, corriente y resistencia; frecuentemente se enuncia simplemente por E (tensión) = I (corriente) × R (resistencia).

organic lining Brake friction material made from nonmetallic fibers bonded together in a composite material.

forro orgánico Material de fricción del freno hecho de fibras no metálicas adheridas en un material compuesto.

overload spring Spring in the end of the cable in cable-operated adjusters that lets the cable move without breaking if the pawl or star wheel is jammed.

resorte de sobrecarga Resorte al final del cable de los reguladores operados por cables que permite que éste se mueva sin romperse si el trinquete o la rueda de estrella están agarrotados.

overvoltage protection relay (OVPR) An electrical device used in ABSs that prevents damage to electronic components and circuits due to electrical surges or spikes.

relé de protección contra sobretensión (OVPR) Dispositivo eléctrico usado en los frenos ABS que evita posibles daños en los componentes electrónicos y en los circuitos debido a subidas y bajadas eléctricas momentáneas.

ozone An unstable molecule of oxygen with three atoms instead of the normal two. Ozone oxidizes other elements and compounds by giving up its extra oxygen atom.

ozono Molécula de oxígeno inestable con tres átomos en lugar de dos. El ozono oxida otros elementos y compuestos cediéndoles su átomo de oxígeno extra.

P-metric system The most common modern system to specify passenger car tire sizes.

sistema métrico P Sistema métrico más moderno y común para especificar el tamaño de los neumáticos de los automóviles.

pad hardware Miscellaneous small parts such as antirattle clips and support clips that hold brake pads in place and keep them from rattling.

hardware de cojines Piezas pequeñas surtidas, como grapas antiresonancia y grapas de apoyo que mantienen los cojines de los frenos en su lugar y evitan que suenen.

pad wear indicators Devices that warn the driver when disc brake linings have worn to the point where they need replacement. Wear indicators may be mechanical (audible) or electrical.

indicadores del desgaste de los cojines Dispositivos que advierten al conductor de que los forros de los frenos de disco se han gastado tanto que es necesario cambiarlos. Los indicadores de desgaste pueden ser mecánicos (acústicos) o eléctricos.

parking brake control The pedal or lever used to apply the parking brakes.

control del freno de estacionamiento Pedal o palanca que se usa para aplicar los frenos de estacionamiento.

parking brakes The disc or drum brakes that hold the vehicle stationary after the service brakes stop it.

frenos de estacionamiento Frenos de disco o de tambor que mantienen el vehículo inmóvil después de ser detenido por los frenos de servicio.

pawl A hinged or pivoted component that engages a toothed wheel or rod to provide rotation or movement in one direction while preventing it in the opposite direction.

trinquete Componente articulado o embisagrado que se engrana a una rueda o varilla dentada para ofrecer rotación o movimiento en un sentido mientras que lo impide en el sentido contrario.

pedal position sensor A sensor that measures brake pedal movement and speed of movement; commonly associated with electronic braking systems.

sensor de posición del pedal Sensor que mide el movimiento del pedal del freno y la velocidad del movimiento. Se asocia comúnmente con los sistemas de freno electrónico.

permanent magnet (PM) generator A reluctance sensor. A sensor that generates a voltage signal by moving a conductor through a permanent magnetic field.

generador de magneto permanente (PM) Sensor de reluctancia. Sensor que genera una señal de tensión al mover un conductor a través de un campo magnético permanente.

perpetual energy Energy that can be produced/converted without using additional energy forever.

energía perpetua Energía que puede producirse/convertise sin usar energía adicional.

phenolic plastic Plastic made primarily from phenol, a compound derived from benzene and also called carbolic acid.

plástico fenólico Plástico hecho principalmente de fenol, compuesto derivado del benceno, llamado también ácido carbólico.

phosgene A poisonous gas that is formed when certain other gases are exposed to flame; also known as mustard gas, the principal poison gas used in World War I.

fosgeno Gas venenoso que se forma al exponer ciertos gases a una llama; conocido también como gas mostaza, fue el gas más usado en la Primera Guerra Mundial.

piston stop A metal part on a brake backing plate that keeps the wheel cylinder pistons from moving completely out of the cylinder bore.

tope de pistón Parte metálica de la placa de refuerzo del freno que impide que los pistones del cilindro de la rueda se salgan del hueco del cilindro.

polyglycol A mixture of several alcohols. Polyalkylene-glycol-ether brake fluids that meet specifications for DOT 3 and DOT 4 brake fluids.

poliglicol Mezcla de varios alcoholes. Líquidos de frenos de polialquilenglicoléter que están dentro de las especificaciones DOT 3 y DOT 4 para líquidos de frenos.

positive wheel slip The excessive wheel spin that occurs during acceleration as a wheel loses traction and spins on the pavement.

deslizamiento positivo de la rueda Giro excesivo de la rueda que se produce durante una aceleración cuando una rueda pierde tracción y gira sobre el pavimento.

POST Acronym for the power on system test used by the microprocessor at key on.

POST Siglas de la potencia en una prueba del sistema usadas por el microprocesador al conectar.

potentiometer A variable resistor that acts as a voltage divider to produce a continuously variable output signal proportional to a mechanical position.

potenciómetro Resistor variable que actúa como separador de tensiones para producir una señal de salida que varía continuamente, proporcional a la posición mecánica.

PowerMaster A self-contained hydraulic power brake system with its own hydraulic reservoir and independent electric pump; developed by General Motors for cars and trucks that could not economically use vacuum-assisted power brakes.

PowerMaster Sistema de frenos hidráulico autónomo con su propia reserva hidráulica y bomba eléctrica independiente; producido por General Motors para coches y camiones que no podían usar económicamente frenos de potencia asistidos por vacío.

pressure Force exerted on a given unit of surface area. Pressure equals force divided by area and is measured in pounds per square inch (psi) or kilopascals (kPa).

presión Fuerza ejercida sobre una unidad de superficie determinada. La presión es igual a la fuerza dividida por el área y se mide en libras por pulgadas cuadradas (psi) o kilopascales (kPa).

pressure differential The difference between two pressures on two surfaces or in two separate areas. The pressures can be either pneumatic (air) or hydraulic.

diferencial de presión Diferencia entre dos presiones en dos superficies o áreas distintas. La presión puede ser neumática (aire) o hidráulica.

pressure differential valve A hydraulic valve that reacts to a difference in pressure between the halves of a split brake system. When a pressure differential exists, the valve moves a plunger to close the brake warning lamp switch.

válvula de diferencial de presión Válvula hidráulica que reacciona ante una diferencia de presión entre las dos partes de un sistema de frenos dividido. Cuando hay una diferencia de presión, la válvula mueve un pistón que cierra el conmutador de la luz indicadora de freno.

pressure-sensitive base A software that measures tire pressure with on-wheel sensors and relays the pressure to an onboard computer.

base de presión sensitiva Software que mide la presión de las llantas con sensores en la rueda y transmite la presión a una computadora del vehículo.

primary shoe The leading shoe in a duo-servo brake. The primary shoe is self-energizing and applies servo action to the secondary shoe to increase its application force. Primary shoes have shorter linings than secondary shoes.

zapata primaria Zapata guía en un servofreno dual. La zapata primaria es autónoma y aplica una servoacción a la zapata secundaria para aumentar su fuerza de aplicación. Las zapatas primarias tienen forros más pequeños que las secundarias.

program The job instructions for a computer.

programa Instrucciones para que un computador realice un trabajo.

programmable read-only memory (PROM) A computer memory integrated circuit chip that can be programmed once to store the computer program and other data.

memoria de sólo lectura programable (PROM) Chip de circuito integrado de la memoria de un computador que se puede programar una vez para almacenar el programa del computador y otros datos.

proportioning valve A hydraulic control valve that controls the pressure applied to rear drum brakes. A proportioning valve decreases the rate of pressure application above its split point as the brake pedal is applied harder.

válvula de dosificación Válvula de control hidráulico que controla la presión aplicada a los frenos de tambor traseros. Una válvula de dosificación disminuye la proporción de aplicación de presión por encima de su punto de separación cuando se presiona más fuerte el pedal del freno.

pull up resistor A fixed resistor in a voltage divider circuit for a sensor input signal. A variable-resistor sensor changes its voltage drop proportionally to the quantity being measured, and the pull up resistor drops the rest of the reference voltage.

resistor de arranque Resistor fijo en un circuito divisor de tensión que alimenta a un sensor de señal de entrada. Un sensor del resistor variable cambia su caída de tensión proporcionalmente a la cantidad que ha medido y el resistor de arranque baja el resto de la tensión de referencia.

pulse width The amount of time that an electromechanical device, such as a solenoid, is energized. Pulse width usually is measured in milliseconds.

anchura entre impulsos Tiempo que emplea un mecanismo electromecánico, como por ejemplo un solenoide, en activarse. La anchura entre impulsos se mide normalmente en milisegundos.

pulse-width modulation (PWM) The characteristic of a continuous on-and-off cycling of a solenoid for a fixed number of times per second. While the frequency of the cycles remains constant, the ratio of on-time to total cycle time varies or is modulated.

modulación de impulsos en altura (PWM) Característica del intercambiador cíclico continuo de un solenoide para un número determinado de veces por segundo. Mientras que la frecuencia de ciclos permanece constante, la relación entre el tiempo en activo y tiempo del ciclo total varía o se modula.

quick take-up master cylinder A dual master cylinder that supplies a large volume of fluid to the front disc brakes on initial brake application, which takes up the clearance of low-drag calipers.

cilindro maestro de tensor rápido Cilindro maestro doble que proporciona una gran cantidad de líquido a los frenos de disco delanteros en la primera frenada, compensando la holgura de los calibres de baja resistencia.

quick take-up valve The part of the quick take-up master cylinder that controls fluid flow between the reservoir and the primary low-pressure chamber.

válvula de tensor rápido Parte del cilindro maestro del tensor rápido que controla el flujo de líquido entre el depósito y la cámara de baja presión primaria.

radial ply tire Tire construction in which the cords in the body plies of the carcass run at an angle of 90 degrees to the steel beads in the inner rim of the carcass. Each cord is parallel to the radius of the tire circle.

neumático de capas radiales Fabricación de neumáticos en los que los cordones de las capas del cuerpo de la carcasa se fijan con un ángulo de 90 grados a las nervaduras de acero en borde interior de la carcasa. Cada cordón es paralelo a los radios del círculo del neumático.

radial runout An out-of-round condition in which the radius of the wheel or tire is not consistent from the wheel center to any point on the rim or the tread.

desgaste radial Pérdida de redondez en la que el radio de la rueda o neumático no es consistente desde el centro de la rueda a cualquier punto de la llanta o de la rodadura.

radio frequency interference (RFI) A narrow band of frequencies within the electromagnetic interference spectrum.

interferencia de radiofrecuencia (RFI) Banda estrecha de frecuencias dentro del espectro de interferencia electromagnética.

random access memory (RAM) Computer read-write memory on which information can be written and from which it can be read.

memoria de acceso aleatorio (RAM) Memoria de lectura y escritura del computador en la cual se puede ingresar información y leerla.

reaction disc (or plate and levers) The components in a vacuum power booster that provide pedal feel or feedback to the driver.

disco de reacción (o placa y palancas) Componentes de un reforzador de vacío que proporcionan sensaciones en el pedal o respuesta al conductor.

read-only memory (ROM) The permanent program memory of a computer. Instructions can be read from ROM but nothing can be written into it and it cannot be changed.

memoria de sólo lectura (ROM) Memoria permanente de programas de un computador. Se pueden leer instrucciones de la ROM pero no es posible escribir ni cambiar nada en ella.

rear antilock brake system (RABS) A two-wheel Kelsey–Hayes ABS used on the rear wheels of Ford light-duty pickup trucks.

sistema antibloqueo de frenos traseros (RABS) Frenos ABS Kelsey–Hayes para dos ruedas utilizado en las ruedas traseras de las camionetas Ford de servicio ligero.

rear wheel antilock (RWAL/RABS) A two-wheel ABS used on the rear wheels of light-duty pickup trucks and some SUVs. One of the best known Kelsey–Hayes systems.

antibloqueo de ruedas traseras (RWAL) Frenos ABS para dos ruedas utilizado en las ruedas traseras de camionetas ligeras y en algunos SUV. Uno de los sistemas de Kelsey–Hayes más conocidos.

reference voltage A fixed voltage supplied to the sensor by a voltage regulator inside the computer, or control module. As the sensor changes, the return voltage is altered and sent back to the computer for use. Most computer control systems operate with a 5-volt reference voltage.

tensión de referencia Tensión fija suministrada al sensor por un regulador de tensión del interior del computador o módulo de control. Al cambiar el sensor, la tensión de vuelta se altera y se envía de vuelta al computador para su uso. La mayoría de los sistemas de control por computador funcionan con una tensión de referencia de 5 voltios.

regenerative braking Converting the energy (heat and mass) of a moving vehicle to stop that vehicle.

frenar regenerador El frenar regenerador está convirtiendo la energía (calor y masa) de un vehículo móvil para parar ese vehículo.

relay An electromagnetic switch that uses a small amount of current in one circuit to open or close a circuit with greater current flow. Relays are used as remotely controlled switches for circuits.

relé Un conmutador electromagnético que toma una pequeña cantidad de corriente de un circuito para abrir o cerrar otro con mayor flujo de corriente. Los relés se utilizan como conmutadores de control remoto para los circuitos.

reluctance sensor A magnetic pulse generator or pickup coil that sends a voltage signal in response to varying reluctance of a magnetic field.

sensor de reluctancia Generador de impulsos magnéticos o bobina captadora que envía una señal de tensión como respuesta a una variación en la reluctancia de un campo magnético.

reluctor A metal tooth ring used to influence the magnetic flux lines of the PM generator.

reluctor Anillo metálico dentado que se usa para influir en las líneas de flujo magnético del generador de PM.

repeatability The ability of a sensor to send the same signal voltage every time it measures the same value or quantity.

repetitividad Capacidad de un sensor de enviar la misma tensión de señal cada vez que mide el mismo valor o cantidad.

replenishing port The rearward port in the master cylinder bore; also called other names.

puerto de abastecimiento Puerto situado más atrás en el hueco del cilindro maestro; también recibe otros nombres.

reservoir Storage tank for the master cylinder.

depósito Tanque de almacenamiento para el cilindro maestro.

residual pressure check valve A check valve used in the outlet port for drum brakes in some master cylinders. The valve retains a slight pressure in the lines to the drum brakes when the brakes are not applied, which holds the wheel cylinder cup seals against the cylinder walls. Residual pressure check valves have been replaced in many late-model systems by cup expanders in the wheel cylinders.

válvula de control de presión residual Válvula de control usada en el puerto de salida para frenos de tambor en algunos cilindros maestros. La válvula retiene una ligera presión en las líneas que van a los frenos de tambor cuando no se están utilizando, lo que sujeta las cubetas de obturación del cilindro de la rueda contra las paredes del cilindro. Las válvulas de control de presión residual han sido reemplazadas en muchos sistemas de últimos modelos por cubetas de expansión en los cilindros de las ruedas.

return spring A strong spring that retracts a drum brake shoe when hydraulic pressure is released

resorte de vuelta Resorte fuerte que retrae una zapata de freno de tambor cuando se libera presión hidráulica.

reverse braking A condition in which the trailer brakes attempt to apply when the vehicle/trailer is in reverse; a condition that requires devices for preventing trailer braking in reverse.

frenado de reversa Condición que los frenos del trailer tratan de aplicar cuando el vehículo/trailer va en reversa; una condición que requiere de unos dispositivos para prevenir que el trailer frene en reversa.

riveted lining Brake lining attached to the pad or shoe by copper or aluminum rivets.

forro remachado Forro de freno unido a la pastilla o zapata por remaches de cobre o de aluminio.

rolling circumference The distance around the outside of an inflated tire mounted and loaded by the vehicle's weight. It increases with increased inflation and tire size.

circunferencia del balanceo La distancia alrededor del exterior de un neumático inflado montado y cargado por el peso del vehículo. Aumenta con tamaño creciente de la inflación y del neumático.

rolling resistance The resistance of the vehicle's weight and mass and the friction of all moving components to the movement of the vehicle.

coeficiente de resistencia a la rodadura   Resistencia del peso y masa de un vehículo y de la fricción de todos los componentes movibles al movimiento del vehículo.

root cause   The event or events that initially caused the fault.

causa de la raíz   Una causa de la raíz es el acontecimiento o los acontecimientos que causaron inicialmente la avería.

rotor   The rotating part of a disc brake that is mounted on the wheel hub and contacted by the pads to develop friction to stop the car. Also called a disc.

rotor   Parte giratoria de un freno de disco que va montada en el buje de la rueda y entra en contacto con las pastillas para causar el rozamiento que detiene el vehículo. También llamado disco.

screw end nut   A common kind of caliper-actuated parking brake.

tornillo y tuerca   Tipo común de freno de estacionamiento accionado por calibre.

scrub radius   The distance from the tire contact patch centerline to the point where the steering axis intersects the road.

radio de fricción   Distancia desde la línea central de la banda de rodadura del neumático hasta el punto en que el eje de dirección corta la vía.

seamless   Steel tubing with absolutely no seam where a leak may occur and no restrictions internally to cause fluid flow to be restricted.

sin costura   Tubo de acero sin ninguna costura donde puede darse una pérdida sin restricciones internas que hagan que disminuya el flujo de líquido.

secondary shoe   The trailing shoe in a duo-servo brake. The secondary shoe receives servo action from the primary shoe to increase its application force. Secondary shoes provide the greater braking force in a duo-servo brake and have longer linings than primary shoes.

zapata secundaria   Zapata de remolque en un servofreno dual. La zapata secundaria recibe acción asistida de la primaria para aumentar su fuerza de aplicación. Las zapatas secundarias proporcionan mayor potencia de frenado en un servofreno dual y tienen forros más grandes que las primarias.

section width   The width of a tire across the widest point of its cross section, usually measured in millimeters.

anchura de sección   Anchura de un neumático en el punto más amplio de su sección transversal; se suele medir en milímetros.

select low   The program used by the microprocessor to determine which wheel is beginning to lock up.

seleccionar baja   Programa que utiliza el microprocesador para determinar la rueda que está empezando a bloquearse.

self-adjusters   A cable, lever, screw, strut, or other linkage part that provides automatic shoe adjustment and proper lining-to-drum clearance as a drum brake lining wears.

autoreguladores   Cable, palanca, tornillo, puntal u otro mecanismo de unión que proporciona un ajuste automático a la zapata y la holgura adecuada entre forro y tambor cuando se desgasta el forro de un freno de tambor.

self-energizing operation   The action of a drum brake shoe when drum rotation increases the application force of the shoe by wedging it tightly against the drum.

operación autoactivada   Acción de una zapata de freno de tambor cuando la rotación del tambor aumenta la fuerza de aplicación de la zapata al ajustarse como una cuña contra el tambor.

semimetallic lining   Brake friction material made from a mixture of organic or synthetic fibers and certain metals; these linings do not contain asbestos.

forro semimetálico   Material de fricción del freno hecho con una mezcla de fibras orgánicas o sintéticas y ciertos metales; estos forros no contienen amianto.

sensor   Any device that sends an input signal to a computer.

sensor   Dispositivo que envía una señal de entrada a un computador.

service brakes   The disc or drum brakes operated by the driver to stop the vehicle.

frenos de servicio   Frenos de disco o de tambor sobre los que actúa el conductor para detener el vehículo.

servo action   The operation of a drum brake that uses the self-energizing operation of one shoe to apply mechanical force to the other shoe to assist its application. Broadly, servo action is any mechanical multiplication of force.

servoacción   Acción de un freno de tambor en el que una zapata actúa de forma autónoma al aplicar una fuerza mecánica a la otra zapata y ayudarla en su funcionamiento. Más ampliamente, una servoacción es cualquier multiplicación mecánica de una fuerza.

setback   A difference in wheelbase from one side of a vehicle to the other.

retroceso   Diferencia en la distancia entre ejes en un lado y otro del vehículo.

shoe anchor   The large pin, or post, or block against which a drum brake shoe pivots or develops leverage.

anclaje de zapata   Pasador grande, o poste, o bloque contra el que la zapata de un freno de tambor pivota o desarrolla transmisión por palancas.

signals   Electronic data transmitted to and from a vehicle computer. Signals may be data from a sensor or a command to an actuator.

señales   Datos electrónicos que se transmiten desde y hacia la computadora del vehículo. Las señales pueden ser datos de un sensor o una orden a un servo-motor.

silica   A natural element used to make silicone and other man-made materials. Silicone can be used with other chemicals to make a brake fluid with a high boiling point and is classified as DOT 5.

silicona   Un elemento natural usado para hacer el silicón y otros materiales artificiales. El silicón se puede utilizar con otros productos químicos para hacer el líquido de frenos con un alto punto de ebullición y que se clasifique como DOT 5.

sliding caliper   A caliper that is mounted to its support on two fixed sliding surfaces or ways. The caliper slides on the rigid ways and does not have the flexibility of a floating caliper.

calibre de desplazamiento   Calibre que se monta en el soporte sobre dos superficies o vías fijas de deslizamiento. El calibre se desliza por vías rígidas y no tiene la flexibilidad de un calibre flotante.

slope   The numerical ratio or proportion of rear drum brake pressure to full system pressure that is applied through a proportioning valve. If half of the system pressure is applied to the rear brakes, the slope is 1:2 or 50 percent.

atenuación diferencial   Razón numérica o proporción entre la presión del freno de tambor trasero y la presión total del sistema que se aplica a través de una válvula de dosificación. Si la mitad de la presión del sistema se aplica a los frenos traseros, la atenuación diferencial es 1:2 ó del 50 por ciento.

software The various programs in RAM and ROM that provide a microprocessor with memory and operating instructions.

software Diferentes programas en la RAM y la ROM que proporcionan memoria e instrucciones de operación a un microprocesador.

solenoid An electromagnetic device similar in operation to a relay, but movement of the armature or iron core changes electrical energy into mechanical energy.

solenoide Dispositivo electromagnético similar en funcionamiento a un relé, pero en el que el movimiento de la armadura o del núcleo de hierro transforman la energía eléctrica en mecánica.

solenoid valve A mechanical valve operated by a solenoid to control the flow of liquid or gas. Solenoid valves are used extensively in electronically controlled brake, steering, cruise control, and suspension systems.

válvula de solenoide Válvula mecánica que un solenoide hace funcionar para controlar el flujo del líquido o gas. Las válvulas de solenoide se usan mucho en sistemas de frenado, dirección, control de crucero y suspensión controlados electrónicamente.

solid rotor A rotor that is a solid piece of metal with a friction surface on each side.

rotor sólido Rotor formado por una pieza sólida de metal con una superficie de fricción a cada lado.

split point The pressure at which a proportioning valve closes during brake application and reduces the rate at which further pressure is applied to rear drum brakes.

punto de separación Presión a la que la válvula de dosificación se cierra durante la aplicación de los frenos y se reduce la relación con la que se aplica más presión a los frenos de tambor traseros.

spool valve A cylindrical sliding valve that uses lands and valleys around its circumference to control the flow of hydraulic fluid through the valve body.

válvula de carrete Válvula cilíndrica de deslizamiento que usa partes planas y hundimientos en su circunferencia para controlar el flujo del líquido hidráulico a través del cuerpo de la válvula.

square-cut piston seal A fixed seal for a caliper piston that has a square cross section.

junta cuadrada de pistón Junta fija para un pistón de calibre con sección transversal cuadrada.

star wheel A small wheel that is part of a drum brake adjusting link. Turning the star wheel lengthens or shortens the adjuster link to position the shoes for proper lining-to-drum clearance.

rueda en estrella Rueda pequeña que forma parte de un acoplamiento de ajuste de un freno de tambor. Al hacer girar la rueda en estrella se alarga o acorta el acoplamiento para poner en posición las zapatas y conseguir una tolerancia conveniente entre forro y tambor.

static friction Friction between two stationary objects or surfaces.

fricción estática Fricción entre dos objetos o superficies estáticas.

steering axis inclination (SAI) The angle formed by the steering axis of a front wheel and a vertical line through the wheel when viewed from the front with the wheels straight ahead.

inclinación del eje de dirección (SAI) Ángulo formado por el eje de dirección de una rueda delantera y una línea vertical que pasa por la rueda cuando se mira desde la parte delantera con las ruedas directamente hacia delante.

steering knuckle The outboard part of the front suspension that pivots on the ball joints and lets the wheels turn for steering control.

charnela de dirección Parte exterior de la suspensión delantera que pivota en las juntas de bola y permite que las ruedas giren para controlar la dirección.

stroking seal A seal, similar to a lip seal, that is installed at the rear of a caliper piston. Unlike a fixed seal, which is installed in a groove in the caliper bore, a stroking seal is installed on the piston and moves with it.

junta del pistón Junta similar a la del borde, que se instala en la parte trasera de un pistón de calibre. A diferencia de la junta fija que se instala en una hendidura de la superficie interior del calibre, la junta del pistón se instala en el pistón y se mueve con éste.

swept area The total area of the brake drum or rotor that contacts the friction surface of the brake lining.

zona barrida Área total del freno de tambor o del rotor que toca la superficie de fricción del forro del freno.

synthetic lining Brake friction materials made from nonorganic, nonmetallic, and nonasbestos materials; typically fiberglass and aramid fibers.

forro sintético Materiales de fricción de frenos hechos de materiales no orgánicos, no metálicos y sin amianto; por lo común, fibra de vidrio y fibras de aramida.

table The outer surface of a brake shoe to which the lining is attached.

tabla Superficie exterior de una zapata de freno a la que se une el forro.

tandem Two or more devices placed one behind the other in line.

en serie Dos o más dispositivos colocados en línea uno detrás de otro.

tandem booster A power brake vacuum booster with two small diaphragms in tandem to provide additive vacuum force.

reforzador en serie Reforzador de vacío para frenos con dos pequeños diafragmas en serie que proporcionan mas fuerza al vacío.

tapered roller bearing A specific kind of bearing that is based on tapered steel rollers held together in a cage.

cojinete de bolillas cónicas Clase específica de cojinete que se basa en bolillas cónicas de acero encerradas en una jaula.

tensile force The moving force that slides or pulls an object over a surface.

fuerza de tracción Fuerza de movimiento que desliza o arrastra un objeto sobre una superficie.

tetrachloroethylene A chlorinated cleaning solvent with four chlorine atoms per molecule; often used in aerosol brake cleaner.

tetracloroetileno Solvente limpiador clorado con cuatro átomos de cloro por molécula; se usa frecuentemente en el limpiador de frenos en aerosol.

thermal energy The energy of heat.

energía térmica Energía del calor.

thermistor A variable resistor whose resistance changes when temperature changes.

termistor Resistor variable cuya resistencia se ve modificada al cambiar la temperatura.

thrust angle The angle between the geometric centerline and the thrust line of a vehicle.

ángulo de empuje El ángulo formado por la línea geométrica central y la línea de empuje de un vehículo.

thrust line The bisector of total toe on the rear wheels, or the direction in which the rear wheels are pointing.

línea de empuje Bisectriz de la separación total de las ruedas traseras, o la dirección en que señalan dichas ruedas.

tire inflation monitoring system (TIMS) A system that uses the monitoring of wheel speed to warn the vehicle operator of low tire pressure.

sistema de control de la inflación de la llanta (TIMS) Sistema que utiliza el control de velocidad de la rueda para advertir al conductor del vehículo sobre la baja presión de la llanta.

tire load range The load-carrying capacity of a tire, expressed by the letters A through L. The load range letters replace the older method of rating tire strength by the number of plies used in its construction.

límites de carga de neumáticos Capacidad de carga de un neumático, expresada por letras de la A a la L. Las letras de los límites de carga reemplazan el método anterior de clasificar la resistencia de los neumáticos mediante el número de capas usadas en su construcción.

tire pressure monitoring system (TPMS) A system of devices and software used to alert the driver of underinflated tire(s).

sistema de supervisión de la presión del neumático (TPMS) Un sistema de los dispositivos y del software que alerta el conductor de neumáticos inflados inferiores.

toe angle The difference in the distance between the centerlines of the tires on either axle (front or rear) measured at the front and rear of the tires and at spindle height.

ángulo de separación Diferencia de la distancia entre las líneas centrales de los neumáticos en cada eje (delantero o trasero) medida en las partes delantera y trasera de los neumáticos y a la altura del husillo.

toe-out on turns (turning radius) The difference between the angles of the front wheels in a turn.

divergencia en los giros (radio de giro) Diferencia entre los ángulos de las ruedas delanteras al girar.

traction control system (TCS) A system that attempts to control wheel spin during acceleration on slick road surfaces.

sistema de control de tracción (TCS) Sistema que intenta controlar el giro de la rueda en la aceleración sobre pavimentos resbaladizos.

trailer breakaway condition A condition created when the trailer unintentionally disconnects from the towing vehicle; a condition requiring specific DOT safety devices.

condición de desenganche del trailer Condición creada cuando el trailer accidentalmente se desconecta de la grúa; una condición que requiere dispositivos de seguridad DOT específicos.

trailing shoe The second shoe in the direction of drum rotation in a leading-trailing brake. When the vehicle is going forward, the rear shoe is the trailing shoe, but the trailing shoe can be the front or the rear shoe depending on whether the drum is rotating forward or in reverse and whether the wheel cylinder is at the top or the bottom of the backing plate. The trailing shoe is non-self-energizing, and drum rotation works against shoe application.

zapata de remolque Segunda zapata en la dirección de giro del tambor en un freno de tracción-remolque. Cuando el vehículo avanza, la zapata trasera es la remolcada, pero la zapata de remolque puede ser la delantera o la trasera dependiendo de si el tambor está girando hacia delante o hacia atrás y de si el cilindro de la rueda está en la parte superior de la placa de refuerzo o en la inferior. La zapata de remolque no es autónoma, y el giro del tambor funciona contra la aplicación de la zapata.

transistor A three-element semiconductor device of NPN or PNP materials that transfers electrical signals across a resistance.

transistor Dispositivo semiconductor de tres elementos de materiales NPN o PNP que pasa señales eléctricas a través de una resistencia.

tread The layer of rubber on a tire that contacts the road and contains a distinctive pattern to provide traction.

rodadura Capa de goma de un neumático que está en contacto con la carretera y contiene un patrón distintivo que favorece la tracción.

tread contact patch The area of the tire tread that contacts the road; determined by tire section width, diameter, and inflation pressure.

banda de rodadura Superficie de la rodadura del neumático que se pone en contacto con la carretera; viene determinada por la anchura de la sección del neumático, el diámetro y la presión de inflado.

tread wear indicator A continuous bar that appears across a tire tread when the tread wears down to the last $\frac{2}{32}$ ($\frac{1}{16}$) inch. When a tread wear indicator appears across two or more adjacent grooves, the tire should be replaced.

indicador de desgaste de la rodadura Barra continua que aparece transversalmente en la rodadura del neumático cuando ésta se desgasta más de $\frac{2}{32}$ ($\frac{1}{16}$) pulgadas. Cuando aparece el indicador de desgaste de rodadura cruzando dos o más hendiduras adyacentes, se debe cambiar el neumático.

trichloroethane A chlorinated cleaning solvent often used in aerosol brake cleaner; less toxic than trichloroethylene.

tricloroetano Solvente clorado limpiador usado a menudo en limpiadores de frenos en aerosol; menos tóxico que el tricloroetileno.

trichloroethylene A chlorinated toxic cleaning solvent often used in aerosol brake cleaner and as an insecticide fumigant.

tricloroetileno Solvente clorado limpiador, tóxico, usado frecuentemente como limpiador de frenos en aerosol y como fumigador insecticida.

unidirectional rotor A rotor with cooling fins that are curved or formed at an angle to the hub center to increase cooling airflow. Because the fins work properly only when the rotor turns in one direction, unidirectional rotors cannot be interchanged from right to left on the car.

rotor unidireccional Rotor con álabes refrigerantes curvadas o formando un ángulo con el centro del buje para aumentar el flujo de aire de refrigeración. Como los álabes sólo funcionan adecuadamente cuando el rotor gira en una sola dirección, los rotores unidireccionales no se pueden intercambiar entre la derecha y la izquierda del auto.

unidirectional tread pattern A tire tread pattern that can be rotated only in one direction. Thus, left- and right-side tires with unidirectional tread cannot be interchanged.

patrón de rodadura unidireccional Patrón de rodadura de un neumático que se puede girar en una única dirección. Es decir, los neumáticos de la izquierda y de la derecha con rodadura unidireccional no se pueden intercambiar.

uniform tire quality grading (UTQG) indicators Letters and numbers molded into the sidewall of a tire to indicate relative tread life, wet weather traction, and heat resistance.

indicadores del grado de calidad de neumáticos uniformes (UTQG) Letras y números moldeados en los lados de un neumático para indicar la duración relativa de un neumático, la tracción en tiempo lluvioso y la resistencia al calor.

vacuum In automotive service, vacuum is generally considered to be air pressure lower than atmospheric pressure. A true vacuum is a complete absence of air.

vacío En relación con los automóviles, el vacío se suele considerar como una presión de aire menor que la atmosférica. Un vacío verdadero es la completa ausencia de aire.

vacuum suspended A term that describes a power brake vacuum booster in which vacuum is present on both sides of the diaphragm when the brakes are released. The most common kind of vacuum booster.

de suspensión de vacío Término que describe un reforzador de vacío para frenos en el que existe vacío a ambos lados del diafragma cuando se sueltan los frenos. El tipo más común de reforzador de vacío.

valleys The annular grooves or recessed areas between the lands of a valve spool.

hundimientos Hendiduras anulares o áreas ahuecadas entre las partes de una válvula de carrete.

vent port The forward port in the master cylinder bore; also called other names.

válvula de ventilación Válvula delantera de la parte interior del cilindro maestro; también recibe otros nombres.

ventilated rotor A rotor that has cooling fins cast between the braking surfaces to increase the cooling area of the rotor.

rotor ventilado Rotor que tiene álabes refrigerantes entre las superficies de frenado para aumentar la superficie refrigerante del rotor.

volt The unit used to measure the amount of electrical force or energy.

voltio Unidad usada para medir la cantidad de fuerza o energía eléctrica.

voltage The electromotive force that causes current to flow. The potential force that exists between two points when one is positively charged and the other is negatively charged.

tensión Fuerza electromotriz que hace que la corriente fluya. La fuerza potencial que existe entre dos puntos cuando uno está cargado positivamente y el otro negativamente.

water fade Brake fade that occurs when water is trapped between the brake linings and the drum or rotor and the coefficient of friction is reduced.

pérdida de agua Pérdida de frenado debido a que el agua queda atrapada entre los forros de los frenos y el tambor o el rotor, reduciéndose el coeficiente de rozamiento.

ways Machined surfaces on the caliper support on which a sliding caliper slides.

conductos Superficies talladas en el soporte del calibre por las que se desliza un calibre de desplazamiento.

web The inner part of a brake shoe that is perpendicular to the table and to which all of the springs and other linkage parts attach.

membrana Parte interior de una zapata de freno perpendicular a la tabla y en la que se fijan todos los resortes y otras partes de acoplamiento.

weight The measure of the Earth's gravitational force or pull on an object.

peso Medida de la fuerza de gravedad de la Tierra o de la atracción sobre un objeto.

wheel cylinder The hydraulic slave cylinder mounted on the backing plate of a drum brake assembly. The wheel cylinders convert hydraulic pressure from the master cylinder to mechanical force that applies the brake shoes.

cilindro de la rueda Cilindro hidráulico auxiliar montado en la placa de refuerzo de un conjunto de frenos de tambor. Los cilindros de la rueda convierten la presión hidráulica del cilindro maestro en la fuerza mecánica que se aplica a las zapatas de freno.

wheel offset The distance between the centerline of the rim and the mounting plane of the wheel.

desajuste de la rueda Distancia entre la línea central de la llanta y el plano de montaje de la rueda.

wheel speed base A software program in an ABS used to determine inflation of a tire.

base de velocidad de la rueda Programa de software en un ABS que se usa para determinar la cantidad de aire de una llanta.

yaw Swinging motion to the left or to the right of the vertical centerline or rotation around the vertical centerline.

derrape Movimiento de balanceo de izquierda a derecha de la línea central vertical o giro alrededor de dicha línea.

Index

Note: Page numbers in bold print reference non-text material.

A

ABS. *See* Antilock brake system
Accumulator, 149, 259. *See also* Pumps and accumulators
Active brake systems, 272–273
Active braking, 11
Actuator, 11, 248–249, **250**
Adjustable pedal systems (APS), 71, **71**
Adsorption, 27
Air systems, 134–135, **134**
Alignment
  wheel performance, 58–59, **58**
Ampere (A), 36
Analog signals, 248, **249**
Antilock brake systems (ABSs), 1, 8, 10–11, 252–274
  brands, 254
  components, 254–260
    controllers, 255
    lamps and communications, 259–260
    magnetoresistive wheel speed sensors, 257–259, **257–258**
    microprocessor, 255
    motion sensors, 259
    pumps and accumulators, 259
    wheel sensors, 255–257, **256**
  Continental Teves Mark 20 and Mark 20E, 264–269, **264–269**
  future braking systems, 273–274
    regenerative braking, 274
  system integration, 261–264, **262–263**
  traction control systems, 260–261
  types and general operations, 252–254
    four wheel ABS, 254
    integrated and nonintegrated ABS, 252, **253**
    real wheel ABS (RWAL) or RABS, 253–254, **253**
APS. *See* Adjustable pedal system
Aramid fibers, 176
ARC. *See* Automatic ride control
Arcing the brake shoe, 209
Asbestos
  in brake linings, 25, 174–175
Aspect ratio, 43
Atmospheric pressure, 132, **133**
Atmospheric suspended vacuum booster, 139
Automatic parking brake release, 232–233, **234**
Automatic ride control (ARC), 11, 262
Automotive Friction Material Edge Code, 176–177, **177**
Auxiliary drum parking brakes, 242, **242**
Auxiliary vacuum pumps, 147–149, **148**

B

BA. *See* Brake assist
Backing plate, 203, 204, **204**, 210, **211**
Ball-and-ramp caliper-actuated parking brakes, 242, 243–244, **244**

Ball bearings, 50, 51, **52**
Banjo fittings, 104, **105**
Bearing cage, 173
Bearing cone, 173
Bearing cup, 173
Belted bias ply tires, 43, **43**
Bendix, 254
Bias ply tire, 43, **43**
Bimetallic aluminum drums, 207, **207**
Binders, 175
Boiling point, brake fluid, 64–65, **64 table**, **65**
Bonded linings, 179
Bosch, 254
BPMV. *See* Brake pressure modulator valve
Brake assist (BA), vacuum booster with, 155
Brake caliper, 182
Brake fade, 26–27, 163
  gas fade, 27
  lining fade, 26
  mechanical fade, 26–27
Brake fluid. *See* Hydraulic brake fluid
Brake friction materials, 174–176, 209
  attachment, 178–180, **178**
  selection, 176–178, **177 Table**
  types
    ceramic brake pads, 176
    metallic linings, 175–176
    organic linings, 174–175
    semimetallic linings, 175
    synthetic linings, 176
Brake lines, 8, **8**, 95–106, **97**
  fittings, 101–104, **102–105**
  hoses, 99–101, **99–101**
  line and fitting precautions, 105–106
  tubes or pipes, 96–99, **97–99**
Brake pad indicators, 120
Brake pads, 174, **174**
Brake pedal and pushrod, 68–71, **69–71**
  adjustable pedal systems, 71, **71**
  brake linkage free play, 69, **70**
Brake pedal feel, 143–144
Brake pressure modulator valve (BPMV), 269
Brake shoes, 207–210, **208–210**
Brake switch, 258–259, **258**
Brake system
  braking dynamics, 21–22, **21**
  electric, 12
  electrical principles, 36–37
    amperage, voltage, and resistance, 36
    Ohm's law, 37, **37**
  electronic, 10–11

Brake system (*cont.*)
energy, 18–21
inertia and momentum, 20–21
kinetic energy, mass, weight, and speed, 18–20, **19–20**
friction principles, 22–28
brake fade, 26–27
brake friction materials, 27–28, **27–28**
coefficient of friction, 23–26
friction and pressure, 22–23, **23**
friction and surface area, 23, **23**
kinetic and static friction, 22, **22**
functions, 1
hydraulic pressure, 17–18, **18**
hydraulic principles, 31–36
hydraulic surge, 12
operation, 17–18
overview, 1–11, **2**
antilock brake systems, 10–11
hydraulic systems, 6–9, **6–9**, 12
leverage and brake pedal design, 1–2, **3**
parking brakes, 10
power boosters, 10, **10**
service brake design, 3–5, **4–5**
vacuum and air pressure principles, 36
Brake warning system. *See* Electrical warning system
Braking dynamics, 21–22, **21**
Breakaway condition for trailers, 12

C
CAB. *See* Controller antilock brake
CAN. *See* Controller area network
Cables, parking brakes, 234–237, **235–237**
Caliper, 4
parts and operation, 182–188, **183–184**
body, 183–184, **184**
dust boots, 187–188, **188**
hydraulic passages and lines, 184–185, **185**
pistons, 8–9, **9**, 185–186, **185**
piston seals, 186–187, **187–188**
support, 183
Caliper-actuated parking brakes
screw-and-nut operation, 242, 243, **243**
ball-and-ramp operation, 243–244, **244**
Camber, 52, **53**
Cam-ground lining, 209, **209**
Carcass, tire, 42
Casing, tire, 42
Caster, 52, **53**
Center high-mounted stoplamp (CHMSL), 122, **127**
Central-valve master cylinders, 89–90, **90**
Ceramic brake pads, 28, 176
Check valve, 135
CHMSL. *See* Center high-mounted stoplamp
Class 2 wiring, 272

Coefficient of friction, 23–26, **24, 26**
heat dissipation, 26
materials, 25
surface finish, 25
temperature, 25
Cold inflation pressure, 45, **45**
Combination valve, 118–119, **118–119**
Commands, 250
Communications, 260
Composite drums, 206–207, **207**
Composite rotor, 170–172, **172**
Compression fittings, 104, **105**
Conduit, 237
Continental Teves system, 240–241, **241**, 264–269
Mark 20, 264–267, **264–267**
Mark 20E, 268–269, **268–269**
Control area network (CAN), 250
Controller, 249, **250**
Controller antilock brake (CAB) module, 255
Cup expander, 82, **82**, 212, **212**
Cup seal, 76
Curing agents, 175

D
Delco-Bosch, 254
Delphi DBC-7 ABS, 269–272, **270–271**
Diagonally split hydraulic system, 84–86, **84**
Diaphragm, 72, 138
suspension, 139–140
Digital signals, 248, **249**
Directional control, 252
Directional stability, 252
Disc brakes, 4–6, **5**, 163–195
advantages and disadvantages, 163–169
fade resistance, 163–165, **164–165**
parking brake, 168–169
self-adjustment, 165, **166**
brake servo action, pulling, and grabbing, 166–168, **167**
caliper parts and operation, 182–188, **183–184**
body, 183–184, **184**
dust boots, 187–188, **188**
hydraulic passages and lines, 184–185, **185**
piston seals, 186–187, **187–188**
pistons, 185–186, **185**
construction, 169–182
brake pad wear indicators, 180–182, **181–182**
brake pads, 174, **174**
composite rotors, 170–172, **172**
fixed and floating rotors, 170, **171**
friction materials, 174–180, **174, 177 table, 178**
pad-to-caliper attachment, 180, **180**
rotor hubs and wheel bearings, 173–174, **173**
rotors, hubs, and bearings, 169–170, **170–171**
solid and ventilated rotors, 172–173, **172**
friction material attachment, 178–180, **178**

friction material selection, 176–178, **177 table**

noise, 168

operation, **168,** 169

pad-to-caliper attachment, 180, **180**

    brake pad wear indicators, 180–182, **181–182**

performance disc brakes, 193, 195

rear wheel disc brakes, 193, **193–194**

types, 188–193

    fixed caliper, 188–190, **189–190**

    floating caliper, 190–192, **191**

    sliding caliper, 192–193, **192**

DOT 5 silicone fluid, 66

Double flare, 102–103, **103**

DRP. *See* Dynamic rear proportioning

Drum brakes, 3–4, **4,** 199–227

advantages and disadvantages, 199–203

    fade resistance, 201–203, **202–203**

    lack of noise, 200

    parking brake operation, 200

    pulling and grabbing, 199–200

    self-adjustment, 200–201

    self-energizing and servo action, 199, **200–201**

backing plate, 210, **211**

brake drums and hubs, 204–207, **205–207**

    bimetallic aluminum drums, 207, **207**

    solid cast-iron drums, 206, **206**

    steel and iron drums, 206–207, **207**

brake shoes and linings, 207–210, **208–210**

construction, 203–219, **203**

designs, 219–224

    duo-servo brakes, 219, **219,** 223–224

    leading-trailing brakes, 219, **220,** 221–223, **223**

    self-energizing and servo actions, 219–221, **220–222**

friction materials, 209

lining-to-drum fit, 209–210, **209–210**

parking brake linkage, 218–219

return and hold-down springs, 212–213, **212–213**

self-adjusters, 214–218

    duo-servo star wheel adjusters, 214–216, **214–216**

    leading-trailing-shoe cam adjusters, 217–218, **218**

    leading-trailing-shoe ratchet adjusters, 216–217, **217**

    leading-trailing-shoe star wheel adjusters, 216

    precautions, 218

Drums

    bimetallic aluminum, 207, **207**

    solid cast-iron, 206, **206**

    steel and iron, 206–207, **207**

Drum web, 204

Dual-piston master cylinder, 72–84, **72, 85–86**

construction, 74–77, **75–77**

operation, 77–81, **78–81**

ports, 74, **74**

reservoir, 72–74, **73**

residual pressure check valve, 81–84, **81–83**

Duo-servo brakes, 219, **220,** 223–224

Duo-servo star wheel adjusters, 214–216, **214–216**

Dust boots, caliper, 187–188, **188**

Dynamic rear proportioning (DRP), 271

    valve, 259

E

EBCM. *See* Electronic brake control module

EBTCM. *See* Electronic brake traction control module

Eccentric shaft, 242

EHB. *See* Electro-hydraulic brake

Electric braking system, 12, 247–274. *See also* Antilock brake system

components, 247–251

    actuators, 248–249, **250**

    commands, 250

    control area network (CAN), 250

    controller, 249, **250**

    hydraulic modulator, 250, **251**

    multiplexing, 250, **251**

    sensors, 247, **248**

    signals, 248, **249**

Electrical parking brake systems, 239–241, **240–241**

Electrical warning system, 120–126

    brake pad indicators, 120

    master cylinder fluid level switch, 120–121, **121**

    parking brake switch, 120

    stoplamp bulbs, 124–126, **124–127**

    stoplamp switch and circuit, 122, **122**

    stoplamps, 122–124, **123–124**

Electro-hydraulic brake (EHB), 157–159, **158–159**

Electrohydraulic unit, 255

Electro-Lok, 239, **240**

Electronic brake control module (EBCM), 255

Electronic brake traction control module (EBTCM), 269

Electronically variable brake proportioning (EVBP), 268

Energy, 28–29

    perpetual, 29

Enhanced traction system (ETS), 271

Equalizer, 235, **235,** 238

Eyeball, 257

F

Fade resistance, 163–165, **164**

    disc brake, 201, **202**

    drum brake, 201–203

Failure warning lamp switch, 115–118, **115–118**

Fast-fill and quick take-up master cylinders, 86–89

Federal Motor Vehicle Safety Standards (FMVSS), 4

Fillers, 175

Fittings, brake, 101–104, **102–105**

Fixed caliper brake, 188–190, **189–190**

Fixed rotor, 170, **171**

Fixed seals, 186, **187**

Floating calipers, 190–192, **191**

Floating rotors, 170, **171**
Flux lines, 256
FMVSS. *See* Federal Motor Vehicle Safety Standards
Force, 32
Force multiplication, 131
Four-wheel antilock systems, 254
Free play, 69–71
Frequency, 248
Friction, 3
    principles, 22–28
        coefficient of, 23–26
        kinetic and static friction, 22, **22**
        and pressure, 22–23, **23**
        and surface area, 23, **23**
Friction materials, brake, 27–28, **27–28.** *See also* Brake friction
        materials
    ceramic, 28
    fully metallic, 28
    nonmetallic, 28
    semimetallic, 28
Friction modifiers, 175
Front-to-rear split hydraulic system, 84–86, **85**
Fulcrum, 2
Fully metallic brake lining materials, 28

G
Gas fade, 27, 165, 202
Geometric centerline, 54
Gross vehicle weight rating (GVWR), 41
GVWR. *See* Gross vehicle weight rating

H
Heat dissipation, 26
Height-sensing proportioning valve, 113–114, **113–115**
High-pressure pumps and accumulators, 259
Hold-down springs, 213, **213**
Hoses, brake, 99–101, **99–101**
HSMO. *See* Hydraulic system mineral oil
Hubs, rotor, 173–174
Hybrid vehicles, 273–274
Hydraulic advantage, 131
Hydraulic brake fluid, 63–68
    boiling point, 64–65, **64 table**, **65**
    contaminated fluid problems, 68
    DOT 5 silicone fluid, 66
    fluid compatibility, 66–67
    hydraulic system mineral oil fluids, 66
    precautions, 67–68
    requirements, 66
    storage and handling, 67
    synthetic fluids, 66
Hydraulic channels, 269
    four channel, 269
    three channel, 269
Hydraulic modulator, 250, **251**

Hydraulic power brakes, 149–157, **150, 152–157**
    hydro-boost operation, 151–155, **152–154**
    hydro-boost principles, 149, **150**
    Powermaster general operation, 155–157, **156–157**
    vacuum booster with brake assist, 155
Hydraulic pressure, 17–18, **18**
    control without valves, 119–120
Hydraulic principles, 31–36
    and brake system engineering, 36
    fluids apply pressure to transmit and increase force, 32–35,
        32–35
    fluids cannot be compressed, 31, **31**
    fluids can transmit movement, 31–32, **31**
Hydraulic surge, 12
Hydraulic system mineral oil (HSMO), 66
Hydraulic systems, 6–9
    lines and hoses, 8, **8**
    master cylinder, 6, **6–7**
    pressure control valves, 8, **9**
    split systems, 6–7, **7**
    wheel cylinders and caliper pistons, 8–9, **9**
Hydro-boost, 149
    operation, 151–155, **152–154**
        brakes being released, 151, **152**
        brakes holding, 151
        brakes not applied, 151, **152**
        moderate brake application, 151, **153**
        PowerMaster, 155–157, **156–157**
        reserve brake application, 151–155, **154**
    principles, 149, **150**

I
Inertia, 20–21
Intake manifold systems, 135, **135**
Integrated ABS, 252, **253**
Intermediate lever, 238
International Standards Organization (ISO), 95
Inverted double flare, 98, **98**, **104**
ISO. *See* International Standards Organization
ISO flare, 98, **98**
    fittings, 103–104, **104**
Isolation/dump valve, 253

K
Kelsey–Hayes, 254
Kinetic energy, 18–20, **20**
Kinetic friction, 22, **22**

L
Lamps and communications, 259–260
    communications, 260
    warning lamps, 260
Lands, 151
Lateral accelerometer, 259
Lathe-cut seals, 186
Laws of motion, 29–30, **30**

Leading shoe, 219
Leading-trailing brakes, 219, **219**, 221–223, **223**
Leading-trailing-shoe cam adjusters, 217–218, **218**
Leading-trailing-shoe ratchet adjusters, 216–217, **217**
Leading-trailing-shoe star wheel adjusters, 216
Leverage, 131
    and brake pedal design, 2, **3**
Levers, parking brakes, 232, 238
Lining, brake shoe, 207–210, **208**
Lining fade, 26, 202
Lining-to-drum fit, 209–210, **209–210**
Linkage, parking brake, 218–219
Lockup, wheel, 10
Loughead (Lockheed), Malcolm, 6
Low-drag caliper seals, 186–187
Low-pressure return pumps, 259

M
Magnetoresistive wheel speed sensors, 257–259, **257–258**
    brake switch, 258–259, **258**
Mark 20, 264–267, **264–267**
Mark 20E, 268–269, **268–269**
Mass, 18–19, **19**
Master cylinder fluid level switch, 120–121, **121**
Master cylinders, 6, **6–7**
    central valve, 89–90, **90**
    dual-piston, 72–84, **85–86**
    fast fill and quick take-up, 86–89, **87**
Materials, 25
Mechanical advantage, 33, 131
Mechanical fade, 26–27, 201
Metallic linings, 175–176
Metering valve, 106–108, **107–109**
Microprocessor inputs, 255
    permanent magnet generators, 255, **255**
Mold-bonded linings, 180
Momentum, 20–21
Motion, laws of, 29–30, **30**
Motion sensors, 259
Multiplexing, 250

N
National Highway Transportation and Safety Administration
    (NHTSA), 47
Negative wheel slip, 10, 252
Network, 250
Newton's laws of motion, 29–30, **30**
NHTSA. *See* National Highway Transportation and Safety
    Administration
Noise, disc brake, 168, **168**
Nonintegrated ABS, 252
Nonmetallic brake lining materials, 28

O
Ohm (Ω), 36
Ohm, Georg Simon, 37

Ohm's law, 37, **37**
Organic disc pad linings, 174–175
O-rings, 72
Overload spring, 216

P
Pad hardware, 180, **180**
Pad-to-caliper attachment, 180
    brake pad wear indicators, 180–182, **181–182**
Pad wear indicators, 180–182, **181–182**
Parking brakes, 1, 10, 229–246, **230–231**
    controls—levers and pedals, 231–234, **231, 233–234**
        automatic parking brake release, 232–233, **234**
        levers, 232
        pedals, 232, **233**
    electrical, 239–241, **240–241**
    linkage, 218–219, 234–238
        cables, 234–237, **235–237**
        equalizers and adjusters, 238
        levers, 238
        rods, 237
    operation, 229–231
        disc brakes, 168–169
        drum brakes, 200
    rear disc, 242–244
        auxiliary drum parking brakes, 242, **242**
        caliper-actuated parking brakes, 242–244, **243–244**
    rear drum, 238–239, **238–239**
    warning lamps, 234
Parking brake switch, 120
Pascal, Blaise, 35
Pawl, 215
Pedal force, 131
Pedals, 232, **233**
Performance
    brakes, 193, 195
    tires, 57
    wheel, 57–58
    wheel alignment, 58–59, **58**
Permanent magnet (PM) generators, 255, **255**
Perpetual energy, 29
Phenolic plastic, 185
Piston seals, calipers, 186–187, **187–188**
Piston stops, 210
Plate-and-lever boosters, 144–145, **145**
PM. *See* Permanent magnet
P-metric system, 44
Polyglycol, 64
Positive wheel slip, 260
POST. *See* Power on System test
Power boosters, 10, **10**, 132
Power brake systems, 131–159
    electro-hydraulic brake (EHB), 157–159, **158–159**
    hydraulically-assisted power brakes, 149–157, **150, 152–157**
    hydro-boost operation, 151–155, **152–154**
    increase in braking power, 131–132

Power brake systems (*cont.*)
    tandem, boosters, 145–149, **146–148**
    vacuum and air systems for power boosters, 134–137, **134–137**
    vacuum power boosters, 137–145, **138–145**
    vacuum principles, 132–134, **133**
PowerMaster system
    master cylinder, 156, **156**
    operation, 155–157, **155–157**
Power on System Test (POST) program, 256, 260
Pressure, 6
Pressure control valves, 8, **9**
Pressure differential, 132
Pressure differential valve, 115–118, **115–118**
Pressure-sensor base (PSB), 47
Primary shoe, 221, **222**
Proportioning valve, 110–113, **110–113**
    height-sensing, 113–114, **113–115**
PSB. *See* Pressure-sensor base
Pumps and accumulators, 259
    dynamic rear proportioning (DRP) valve, 259
    high-pressure, 259
    low pressure-return pumps, 259
Pushrod. *See* Brake pedal and pushrod

Q
Quick take-up master cylinder, 86–89, **87**
Quick take-up valve, 87, **87**

R
RABS. *See* Rear antilock brake system
Radial ply tire, 42, 43, **43**
Raymond Company, 28
Reaction disc, 143
Reaction-disc booster, 144, **144**
Reaction plate-and-lever booster, **145**
Reaction plate and levers, 143
Rear antilock brake system (RABS), 253–254
Rear wheel antilock (RWAL) system, 253–254, **253**
Rear wheel disc brakes, 193, **193–194**
Regenerative braking, 274
Reluctor, 256
Replenishing port, 74
Reservoir, 72–74
Residual pressure check valve, 81–84, **81**
Return springs, 212–213, **212–213**
Reverse braking, 12–14, **13–14**
Rim width, 48, **49**
Riveted linings, 179
Rods, parking brakes, 237
Rotor, disc brake, 169–173, **170**
    composite, 170–172, **172**
    fixed, 170, **171**
    floating, 170, **171**
    solid, 172–173, **172**
    unidirectional, 173
    ventilated, 172–173, **172**

Run-flat tires, 47
RWAL. *See* Rear wheel antilock

S
SAE flare fittings, 102–103, **103**
SAI. *See* Steering axis inclination
Screw-and-nut caliper-actuated parking brakes, 242, 243, **243**
Scrub radius, 49, **50**
Seamless tubing, 96
Secondary shoe, 221, **222**
Section width, 43
Self-adjusters, 214–218
    duo-servo star wheel adjusters, 214–216, **214–216**
    leading-trailing-shoe cam adjusters, 217–218, **218**
    leading-trailing-shoe ratchet adjusters, 216–217, **217**
    leading-trailing-shoe star wheel adjusters, 216
    precautions, 218
Self-adjustment
    disc brake, 165, **166**
    drum brake, 200–201
Self-energizing operation, 166, **167**
    and brake design, 219–221
    drum brakes, 199, **200–201**
Semimetallic linings, 28, 176
Sensors, 247, **247**
Service brake design, 1, 3–5
    disc brakes, 4–5, **5**
    drum brakes, 3–4, **4**
Servo action
    and brake design, 219–221
    disc brakes, 166–168, **167**
    drum brakes, 199, **201**
Shoe anchors, 210
Signals, 248, **249**
Sliding calipers, 192–193, **192**
Slope, 112
Solenoid valves, 248, **250**
    reversing, 13–14, **14**
Solid cast-iron drums, 206, **206**
Solid rotor, 172–173, **172**
Speed, 18–20, **20**
    wheel sensors, 11
Split point, 111
Split system, 6–7, **7**
Spool valve, 149
Square-cut piston seal, 186
Star wheel, 215
Static friction, 22, **22**
Steel and iron drums, 206–207, **207**
Steering axis inclination (SAI), 53–54, **54**
Steering knuckle, 183
Stoplamp bulbs, 124–126, **124–127**
Stoplamp switch and circuit, 122, **122**
Stoplamps, 122–124, **122–124**
Straight roller bearings, 50, 51, **51**

Stroking seals, 187, **188**
Sumitomo ABS, 254
Surface finish, 25
Swept area
    disc brake, 163–164, 201, **202**
    drum brake, 201
Synthetic fluids, 66
Synthetic linings, 176
System integration, ABS, 261–264, **262–263**

T
Table, brake shoe, 207, **208**
Tandem, 145
Tandem boosters, 145–149, **146–148**
    auxiliary vacuum pumps, 147–149, **148**
Tapered roller bearing, 50–51, **51**, 173
TCS. *See* Traction control system
Temperature, 25
Tensile force, 24
Teves. *See* Continental Teves
Thermal energy, 22
Thrust angle, 54, **55**
Thrust line, 54
TIMS. *See* Tire inflation monitoring system
Tire inflation monitoring system (TIMS), 271
Tire load range, 44–45
Tire pressure monitoring system (TPMS), 47
Tires, 41–48
    carmaker's recommendations, 41–42, **42**
    construction, 42–43, **42–43**
    effect on braking, 47–48, **47**
    inflation monitoring, 47–48
    performance, 57
    run-flat, 47
    size, 43–44, **45**
    specifications, 44–46
    tread design, 46, **47**
Toe angle, 53, **53**
Toe-out on turns, 54, **55**
TPMS. *See* Tire pressure monitoring system
Traction control systems (TCSs), 11, 260–261
    wheel slip, 260
    wheel spin control strategies, 260–261
Trailer brakes, 11–14
    breakaway condition, 12
    electric brakes, 12
    hydraulic surge, 12
    reverse braking, 12–14, **13**
Trailing shoe, 221
Tread, tire, 42
Tread contact patch, 56
Tread wear indicator, 46
Tubing, 96–99, **97**
    installation, 97–99, **98–99**
    sizes, 97, **97**
Turning radius, 54, **55**

U
Unidirectional rotor, 173
Unidirectional tread pattern, 46
Uniform tire quality grading (UTQG) indicators, 46
UTQG. *See* Uniform tire quality grading

V
Vacuum, 1, 36, 132
Vacuum and air systems for power boosters, 134–137
    air systems, 134–135 **134**
    intake manifold systems, 135, **135**
    vacuum check valves, 135–137, **136–137**
Vacuum booster, 137–145
    brake assist, 155
    brake pedal feel, 143–144
    construction, 138–139, **139**
    diaphragm suspension, 139–140
    operation, 141–143
        brakes being released, 142–143
        brakes holding, 142, **143**
        brakes not applied, 141, **141**
        full brake application, 142
        moderate brake application, 141, **142**
    plate-and-lever booster, 144–145, **145**
    reaction-disc booster, 144, **144**
    types of vacuum boosters, 140–143, **140–143**
        general types, 140–141, **140**
        operation, 141–143, **141–143**
Vacuum check valves, 135–137, **136–137**
Vacuum principles, 132–134, **133**
Vacuum reservoir, 135, **135**
Vacuum suspended vacuum booster, 139
Valleys, 151
Valves, 106–120
    combination valve, 118–119, **118–119**
    height-sensing proportioning valve, 113–114, **113–115**
    hydraulic pressure control without valves, 119–120
    metering valve, 106–108, **107–109**
    pressure differential valve, 115–118, **115–118**
    proportioning valve, 110–113, **110–113**
Vent port, 74
Ventilated rotor, 172–173, **172**
Volt (V), 36
Voltage, 36

W
Warning lamps, 120–126, 260
Water fade, 164, 202
Web, brake shoe, 207
Weight, 18–19, **19**
Wheel
    performance, 57–58
Wheel alignment, 52–56
    camber, 52, **53**
    caster, 52, **53**
    effects on braking, 56–57

Wheel alignment (*cont.*)
    performance, 57–58, **58**
    setback, 55–56, **55**
    steering axis inclination, 53–54, **54**
    thrust angle, 54, **55**
    toe angle, 53, **53**
    toe-out on turns, 54, **55**
Wheel bearings, 50–51, **51–52**, 173–174, **173**
    effects on braking, 51
Wheel cylinders, 8–9, **9**, 210–212, **211–212**
Wheel effects on braking, 48–50
    rim width, 48, **49**
    wheel offset, 49, **49**
Wheel offset, 49–50, **50**

Wheel sensors, 255–257, **256**
    magnetoresistive sensors, 257–259, **257–258**
    speed, 11
Wheel slip, 260
    negative, 252
    positive, 260
Wheel-speed base (WSB), 47
Wheel speed sensors, 11
Wheel spin control strategies, 260–261
WSB. *See* Wheel-speed base
Work, 28–29

Y
Yaw, 11, 259